# Principles of
# Highway Engineering and
# Traffic Analysis

# Principles of
# Highway Engineering and
# Traffic Analysis
## SI Version

## Fifth Edition

## Fred L. Mannering
*Purdue University*

## Scott S. Washburn
*University of Florida*

# WILEY

# Preface

## INTRODUCTION

The first four editions of *Principles of Highway Engineering and Traffic Analysis* sought to redefine how entry-level transportation engineering courses are taught. When the first edition was published over two decades ago, there was a need for an entry-level transportation engineering book that focused exclusively on highway transportation and provided the depth of coverage needed to serve as a basis for future transportation courses as well as the material needed to answer questions likely to appear on the Fundamentals of Engineering (FE) and/or Principles and Practice of Engineering (PE) exams in civil engineering. The subsequent use of the various editions of this book, over the years, at some of the largest and most prestigious schools in the U.S. and throughout the world suggests that a vision of a concise, highly focused, and well written entry-level book is shared by many educators.

## APPROACH

This fifth edition of *Principles of Highway Engineering and Traffic Analysis* continues the spirit of the previous four editions by again focusing exclusively on highway transportation and providing the depth of coverage necessary to solve the highway-related problems that are most likely to be encountered in engineering practice. The focus on highway transportation is a natural one given the dominance of highway transportation for people and freight movement in the U.S. and throughout the world. While the focus on highway transportation is easily accomplished, identifying the highway-related problems most likely to be encountered in practice and providing an appropriate depth of coverage of them is a more challenging task. Using the first four editions as a basis, along with the comments of other instructors and students who have used previous editions of the book, topics that are fundamental to highway engineering and traffic analysis have been carefully selected. The material provided in this book ensures that students learn the fundamentals needed to undertake upper-level transportation courses, enter transportation employment with a basic knowledge of highway engineering and traffic analysis, and have the knowledge necessary to answer transportation-related questions on the civil engineering FE and PE exams.

## MATHEMATICAL RIGOR

Within the basic philosophical approach described above, this book addresses the concern of some that traditional highway transportation courses are not as mathematically challenging or rigorous as other entry-level civil engineering courses, and that this may affect student interest relative to other civil-engineering fields of study. This concern is not easily addressed because there is a dichotomy with regard to mathematical rigor in highway transportation, with relatively simple mathematics used in practice-oriented material and complex mathematics used in research. Thus it is common for instructors to either insult students' mathematical knowledge or vastly exceed it. This book strives for that elusive middle ground of mathematical rigor that matches junior and senior engineering students' mathematical abilities.

## CHAPTER TOPICS AND ORGANIZATION

The fifth edition of *Principles of Highway Engineering and Traffic Analysis* has evolved from nearly three decades of teaching introductory transportation engineering classes at the University of Washington, University of Florida, Purdue University, and Pennsylvania State University, feedback from users of the first four editions, and experiences in teaching civil-engineering licensure exam review courses. The book's material and presentation style (which is characterized by the liberal use of example problems) are largely responsible for transforming much-maligned introductory transportation engineering courses into courses that students consistently rate among the best civil engineering courses.

The book begins with a short introductory chapter that stresses the significance of highway transportation to the social and economic underpinnings of society. This chapter provides students a basic overview of the problems facing the field of highway engineering and traffic analysis. The chapters that follow are arranged in sequences that focus on highway engineering (Chapters 2, 3, and 4) and traffic analysis (Chapters 5, 6, 7, and 8).

Chapter 2 introduces the basic elements of road vehicle performance. This chapter represents a major departure from the vehicle performance material presented in all other transportation and highway engineering books, in that it is far more involved and detailed. The additional level of detail is justified on two grounds. First, because students own and drive automobiles, they have a basic interest that can be linked to their freshman and sophomore coursework in physics, statics, and dynamics. Traditionally, the absence of such a link has been a common criticism of introductory transportation and highway engineering courses. Second, it is important that engineering students understand the principles involved in vehicle technologies and the effect that continuing advances in vehicle technologies will have on engineering practice.

Chapter 3 presents current design practices for the geometric alignment of highways. This chapter provides details on vertical-curve design and the basic elements of horizontal-curve design. This edition of the book has been updated to the latest design guidelines (*Policy on Geometric Design of Highways and Streets*, American Association of State Highway and Transportation Officials, Washington, DC, 2011).

Chapter 4 provides a detailed overview of traditional pavement design, covering both flexible and rigid pavements in a thorough and consistent manner. The material in this chapter also links well with the geotechnical and materials courses that are likely to be part of the student's curriculum. This edition of the book is significantly revised with new examples and new sections on pavement distresses and mechanistic-empirical approaches to pavement design.

Chapter 5 presents the fundamentals of traffic flow and queuing theory, which provide the basic tools of traffic analysis. Relationships and models of basic traffic-stream parameters are introduced, as well as queuing analysis models for deterministic and stochastic processes. Considerable effort was expended to make the material in this chapter accessible to junior and senior engineering students.

Chapter 6 presents some of the current methods used to assess highway levels of performance. Fundamentals and concepts are discussed along with the complexities involved in measuring and/or calculating highway level of service. This edition of the book has been updated to the latest analysis standards (*Highway Capacity Manual* 2010, Transportation Research Board, National Academy of Sciences, Washington, DC).

Chapter 7 introduces the basic elements of traffic control at a signalized intersection and applies the traffic analysis tools introduced in Chapter 5 to signalized intersections. The chapter focuses on pretimed, isolated signals, but also introduces the reader to the fundamentals of actuated and coordinated signal systems. Both theoretical and practical elements associated with traffic signal timing are presented. This edition of the book has been updated to the latest analysis standards (*Highway Capacity Manual* 2010, Transportation Research Board, National Academy of Sciences, Washington, DC).

Chapter 8, the final chapter, provides an overview of travel demand and traffic forecasting. This chapter concentrates on a theoretically and mathematically consistent approach to travel demand and traffic forecasting that closely follows the approach most commonly used in practice and contains a section on the traditional four-step travel-demand forecasting process. This chapter provides the student with an important understanding of the current state of travel demand and traffic forecasting, and some critical insight into the deficiencies of forecasting methods currently used.

## NEW AND REVISED PROBLEMS

This edition includes many new examples and end-of-chapter problems. Additionally, many end-of-chapter problems carried over from the previous edition have been revised. Users of the book will find the end-of-chapter problems to be extremely useful in supporting the material presented in the book. These problems are precise and challenging, a combination rarely found in transportation/highway engineering books.

## NEW TO THIS EDITION

In this edition we have once again made several enhancements to the content and visual presentation, based on suggestions from instructors. Some new features in this edition of the book include:

**SI units**. The third and fourth editions of the book included both U.S. customary and SI units. For the benefit of our international users, this new SI edition includes only metric units.

**Revised Chapter 3.** Chapter 3 has been revised to include the latest information from the recently published *A Policy on Geometric Design of Highways and Streets*, American Association of State Highway and Transportation Officials, Washington, DC, 2011.

**Significantly revised Chapter 4.** Chapter 4 has been substantially revised to be more tightly focused with new examples and revised sections.

**Mechanistic-empirical approach to pavement design.** In addition to traditional pavement design procedures, Chapter 4 now provides an introduction to the newer mechanistic-empirical pavement design procedures.

**Revised Chapter 6.** Chapter 6 has been revised to include the latest information from the recently published *Highway Capacity Manual* 2010, Transportation Research Board, National Academy of Sciences, Washington, D.C.

**Revised Chapter 7.** Chapter 7 has been revised to include the latest information from the recently published *Highway Capacity Manual* 2010, Transportation Research Board, National Academy of Sciences, Washington, D.C.

## WEBSITE

The website for this book is www.wiley.com/college/mannering and contains the following resources:

**Solutions Manual.** Complete solutions to all the problems in the book.

**Lecture Slides.** Lecture slides developed by the authors, which also include all of the figures and tables from the text.

**In-Class Design Problems.** Design problems developed by the authors for in-class use by students in a cooperative-learning context. The problems support the material presented in the chapters and the end-of-chapter problems.

**Sample Exams.** Sample midterm and final exams are provided to give instructors class-proven ideas relating to successful exam format and problems.

Visit the Instruction Companion Site section of the book website to register for a password to download these resources.

Fred L. Mannering
Scott S. Washburn

# Contents

# Chapter 1

# Introduction to Highway Engineering and Traffic Analysis

## 1.1 INTRODUCTION

Highways have played a key role in the development and sustainability of human civilization from ancient times to the present. Today, in the U.S. and throughout the world, highways continue to dominate the transportation system — providing critical access for the acquisition of natural resources, industrial production, retail marketing and population mobility. The influence of highway transportation on the economic, social and political fabric of nations is far-reaching and, as a consequence, highways have been studied for decades as a cultural, political, and economic phenomenon. While industrial needs and economic forces have clearly played an important part in shaping highway networks, societies' fundamental desire for access to activities and affordable land has generated significant highway demand, which has helped define and shape highway networks.

In the twenty-first century, the role of highways in the transportation system continues to evolve. In most nations, the enormous investment in highway-transportation infrastructure that occurred in the middle of the last century, which included the construction of the U.S. interstate highway system (the largest infrastructure project in human history), has now given way to infrastructure maintenance and rehabilitation, improvements in operational efficiency, various traffic-congestion relief measures, energy conservation, improved safety and environmental mitigation. This shift has forced a new emphasis in highway engineering and traffic analysis — one that requires a new skill set and a deeper understanding of the impact of highway decisions.

## 1.2 HIGHWAYS AND THE ECONOMY

It is difficult to overstate the influence that highway transportation has on the economy of nations. Highway systems have a direct effect on industries that supply vehicles and equipment to support highway transportation and the industries that are involved in highway construction and maintenance. Highway systems are also vital to manufacturing and retail supply chains and distribution systems, and serve as regional and national economic engines.

### 1.2.1 The Highway Economy

In the U.S., more than 15% of average household income is spent on highway vehicle purchases, maintenance, and other vehicle expenditures. As a consequence, the

1

industries providing vehicles and vehicle services for highway transportation have an enormous economic influence. In the U.S. alone, in the light-vehicle market (cars, vans, pickup trucks, and so on), as many as 16 million or more new vehicles can be sold annually (depending on economic conditions), which translates to over 400 billion dollars in sales and more than a million jobs in manufacturing and manufacturing-supplier industries. Add to this the additional employment associated with vehicle maintenance and servicing, and more than five million U.S. jobs can be tied directly to highway vehicles. The influence of the highway economy extends further to the heavy-vehicle sector as well, with more than 1.3 million jobs and billions of dollars expended annually by the trucking, freight movement, and other industries in conducting operations and expanding, replacing, and maintaining their fleets of commercial vehicles in the U.S.

The direct influence that highways have also extends to the construction and maintenance of highways, with over 200 billion dollars in annual expenditures in the U.S. alone. This too has an enormous impact on employment and other aspects of the economy.

### 1.2.2   Supply Chains

The survival of modern economies is predicated on efficient and reliable supply chains. Industries have become increasingly dependent on their supply chains to reduce costs and remain competitive. As an example, most manufacturing industries today rely on just-in-time (JIT) delivery to reduce inventory-related costs, which can be a substantial percentage of total costs in many industries. The idea of JIT delivery is that the materials required for production are supplied just before they are needed. While such a strategy significantly reduces inventory costs, it requires a very high degree of certainty that the required materials will be delivered on time. If not, the entire production process could be adversely affected and costs could rise dramatically.

In retail applications, effective supply chains can significantly reduce consumer costs and ensure that a sufficient quantity of goods is available to satisfy consumer demand. The ability of highways to provide reliable service for JIT inventory control and other supply-chain-related industrial and retail applications has made highways critical to the function of modern economies.

### 1.2.3   Economic Development

It has long been recognized that highway construction and improvements to the highway network can positively influence economic development. Such improvements can increase accessibility and thus attract new industries and spur local economies. To be sure, measuring the economic-development impacts of specific highway projects is not an easy task because such measurements must be made in the context of regional and national economic trends. Still, the effect that highways can have on economic development is yet another example of the far-reaching economic influences of highway transportation.

## 1.3   HIGHWAYS, ENERGY, AND THE ENVIRONMENT

As energy demands and supplies vary, and nations become increasingly concerned about environmental impacts, the role that highway transportation plays has come under close scrutiny. As a primary consumer of fossil fuels and a major contributor to air-borne pollution, highway transportation is an obvious target for energy conservation and environmental impact mitigation efforts.

In the U.S., highway transportation is responsible for roughly 60 percent of all petroleum consumption. This translates into about 12 million barrels of oil a day. In light of the limitations of oil reserves, this is an astonishing rate of consumption. In terms of emission impacts, highway transportation is responsible for roughly 25 percent of U.S. greenhouse gas emissions (including over 30 percent of carbon dioxide emissions). Highway transportation's contribution to other pollutants is also substantial. Highway travel is responsible for about 35 percent of all nitrous oxide emissions (NOx) and 25 percent of volatile organic compound emissions (VOC), both major contributors to the formation of ozone. Highway travel also contributes more than 50 percent of all carbon monoxide (CO) emissions in the U.S. and is a major source of fine particulate matter (2.5 micrometers or smaller, $PM_{2.5}$), which is a known carcinogen.

Given these numbers, the energy and environmental impacts of highway transportation are clearly substantial, and an important consideration in the design and maintenance of highway facilities and the development and implementation of policies affecting highway transportation.

## 1.4   HIGHWAYS AS PART OF THE TRANSPORTATION SYSTEM

It is important to keep in mind that highway transportation is part of a larger transportation system that includes air, rail, water and pipeline transportation. In this system, highways are the dominant mode of most passenger and freight movements. For passenger travel, highways account for about 90 percent of all passenger-kilometers. On the freight side, commercial trucks account for about 37 percent of the freight tonne-kilometers and, because commercial trucks transport higher-valued goods than other modes of transportation (with the exception of air transportation), nearly 80 percent of the dollar value of all goods is transported by commercial trucks.

While highways play a dominant role in both passenger and freight movement, in many applications there are critical interfaces among the various transportation modes. For example, many air, rail, water and pipeline freight movements involve highway transportation at some point for their initial collection and final distribution. Interfaces between modes, such as those at water ports, airports and rail terminals, create interesting transportation problems but, if handled correctly, can greatly improve the efficiency of the overall transportation system.

## 1.5   HIGHWAY TRANSPORTATION AND THE HUMAN ELEMENT

Within the highway transportation system, passenger options include single-occupant private vehicles, multi-occupant private vehicles, and public transportation modes (such as bus). It is critical to develop a basic understanding of the effect that highway-related projects and policies may have on the individual highway modes of travel (single-occupant private vehicles, bus and so on) because the distribution of travel among modes will strongly influence overall highway-system performance. In addition, highway safety and the changing demographics of highway users are important considerations.

## 1.5.1    Passenger Transportation Modes and Traffic Congestion

Of the available urban transportation modes (such as bus, commuter train, subway, private vehicle, and others), private vehicles (and single-occupant private vehicles in particular) offer an unequaled level of mobility. The single-occupant private vehicle has been such a dominant choice that travelers have been willing to pay substantial capital and operating costs, confront high levels of congestion, and struggle with parking-related problems just to have the flexibility in travel departure time and destination choices that is uniquely provided by private vehicles. In the last 50 years, the percentage of trips taken in private vehicles has risen from slightly less than 70 percent to over 90 percent (public transit and other modes make up the balance). Over this same period, the average private-vehicle occupancy has dropped from 1.22 to 1.09 persons per vehicle, reflecting the fact that the single-occupant vehicle has become an increasingly dominant mode of travel.

Traffic congestion that has arisen as a result of extensive private-vehicle use and low-vehicle occupancy presents a perplexing problem. The high cost of new highway construction (including monetary, environmental and social costs) often makes building new highways or adding additional highway capacity an unattractive option. Trying to manage the demand for highways also has its problems. For example, programs aimed at reducing congestion by encouraging travelers to take alternate modes of transportation (bus-fare incentives, increases in private-vehicle parking fees, tolls and traffic-congestion pricing, rail- and bus-transit incentives) or increasing vehicle occupancy (high-occupancy vehicle lanes and employer-based ridesharing programs) can be considered viable options. However, such programs have the adverse effect of directing people toward travel modes that inherently provide lower levels of mobility because no other mode offers the departure-time and destination-choice flexibility provided by private, single-occupant vehicles. Managing traffic congestion is an extremely complex problem with significant economic, social, environmental, and political implications.

## 1.5.2    Highway Safety

The mobility and opportunities that highway infrastructure provides also have a human cost. Although safety has always been a primary consideration in highway design and operation, highways continue to exact a terrible toll in loss of life, injuries, property damage, and reduced productivity as a result of vehicle accidents. Highway safety involves technical and behavioral components and the complexities of the human/machine interface. Because of the high costs of highway accidents, efforts to improve highway safety have been intensified dramatically in recent decades. This has resulted in the implementation of new highway-design guidelines and countermeasures (some technical and some behavioral) aimed at reducing the frequency and severity of highway accidents. Fortunately, efforts to improve highway design (such as more stringent design guidelines, breakaway signs, and so on), vehicle occupant protection (safety belts, padded dashboards, collapsible steering columns, driver- and passenger-side airbags, improved bumper design), as well as advances in vehicle technologies (antilock braking, traction control systems, electronic stability control) and new accident countermeasures (campaigns to reduce drunk driving), have gradually managed to reduce the fatality rate — the number of fatalities per kilometer driven. However, in spite of continuing efforts and unprecedented advancements in vehicle safety technologies, the total number of fatalities per year in the U.S. has remained unacceptably high at more than 30,000 per year.

To understand why U.S. highway fatality numbers have not dramatically decreased or why fatality rates (fatalities per distance driven) have not dropped more than they have as a result of all the safety efforts, a number of possible explanations arise: an increase in the overall level of aggressive driving; increasing levels of disrespect for traffic control devices (red-light and stop-sign running being two of the more notable examples); in-vehicle driving distractions (cell phones, eating, talking to passengers); driver impairments (alcohol, drugs, fatigue); and poor driving skills in the younger and older driving populations. Two other phenomena are being observed that may be contributing to the persistently stable number of fatalities. One is that some people drive more aggressively (speeding, following too closely, frequent lane changing) in vehicles with advanced safety features, thus offsetting some or all of the benefits of new safety technologies. Another possibility is that many people are more influenced by style and function than safety features when making vehicle purchase decisions. This is evidenced by the popularity of vehicles such as sport utility vehicles, mini-vans, and pickup trucks, despite their consistently overall lower rankings in certain safety categories, such as roll-over probability, relative to traditional passenger cars. These issues underscore the overall complexity of the highway safety problem and the trade-offs that must be made with regard to cost, safety, and mobility (speed).

### 1.5.3    Demographic Trends

Travelers' commuting patterns (which lead to traffic congestion) are inextricably intertwined with such socioeconomic characteristics as age, income, household size, education, and job type, as well as the distribution of residential, commercial, and industrial developments within the region. Many American metropolitan areas have experienced population declines in central cities accompanied by a growth in suburban areas. One could argue that the population shift from the central cities to the suburbs has been made possible by the increased mobility provided by the major highway projects undertaken during the 1960s and 1970s. This mobility enabled people to improve their quality of life by gaining access to affordable housing and land, while still being able to get to jobs in the central city with acceptable travel times. Conventional wisdom suggested that as overall metropolitan traffic congestion grew (making the suburb-to-city commuting pattern much less attractive), commuters would seek to avoid traffic congestion by reverting to public transport modes and/or once again choosing to reside in the central city. However, a different trend has emerged. Employment centers have developed in the suburbs and now provide a viable alternative to the suburb-to-city commute (the suburb-to-suburb commute). The result is a continuing tendency toward low-density, private-vehicle based development as people seek to retain the high quality of life associated with such development.

Ongoing demographic trends also present engineers with an ever-moving target that further complicates the problem of providing mobility and safety. An example is the rising average age of the U.S. population that has resulted from population cohorts (the baby boom following the Second World War) and advances in medical technology that prolong life. Because older people tend to have slower reaction times, taking longer to respond to driving situations that require action, engineers must confront the possibility of changing highway-design guidelines and practices to accommodate slower reaction times and the potentially higher variance of reaction times among highway users.

## 1.6   HIGHWAYS AND EVOLVING TECHNOLOGIES

As in all fields, technological advances at least offer the promise of solving complex problems. For highways, technologies can be classified into those impacting infrastructure, vehicles, and traffic control.

### 1.6.1   Infrastructure Technologies

Investments in highway infrastructure have been made continuously throughout the 20th and 21st centuries. Such investments have understandably varied over the years in response to need, and political and national priorities. For example, in the U.S., an extraordinary capital investment in highways during the 1960s and 1970s was undertaken by constructing the interstate highway system and upgrading and constructing many other highways. The economic and political climate that permitted such an ambitious construction program has not been replicated before or since. It is difficult to imagine, in today's economic and political environment, that a project of the magnitude of the interstate highway system would ever be seriously considered. This is because of the prohibitive costs associated with land acquisition and construction and the community and environmental impacts that would result.

It is also important to realize that highways are long-lasting investments that require maintenance and rehabilitation at regular intervals. The legacy of a major capital investment in highway infrastructure is the proportionate maintenance and rehabilitation schedule that will follow. Although there are sometimes compelling reasons to defer maintenance and rehabilitation (including the associated construction costs and the impact of the reconstruction on traffic), such deferral can result in unacceptable losses in mobility and safety as well as more costly rehabilitation later.

As a consequence of past capital investments in highway infrastructure and the current high cost of highway construction and rehabilitation, there is a strong emphasis on developing and applying new technologies to more economically construct and extend the life of new facilities and to effectively combat an aging highway infrastructure. Included in this effort are the extensive development and application of new sensing technologies in the emerging field of structural health monitoring. There are also opportunities to extend the life expectancy of new infrastructure with the ongoing nanotechnology advances in material science. Such technological advances are essential elements in the future of highway infrastructure.

### 1.6.2   Vehicle Technologies

Until the 1970s, vehicle technologies evolved slowly and often in response to mild trends in the vehicle market as opposed to an underlying trend toward technological development. Beginning in the 1970s, however, three factors began a cycle of unparalleled advances in vehicle technology that continues to this day: (1) government regulations on air quality, fuel efficiency, and vehicle-occupant safety, (2) energy shortages and fuel-price increases, and (3) intense competition among vehicle manufacturers (foreign and domestic). The aggregate effect of these factors has been vehicle consumers that demand new technology at highly competitive prices. Vehicle manufacturers have found it necessary to reallocate resources and to restructure manufacturing and inventory control processes to meet this demand. In recent years, consumer demand and competition among vehicle manufacturers has resulted in the widespread implementation of new technologies including supplemental restraint systems, anti-lock brake systems, traction control systems, electronic stability control, and a host of other applications of new technologies to improve the safety and comfort in highway vehicles. There is little doubt that the combination of consumer demand and intense competition in the vehicle industry will continue to spur vehicle technological innovations.

Evolving vehicle technologies play a critical role in the highway system. Such technologies directly influence highway design and traffic operations, and are critical considerations in providing high levels of mobility and safety. It is essential that highway engineers understand how vehicle design and technology are interrelated with highway design and operation.

### 1.6.3    Traffic Control Technologies

Intersection traffic signals are a familiar traffic-control technology. At signalized intersections, the trade-off between mobility and safety is brought into sharp focus. Procedures for developing traffic signal control plans (allocating green time to conflicting traffic movements) have made significant advances over the years. Today, signals at critical intersections can be designed to respond quickly to prevailing traffic flows, groups of signals can be coordinated to provide a smooth through-flow of traffic, and, in some cases, computers control entire networks of signals.

In addition to traffic signal controls, numerous safety, navigational, and congestion-mitigation technologies are now reaching the market under the broad heading of Intelligent Transportation Systems (ITS). Such technological efforts offer the potential to significantly reduce traffic congestion and improve safety on highways by providing an unprecedented level of traffic control. There are, however, many obstacles associated with ITS implementation, including system reliability, human response and the human/machine interface. Numerous traffic-control technologies offer the potential for considerable improvement in the efficient use of the highway infrastructure, but one must also recognize the limitations associated with these technologies.

### 1.7    SCOPE OF STUDY

Highway engineering and traffic analysis involve an extremely complex interaction of economic, behavioral, social, political, environmental, and technological factors. This complexity makes highway engineering and traffic analysis far more challenging than typical engineering disciplines that tend to have an overriding focus on only the technical aspects of the problem. To be sure, the technical challenges encountered in highway engineering and traffic analysis easily rival the most complex technical problems encountered in any other engineering discipline. However, it is the economic, behavioral, social, political, and environmental elements that introduce a level of complexity unequalled by any other engineering discipline.

The remaining chapters in this book do not intend to provide a comprehensive assessment of the many factors that influence highway engineering and traffic analysis. Instead, Chapters 2 through 8 seek to provide readers with the fundamental elements and methodological approaches that are used to design and maintain highways and assess their operating performance. This material constitutes the fundamental principles of highway engineering and traffic analysis that are needed to begin to grasp the many complex elements and considerations that come into play during the construction, maintenance, and operation of highways.

# Chapter 2

# Road Vehicle Performance

## 2.1 INTRODUCTION

The performance of road vehicles forms the basis for highway design guidelines and traffic analysis. For example, in highway design, the determination of the length of freeway acceleration and deceleration lanes, maximum highway grades, stopping sight distances, passing sight distances, and numerous accident-prevention devices all rely on a basic understanding of vehicle performance. Similarly, vehicle performance is a major consideration in the selection and design of traffic control devices, the determination of speed limits, and the timing and control of traffic signal systems.

Studying vehicle performance serves two important functions. First, it provides insight into highway design and traffic operations and the compromises that are necessary to accommodate the wide variety of vehicles (from high-powered sports cars to heavily laden trucks) that use highways. Second, it forms a basis from which the impact of advancing vehicle technologies on existing highway design guidelines can be assessed. This second function is particularly important in light of the ongoing unprecedented advances in vehicle technology. Such advances will necessitate more frequent updating of highway design guidelines as well as engineers who have a better understanding of the fundamental principles underlying vehicle performance.

The objective of this chapter is to introduce the basic principles of road vehicle performance. Primary attention will be given to the straight-line performance of vehicles (acceleration, deceleration, top speed, and the ability to ascend grades). Cornering performance of vehicles is overviewed in Chapter 3, but detailed presentations of this material are better suited to more specialized sources [Campbell 1978; Brewer and Rice 1983; Wong 2008].

## 2.2 TRACTIVE EFFORT AND RESISTANCE

Tractive effort (also referred to as thrust) and resistance are the two primary opposing forces that determine the straight-line performance of road vehicles. Tractive effort is simply the force available, at the roadway surface, to perform work and is expressed in newtons (N). Resistance, also expressed in N, is defined as the force impeding vehicle motion. The three major sources of vehicle resistance are (1) aerodynamic resistance, (2) rolling resistance (which originates from the roadway surface–tire interface), and (3) grade or gravitational resistance. To illustrate these forces, consider the vehicle force diagram shown in Fig. 2.1.

**Figure 2.1** Forces acting on a road vehicle.

$R_a$   =   aerodynamic resistance in N,

$R_{rlf}$   =   rolling resistance of the front tires in N,

$R_{rlr}$   =   rolling resistance of the rear tires in N,

$F_f$   =   available tractive effort of the front tires in N,

$F_r$   =   available tractive effort of the rear tires in N,

$W$   =   total vehicle weight in N,

$\theta_g$   =   angle of the grade in degrees,

$m$   =   vehicle mass in kg, and

$a$   =   acceleration in m/s$^2$.

Summing the forces along the vehicle's longitudinal axis provides the basic equation of vehicle motion:

$$F_f + F_r = ma + R_a + R_{rlf} + R_{rlr} + R_g \qquad (2.1)$$

where $R_g$ is the grade resistance and is equal to $W \sin \theta_g$. For exposition purposes it is convenient to let $F$ be the sum of available tractive effort delivered by the front and rear tires ($F_f + F_r$) and similarly to let $R_{rl}$ be the sum of rolling resistance ($R_{rlf} + R_{rlr}$). This notation allows Eq. 2.1 to be written as

$$F = ma + R_a + R_{rl} + R_g \qquad (2.2)$$

Sections 2.3 to 2.8 present a thorough discussion of the components and implications of Eq. 2.2.

## 2.3   AERODYNAMIC RESISTANCE

Aerodynamic resistance is a resistive force that can have significant impacts on vehicle performance. At high speeds, where this component of resistance can become overwhelming, proper vehicle aerodynamic design is essential. Attention to aerodynamic efficiency in design has long been the rule in racing and sports cars. More recently, concerns over fuel efficiency and overall vehicle performance have resulted in more efficient aerodynamic designs in common passenger cars, although not necessarily in pickup trucks or sport utility vehicles (SUVs).

Aerodynamic resistance originates from a number of sources. The primary source (typically accounting for over 85% of total aerodynamic resistance) is the turbulent flow of air around the vehicle body. This turbulence is a function of the shape of the vehicle, particularly the rear portion, which has been shown to be a major source of air turbulence. To a much lesser extent (on the order of 12% of total aerodynamic resistance), the friction of the air passing over the body of the vehicle contributes to resistance. Finally, approximately 3% of the total aerodynamic resistance can be attributed to air flow through vehicle components such as radiators and air vents.

Based on these sources, the equation for determining aerodynamic resistance is

$$R_a = \frac{\rho}{2} C_D A_f V^2 \qquad (2.3)$$

*where*

$R_a$  =  aerodynamic resistance in N,

$\rho$  =  air density in kg/m$^3$,

$C_D$  =  coefficient of drag (unitless),

$A_f$  =  frontal area of the vehicle (projected area of the vehicle in the direction of travel) in m$^2$, and

$V$  =  speed of the vehicle in m/s.

To be truly accurate, for aerodynamic resistance computations, $V$ is actually the speed of the vehicle relative to the prevailing wind speed. To simplify the exposition of concepts soon to be presented, the wind speed is assumed to be equal to zero for all problems and derivations in this book.

Air density is a function of both elevation and temperature, as indicated in Table 2.1. Equation 2.3 indicates that as the air becomes more dense, total aerodynamic resistance increases. The drag coefficient ($C_D$) is a term that implicitly accounts for all three of the aerodynamic resistance sources discussed above. The drag coefficient is measured from empirical data either from wind tunnel experiments or actual field tests in which the vehicle is allowed to decelerate from a known speed with other sources of resistance (rolling and grade) accounted for. Table 2.2 gives some approximation of the range of drag coefficients for different types of road vehicles. Table 2.3 presents drag coefficients for specific automobiles covering the last 40 years.

The general trend toward lower drag coefficients over this period of time reflects the continuing efforts of the automotive industry to improve overall vehicle efficiency by minimizing resistance forces. Table 2.3 also includes some larger personal vehicles, such as pickup trucks and sport utility vehicles, which generally represent the upper range of drag coefficients for automobiles. Also, automobile operating conditions can have a significant effect on drag coefficients. For example, even a small operational change, such as the opening of windows, can increase drag coefficients by 5% or more. More significant operational changes, such as having the top down on a convertible automobile, can increase the drag coefficient by more than 25%. Finally, projected frontal area (approximated as the height of the vehicle multiplied by its width) typically ranges from 1.0 m$^2$ to 2.5 m$^2$ for passenger cars and is also a major factor in determining aerodynamic resistance.

**Table 2.1**  Typical Values of Air Density Under Specified Atmospheric Conditions

| Altitude (m) | Temperature (°C) | Pressure (kPa) | Air density (kg/m$^3$) |
|---|---|---|---|
| 0 | 15.0 | 101.4 | 1.2256 |
| 1500 | 9.7 | 84.4 | 1.0567 |
| 3000 | −4.5 | 70.1 | 0.9096 |

**Table 2.2**  Ranges of Drag Coefficients for Typical Road Vehicles

| Vehicle type | Drag coefficient ($C_D$) |
|---|---|
| Automobile | 0.25–0.55 |
| Bus | 0.5–0.7 |
| Tractor-Trailer | 0.6–1.3 |
| Motorcycle | 0.27–1.8 |

**Table 2.3**  Drag Coefficients of Selected Automobiles

| Vehicle | Drag coefficient ($C_D$) |
|---|---|
| 1967 Chevrolet Corvette | 0.50 |
| 1967 Volkswagen Beetle | 0.46 |
| 1977 Triumph TR7 | 0.40 |
| 1977 Jaguar XJS | 0.36 |
| 1987 Acura Integra | 0.34 |
| 1987 Ford Taurus | 0.32 |
| 1993 Acura Integra | 0.32 |
| 1993 Ford Probe GT | 0.31 |
| 1993 Ford Ranger (truck) | 0.45 |
| 1996 Dodge Viper RT/10 | 0.45 |
| 1997 Lexus LS400 | 0.29 |
| 1997 Infiniti Q45 | 0.29 |
| 2000 Honda Insight (hybrid) | 0.25 |
| 2002 Acura NSX | 0.30 |
| 2002 Lexus LS430 | 0.25 |
| 2003 Dodge Caravan (minivan) | 0.35 |
| 2003 Ford Explorer (SUV) | 0.41 |
| 2003 Dodge Ram (truck) | 0.53 |
| 2003 Hummer H2 | 0.57 |
| 2005 Honda Insight | 0.25 |
| 2011 Mercedes Benz C-class | 0.27 |
| 2011 Honda Element | 0.57 |
| 2011 Mercedes Benz SLS AMG | 0.36 |

Because aerodynamic resistance is proportional to the square of the vehicle's speed, it is clear that such resistance will increase rapidly at higher speeds. The magnitude of this increase can be underscored by considering an expression for the power ($\mathrm{hp}_{R_a}$) required to overcome aerodynamic resistance. With power being the product of force and speed, the multiplication of Eq. 2.3 by speed gives

$$P_{R_a} = \frac{\rho}{2} C_D A_f V^3 \qquad\qquad (2.4)$$

*where*

$P_{R_a}$   = power required to overcome aerodynamic resistance in N-m/s (watts),

Other terms are as defined previously.

Thus, the amount of power required to overcome aerodynamic resistance increases with the cube of speed, indicating, for example, that eight times as much power is required to overcome aerodynamic resistance if the vehicle speed is doubled.

## 2.4   ROLLING RESISTANCE

Rolling resistance refers to the resistance generated from a vehicle's internal mechanical friction and from pneumatic tires and their interaction with the roadway surface. The primary source of this resistance is the deformation of the tire as it passes over the roadway surface. The force needed to overcome this deformation accounts for approximately 90% of the total rolling resistance. Depending on the vehicle's weight and the material composition of the roadway surface, the penetration of the tire into the surface and the corresponding surface compression can also be a significant source of rolling resistance. However, for typical vehicle weights and pavement types, penetration and compression constitute only around 4% of the total rolling resistance. Finally, frictional motion due to the slippage of the tire on the roadway surface and, to a lesser extent, air circulation around the tire and wheel (the fanning effect) are sources accounting for roughly 6% of the total rolling resistance [Taborek 1957].

In considering the sources of rolling resistance, three factors are worthy of note. First, the rigidity of the tire and the roadway surface influence the degree of tire penetration, surface compression, and tire deformation. Hard, smooth, and dry roadway surfaces provide the lowest rolling resistance. Second, tire conditions, including inflation pressure and temperature, can have a substantial impact on rolling resistance. High tire inflation decreases rolling resistance on hard paved surfaces as a result of reduced friction but increases rolling resistance on soft unpaved surfaces due to additional surface penetration. Also, higher tire temperatures make the tire body more flexible, and thus less resistance is encountered during tire deformation. The third and final factor is the vehicle's operating speed, which affects tire deformation. Increasing speed results in additional tire flexing and vibration and thus a higher rolling resistance.

Due to the wide range of factors that determine rolling resistance, a simplifying approximation is used. Studies have shown that overall rolling resistance can be approximated as the product of a friction term (coefficient of rolling resistance) and the weight of the vehicle acting normal to the roadway surface. The coefficient of rolling resistance for road vehicles operating on paved surfaces is approximated as

$$f_{rl} = 0.01\left(1 + \frac{V}{44.73}\right) \tag{2.5}$$

*where*

$f_{rl}$ = coefficient of rolling resistance (unitless), and
$V$ = vehicle speed in m/s.

By inspection of Fig. 2.1, the rolling resistance, in newtons (N), will simply be the coefficient of rolling resistance multiplied by $W \cos \theta_g$, the vehicle weight acting normal to the roadway surface. For most highway applications $\theta_g$ is quite small, so it can be assumed that $\cos \theta_g = 1$, giving the equation for rolling resistance ($R_{rl}$) as

$$R_{rl} = f_{rl}\, W \tag{2.6}$$

From this, the amount of power required to overcome rolling resistance is

$$P_{R_{rl}} = f_{rl} W V \tag{2.7}$$

*where*

$\mathrm{hp}_{R_{rl}}$ = horsepower required to overcome rolling resistance in N-m/s (watts),

$W$ = total vehicle weight in N.

## EXAMPLE 2.1   AERODYNAMIC AND ROLLING RESISTANCE

An 11.0-kN car is driven at sea level ($\rho = 1.2256$ kg/m³) on a level paved surface. The car has $C_D = 0.38$ and 2.0 m² of frontal area. It is known that at maximum speed, 38 kW is being expended to overcome rolling and aerodynamic resistance. Determine the car's maximum speed.

**SOLUTION**

It is known that at maximum speed ($V_m$),

$$\text{available power} = R_a V_m + R_{rl} V_m$$

or

$$\text{available power} = \frac{\rho}{2} C_D A_f V_m^3 + f_{rl} W V_m$$

Substituting, we have

$$38000 = \frac{1.2256}{2}(0.38)(2.0)V_m^3 + 0.01\left(1 + \frac{V_m}{44.73}\right)(11000)V_m$$

or

$$38000 = 0.4657 V_m^3 + 2.459 V_m^2 + 110 V_m$$

Solving for $V_m$ gives

$$V_m = \underline{39.95 \text{ m/s}} \quad \text{or} \quad \underline{143.82 \text{ km/h}}$$

## 2.5 GRADE RESISTANCE

Grade resistance is simply the gravitational force (the component parallel to the roadway) acting on the vehicle. As suggested in Fig. 2.1, the expression for grade resistance ($R_g$) is

$$R_g = W \sin \theta_g \tag{2.8}$$

As in the development of the rolling resistance formula (Eq. 2.6), highway grades are usually very small, so $\sin \theta_g \cong \tan \theta_g$. Rewriting Eq. 2.8, we get

$$R_g \cong W \tan \theta_g = WG \tag{2.9}$$

*where*

$G$ = grade, defined as the vertical rise per some specified horizontal distance (opposite side of the force triangle, Fig. 2.1, divided by the adjacent side) in m/m.

Grades are generally specified as percentages for ease of understanding. Thus a roadway that rises 5 m vertically per 100 m horizontally ($G = 0.05$ and $\theta_g = 2.86°$) is said to have a 5% grade.

## EXAMPLE 2.2  GRADE RESISTANCE

An 8.9-kN car has $C_D = 0.40$, $A_f = 2.0 \text{ m}^2$, and an available tractive effort of 1135 N. If the car is traveling at an elevation of 1500 m ($\rho = 1.0567 \text{ kg/m}^3$) on a paved surface at a speed of 110 km/h, what is the maximum grade that this car could ascend and maintain the 110-km/h speed?

*SOLUTION*

To maintain the speed, the available tractive effort will be exactly equal to the summation of resistances. Thus no tractive effort will remain for vehicle acceleration ($ma = 0$). Therefore, Eq. 2.2 can be written as

$$F = R_a + R_{rl} + R_g$$

For grade resistance (using Eq. 2.9),

$$R_g = WG = 2000G$$

for aerodynamic resistance (using Eq. 2.3),

$$
\begin{aligned}
R_a &= \frac{\rho}{2} C_D A_f V^2 \\
&= \frac{1.0567}{2}(0.4)(2)(110 \times 1000 / 3600)^2 \\
&= 394.63 \text{ N}
\end{aligned}
$$

and for rolling resistance (using Eq. 2.6),

$$
\begin{aligned}
R_r &= f_{rl} W \\
&= 0.01\left(1 + \frac{110 \times 1000 / 3600}{44.73}\right) \times 8900 \\
&= 149.8 \text{ N}
\end{aligned}
$$

Therefore,

$$F = 1135 = 394.63 + 149.8 + 8900G$$
$$G = \underline{0.0664} \quad \text{or a} \quad \underline{6.64} \text{ \% grade}$$

## 2.6 AVAILABLE TRACTIVE EFFORT

With the resistance terms in the basic equation of vehicle motion (Eq. 2.2) discussed, attention can now be directed toward available tractive effort ($F$) as used in Example 2.2. The tractive effort available to overcome resistance and/or to accelerate the vehicle is determined either by the force generated by the vehicle's engine or by some maximum value that will be a function of the vehicle's weight distribution and the characteristics of the roadway surface–tire interface. The basic concepts underlying these two determinants of available tractive effort are presented here.

### 2.6.1 Maximum Tractive Effort

No matter how much force a vehicle's engine makes available at the roadway surface, there is a point beyond which additional force merely results in the spinning of tires and does not overcome resistance or accelerate the vehicle. To explain what determines this point of maximum tractive effort (the limiting value beyond which tire spinning begins), a force and moment-generating diagram is provided in Fig. 2.2.

**Figure 2.2** Vehicle forces and moment-generating distances.

$R_a$ = aerodynamic resistance in N,

$R_{rlf}$ = rolling resistance of the front tires in N,

$R_{rlr}$ = rolling resistance of the rear tires in N,

$F_f$ = available tractive effort of the front tires in N,

$F_r$ = available tractive effort of the rear tires in N,

$W$ = total vehicle weight in N,

$W_f$ = weight of the vehicle on the front axle in N,

$W_r$ = weight of the vehicle on the rear axle in N,

$\theta_g$ = angle of the grade in degrees,

$m$ = vehicle mass in kg,

$a$ = acceleration in m/s$^2$,

$L$ = length of wheelbase,

$h$ = height of the center of gravity above the roadway surface,

$l_f$ = distance from the front axle to the center of gravity, and

$l_r$ = distance from the rear axle to the center of gravity.

To determine the maximum tractive effort that the roadway surface–tire contact can support, it is necessary to examine the normal loads on the axles. The normal load on the rear axle ($W_r$) is given by summing the moments about point $A$ (see Fig. 2.2):

$$W_r = \frac{R_a h + W l_f \cos\theta_g + mah \pm Wh\sin\theta_g}{L} \tag{2.10}$$

In this equation the grade moment ($Wh\sin\theta_g$) is positive for an upward slope and negative for a downward slope. Rearranging terms (assuming $\cos\theta_g = 1$ for the small grades encountered in highway applications) and substituting into Eq. 2.2 gives

$$W_r = \frac{l_f}{L}W + \frac{h}{L}\left(F - R_{rl}\right) \tag{2.11}$$

From basic physics, the maximum tractive effort as determined by the roadway surface–tire interaction will be the normal force multiplied by the coefficient of road adhesion ($\mu$), so for a rear-wheel–drive car

$$F_{max} = \mu W_r \tag{2.12}$$

and substituting Eq. 2.11 into Eq. 2.12,

$$F_{max} = \mu \left[ \frac{l_f}{L} W + \frac{h}{L} \left( F_{max} - R_{rl} \right) \right] \qquad (2.13)$$

$$F_{max} = \frac{\mu W \left( l_f - f_{rl} h \right)/L}{1 - \mu h/L} \qquad (2.14)$$

Similarly, by summing moments about point $B$ (see Fig. 2.2), it can be shown that for a front-wheel–drive vehicle

$$F_{max} = \frac{\mu W \left( l_r + f_{rl} h \right)/L}{1 + \mu h/L} \qquad (2.15)$$

Note that in Eqs. 2.14 and 2.15, because of canceling of units, $h$, $l_f$, $l_r$, and $L$ can be in any unit of length (m, mm, etc.). However, all of these terms must be in the same chosen unit of measure. The units of $F_{max}$ will be the same as the units for $W$ (lb).

---

**EXAMPLE 2.3**   **MAXIMUM TRACTIVE EFFORT**

An 11.0-kN car is designed with a 3.05-m wheelbase. The center of gravity is located 550 mm above the pavement and 1.00 m behind the front axle. If the coefficient of road adhesion is 0.6, what is the maximum tractive effort that can be developed if the car is (a) front-wheel drive and (b) rear-wheel drive?

**SOLUTION**

For the front-wheel–drive case Eq. 2.15 is used:

$$F_{max} = \frac{\mu W \left( l_r + f_{rl} h \right)/L}{1 + \mu h/L}$$

and, from Eq. 2.5, $f_{rl} = 0.01$ because $V = 0$ m/s, so

$$F_{max} = \frac{\left[ 0.6 \times 11000 \times \left( 2.050 + 0.01(0.550) \right) \right]/3.050}{1 + \left( 0.6 \times 0.550 \right)/3.050}$$

$$= \underline{4013.70 \text{ N}}$$

For the rear-wheel–drive case, Eq. 2.14 is used:

$$F_{max} = \frac{\left[ 0.6 \times 11000 \times \left( 1.000 - 0.01(0.550) \right) \right]/3.050}{1 - \left( 0.6 \times 0.550 \right)/3.050}$$

$$= \underline{2413.13 \text{ N}}$$

---

### 2.6.2   Engine-Generated Tractive Effort

The amount of tractive effort generated by the vehicle's engine is a function of a variety of drivetrain design factors. For engine design, critical factors in determining output include the shape of the combustion chamber, the quantity of air drawn into the combustion chamber during the induction phase, the type of fuel used, and fuel intake design. Although a complete description of engine design is beyond the scope of this book, an understanding of how engine output is measured and used is important in the study of vehicle performance. The two most commonly used measures of engine output are torque and power. Torque is the work generated by the engine (the twisting moment) and is expressed in newton-meters (n-m). Power is the rate of engine work, expressed in kilowatts (kW), and is related to the engine's torque by the following equation:

$$P_e = \frac{2\pi M_e n_e}{1000} \tag{2.16}$$

*where*

$P_e$ = engine-generated power in kW,
$M_e$ = engine torque in N-M, and
$n_e$ = engine speed in crankshaft revolutions per second.

Figure 2.3 presents a torque-power diagram for a typical gasoline-powered engine.

**Figure 2.3**  Typical torque-power curve for a gasoline-powered automobile engine.

**EXAMPLE 2.4   ENGINE TORQUE AND POWER**

It is known that an experimental engine has a torque curve of the form $M_e = an_e - bn_e^2$, where $M_e$ is engine torque in N-m, $n_e$ is engine speed in revolutions per second, and $a$ and $b$ are unknown parameters. If the engine develops a maximum torque of 125 N-m at 3200 rev/min (revolutions per minute), what is the engine's maximum power?

*SOLUTION*

At maximum torque, $n_e$ = 53.33 rev/s (3200/60) and

$$\frac{dM_e}{dn_e} = 0 = a - 2bn_e$$

$$a = 2(53.33)b = 106.67b$$

Also, at maximum torque,

$$M_e = an_e - bn_e^2$$
$$125 = a(53.33) - b(53.33)^2$$

Using these two equations to solve for the two unknowns ($a$ and $b$), we find that $b = 0.044$ and $a = 4.692$. Using Eq. 2.16 and $M_e = an_e - bn_e^2$,

$$P_e = \frac{2\pi\left(an_e - bn_e^2\right)n_e}{1000}$$
$$= \frac{2\pi\left(4.692n_e - 0.044n_e^2\right)n_e}{1000}$$

The first derivative of the power equation is used to solve for the engine speed at maximum power:

$$\frac{dP_e}{dn_e} = 0 = (0.00628)\left(9.384n_e - 0.132n_e^2\right)$$
$$n_e = 71.09 \text{ rev/s}$$

so the engine's maximum power is

$$P_e = \frac{2\pi\left(4.692n_e - 0.044n_e^2\right)n_e}{1000}$$
$$= \frac{2\pi\left[4.692(71.09) - 0.044(71.09)^2\right]71.09}{1000}$$
$$= \underline{\underline{49.64 \text{ kW}}}$$

Given the output measures of a vehicle's engine, focus can be directed toward the relationship between engine-generated torque and the tractive effort ultimately delivered to the driving wheels. Unfortunately, the tractive effort needed for acceptable vehicle performance (to provide adequate acceleration characteristics) is greater at lower vehicle speeds, and because maximum engine torque is developed at fairly high engine speeds (crankshaft revolutions), the use of gasoline-powered engines requires some form of gear reduction, as illustrated in Fig. 2.4. This gear reduction provides the mechanical advantage necessary for acceptable vehicle acceleration.

**Figure 2.4** Tractive effort requirements and tractive effort generated by a typical gasoline-powered vehicle.

With gear reductions there are two factors that determine the amount of tractive effort reaching the drive wheels. First, the mechanical efficiency of the drivetrain (the engine and gear reduction devices, including the transmission and differential) must be considered. Typically, 5% to 25% of the tractive effort generated by the engine is lost in gear reduction devices, which corresponds to a mechanical efficiency of the drivetrain ($\eta_d$) of 0.75 to 0.95. Second, the overall gear reduction ratio ($\varepsilon_0$), which includes the gear reductions of the transmission and differential, plays a key role in the determination of tractive effort.

By definition, the overall gear reduction ratio refers to the relationship between the revolutions of the engine's crankshaft and the revolutions of the drive wheels. For example, an overall gear reduction ratio of 4 to 1 ($\varepsilon_0 = 4$) means that the engine's crankshaft turns four revolutions for every one revolution of the drive wheels.

With these terms defined, the engine-generated tractive effort reaching the drive wheels is given as

$$F_e = \frac{M_e \varepsilon_0 \eta_d}{r} \tag{2.17}$$

*where*

$F_e$ = engine-generated tractive effort reaching the drive wheels in N,

$M_e$ = engine torque in N-m,

$\varepsilon_0$ = overall gear reduction ratio,

$\eta_d$ = mechanical efficiency of the drivetrain, and

$r$ = radius of the drive wheels in m.

It follows that the relationship between vehicle speed and engine speed is

$$V = \frac{2\pi r n_e (1-i)}{\varepsilon_0}$$

(2.18)

*where*

$V$ = vehicle speed in m/s,

$n_e$ = engine speed in crankshaft revolutions per second,

$i$ = slippage of the drive axle, generally taken as 2 to 5 percent ($i = 0.02$ to $0.05$) for passenger vehicles, and

Other terms are as defined previously.

To summarize this section, the available tractive effort ($F$ in Eq. 2.2) at any given speed is the lesser of the maximum tractive effort ($F_{max}$) and the engine-generated tractive effort ($F_e$).

## 2.7   VEHICLE ACCELERATION

As defined in the previous section, available tractive effort ($F$) can be used to determine a number of vehicle performance characteristics including vehicle acceleration and top speed. For determining vehicle acceleration, Eq. 2.2 can be applied with an additional term to account for the inertia of the vehicle's rotating parts that must be overcome during acceleration. This term is referred to as the mass factor ($\gamma_m$) and is introduced in Eq. 2.2 as

$$F - \sum R = \gamma_m m a$$

(2.19)

where the mass factor is approximated as

$$\gamma_m = 1.04 + 0.0025\varepsilon_0^2$$

(2.20)

Two measures of vehicle acceleration are worthy of note: the time to accelerate and the distance to accelerate. For both, the force available to accelerate is $F_{net} = F - \Sigma R$. The basic relationship between the force available to accelerate, $F_{net}$, the available tractive effort, $F$ (the lesser of $F_{max}$ and $F_e$), and the summation of resistances is illustrated in Fig. 2.5. In this figure, $F_{net}$ is the vertical distance between the lesser of the $F_{max}$ and the $F_e$ curves and the total resistance curve. So, referring to Fig. 2.5, at speed $V'$, $F_{net}$ will be $F_{max} - \Sigma R$, and at $V''$, $F_{net}$ will be $F_e - \Sigma R$. It follows that when $F_{net} = 0$, the vehicle cannot accelerate and is at its maximum speed for specified conditions (grade, air density, engine torque, and so on). Such was the case for the vehicle described in Example 2.2. When $F_{net}$ is greater than zero (the vehicle is traveling at a speed less than its maximum speed), Eq. 2.19 can be written in differential form as

$$F_{net} = \gamma_m m \frac{dV}{dt} \quad \text{or} \quad dt = \frac{\gamma_m m dV}{F_{net}}$$

**Figure 2.5** Relationship among the forces available to accelerate, available tractive effort, and total vehicle resistance.

and because $F_{net}$ is itself a function of vehicle speed [$F_{net} = f(V)$], integration gives the time to accelerate as

$$t = \gamma_m m \int_{V_1}^{V_2} \frac{dV}{f(V)} \tag{2.21}$$

where $V_1$ is the initial vehicle speed and $V_2$ is the final vehicle speed.

Similarly, it can be shown that the distance to accelerate is

$$d_a = \gamma_m m \int_{V_1}^{V_2} \frac{VdV}{f(V)} \tag{2.22}$$

To solve Eqs. 2.21 and 2.22, numerical integration is necessary because the functional forms of these equations do not lend themselves to closed-form solutions. Such numerical integration is straightforward but requires a computer. Consequently, we do not provide an example of solving these equations.

---

**EXAMPLE 2.5　VEHICLE ACCELERATION**

A car is traveling at 16 km/h on a roadway covered with hard-packed snow (coefficient of road adhesion of 0.20). The car has $C_D = 0.30$, $A_f = 2.0$ m², and $W = 13.3$ kN. The wheelbase is 3.05 m and the center of gravity is 500 mm above the roadway surface and 1.27 m behind the front axle. The air density is 1.0567 kg/m³. The car's engine is producing 130 N-m of torque and is in a gear that gives an overall gear reduction ratio of 4.5 to 1, the wheel radius is 360 mm, and the mechanical efficiency of the driveline is 80%. If the driver needs to accelerate quickly to avoid an accident, what would the acceleration be if the car is (a) front-wheel drive and (b) rear-wheel drive?

**SOLUTION**

We begin by computing the resistances, tractive effort generated by the engine, and mass factor because all of these factors will be the same for both front- and rear-wheel drive.

The aerodynamic resistance is (from Eq. 2.3)

$$R_a = \frac{\rho}{2} C_D A_f V^2$$
$$= \frac{1.0567}{2}(0.3)(2)(16 \times 1000 / 3600)^2$$
$$= 6.26 \text{ N}$$

The rolling resistance is (from Eq. 2.6)

$$R_{rl} = f_{rl} W$$
$$= 0.01 \left( 1 + \frac{16 \times 1000 / 3600}{44.73} \right) \times 13,300$$
$$= 146.22 \text{ N}$$

The engine-generated tractive effort is (from Eq. 2.17)

$$F_e = \frac{M_e \varepsilon_0 \eta_d}{r}$$
$$= \frac{130(4.5)(0.8)}{0.36}$$
$$= 1300 \text{ N}$$

The mass factor is (from Eq. 2.20)

$$\gamma_m = 1.04 + 0.0025 \varepsilon_0^2$$
$$= 1.04 + 0.0025 (4.5)^2$$
$$= 1.091$$

Recall that, to determine acceleration, we need the resistances (already computed) and the available tractive effort, $F$, which is the lesser of $F_e$ or $F_{max}$. For the case of the front-wheel-drive car, Eq. 2.15 can be applied to determine $F_{max}$ :

$$F_{max} = \frac{\mu W (l_r + f_{rl} h)/L}{1 + \mu h / L}$$
$$= \frac{\left[ 0.2 \times 13300 \times (1.780 + 0.011(0.500)) \right]/3.050}{1 + (0.2 \times 0.500)/3.050}$$
$$= 1507.76 \text{ N}$$

Thus for a front-wheel-drive car $F$ = 1300 N (the lesser of 1300 and 1507.76) and the acceleration is (from Eq. 2.19)

$$F - \sum R = \gamma_m ma$$

$$a = \frac{F - \sum R}{\gamma_m m} = \frac{1300 - 152.48}{1.091(13300/9.807)} = \underline{\underline{0.776 \text{ m/s}^2}}$$

For the case of the rear-wheel-drive car, Eq. 2.14 can be applied to determine $F_{max}$ :

$$F_{max} = \frac{\left[0.2 \times 13300 \times (1.270 - 0.011(0.500))\right]/3.050}{1 - (0.2 \times 0.500)/3.050} = 1140.19 \text{ N}$$

Thus for a rear-wheel-drive car $F$ = 1140.19 N (the lesser of 1300 and 1140.19) and the acceleration is (from Eq. 2.19)

$$a = \frac{F - \sum R}{\gamma_m m} = \frac{1140.19 - 152.48}{1.091(13300/9.807)} = \underline{\underline{0.668 \text{ m/s}^2}}$$

| **EXAMPLE 2.6** | **ENGINE-TORQUE AND VEHICLE ACCELERATION** |
|---|---|

A front-wheel drive car is in gear with a gear-reduction ratio 8 to 1 and is traveling at 40 km/h.  The engine torque of the car is given by the equation $M_e = 13n_e - 0.08n_e^2$.  The car has a frontal area of 1.8 m$^2$, $C_D$ of 0.30 and is traveling at sea level (15 degrees C) on a level road.  The wheelbase is 3 meters and the center of gravity is 1 meter behind the front axle and 800 mm above the road surface and the car weighs 12.5 kN. If the car is on a road that is wet with poor pavement, what is the maximum acceleration from 40 km/h (driveline efficiency is 90%, slippage of the drive axle is 2%, wheel radius is 380 mm)?

*SOLUTION*

To begin, the engine speed must be computed so that the torque equation provided in the problem statement can be applied to determine the engine-generated tractive effort. This is done by applying Eq. 2.18 with $i = 0.02$, $\varepsilon_0 = 8$, $r = 0.38$ m, and $V = 11.11$ m/s (40 km/h):

$$V = \frac{2\pi r n_e (1 - i)}{\varepsilon_0} = 40 \times 1000/3600 = \frac{2(3.141)(0.38)n_e(1 - 0.02)}{8}$$

$$n_e = \frac{(40 \times 1000/3600)8}{(2)(3.141)(0.38)(1 - 0.02)} = 38.00 \text{ revolutions per second}$$

Substituting this engine speed into the torque equation provided in the problem statement gives

$$M_e = 13n_e - 0.08n_e^2 = 13(38.00) - 0.08(38.00)^2 = 378.48 \text{ N-m}$$

The engine-generated tractive effort is (from Eq. 2.17) with $\eta_d = 0.9$

$$F_e = \frac{M_e \varepsilon_0 \eta_d}{r} = \frac{378.48(8)(0.9)}{0.380} = 7171.20 \text{ N}$$

As in Example 2.5, the available tractive effort for maximum acceleration, $F$, is the lesser of $F_e$ or $F_{max}$. $F_{max}$ for a front-wheel-drive car is computed using Eq. 2.15. With given values of $l_r = 2$ m, $h = 0.8$ m, $L = 3$ m, $W = 12.5$ kN, $\mu = 0.6$ (from Table 2.4 with poor, wet pavement), and $f_{rl}$ from Eq. 2.5, we find

$$f_{rl} = 0.01\left(1 + \frac{V}{44.73}\right) = 0.01\left(1 + \frac{40 \times 1000/3600}{44.73}\right) = 0.0125$$

Application of Eq. 2.15 yields

$$\begin{aligned}
F_{max} &= \frac{\mu W (l_r + f_{rl} h)/L}{1 + \mu h / L} \\
&= \frac{\left[0.6 \times 12500 \times (2 + 0.0125(0.8))\right]\big/3}{1 + (0.6 \times 0.8)\big/3} \\
&= 4331.90 \text{ N}
\end{aligned}$$

Thus $F = 4331.90$ N (the lesser of 4331.90 and 7171.20). With this, acceleration is determined by applying Eq. 2.19. For input into Eq. 2.19, the rolling resistance (from Eq. 2.6) is

$$\begin{aligned}
R_{rl} &= f_{rl} W \\
&= 0.0125 \times 12500 \\
&= 156.25 \text{ N}
\end{aligned}$$

The aerodynamic resistance with $\rho = 1.2256$ kg/m$^3$ (from Table 2.1), $C_D = 0.30$, $A_f = 1.8$ m$^2$ and $V = 40$ km/h (from Eq. 2.3) is

$$\begin{aligned}
R_a &= \frac{\rho}{2} C_D A_f V^2 \\
&= \frac{1.2256}{2}(0.3)(1.8)(40 \times 1000 / 3600)^2 \\
&= 40.85 \text{ N}
\end{aligned}$$

The mass factor (from Eq. 2.20) is

$$\begin{aligned}
\gamma_m &= 1.04 + 0.0025\varepsilon_0^2 \\
&= 1.04 + 0.0025(8)^2 \\
&= 1.2
\end{aligned}$$

Thus, application of Eq. 2.19 gives the maximum acceleration as

$$F - \sum R = \gamma_m ma$$

$$a = \frac{F - \sum R}{\gamma_m m} = \frac{4331.90 - 156.25 - 40.85}{1.2(12500/9.807)} = \underline{\underline{2.70 \text{ m/s}^2}}$$

Typical values for the coefficient of road adhesion ($\mu$) are shown later, in Table 2.4. However, in determining acceleration (and braking and cornering, as will be shown later in this book) two points are worthy of note. First, it is possible for the coefficient of road adhesion to exceed 1.0. This is because a micro-interaction at the tire-pavement interface results in a "cog-type" effect that, for some high-performance tires that use softer compounds, can increase $\mu$ to above 1.0. This explains why many race cars, particularly drag-racing cars, have initial acceleration rates well in excess of 1 $g$ (9.807 m/s$^2$). Second, at high speed, vehicle aerodynamics can create downward forces that effectively increase $W$ in the preceding equations, and this facilitates greater acceleration. Drag-racing cars and some open-wheel race cars (such as Formula One–style cars) are examples of aerodynamic designs that use air deflectors designed to generate significant downward forces to enhance acceleration, braking, and cornering.

## 2.8   FUEL EFFICIENCY

Given the factors discussed in the preceding sections of this chapter, the elements that determine a vehicle's fuel efficiency are clear. One of the most critical determinants relates to engine design (how the engine-generated tractive effort is produced). Engine designs that increase the quantity of air entering the combustion chamber, improve fuel delivery to the combustion chamber, and decrease internal engine friction lead to improved fuel efficiency. Improvements in other mechanical components, such as decreasing slippage and improving the mechanical efficiency of the transmission and driveshaft, also increase the overall fuel efficiency.

In terms of resistance-reducing options, decreasing overall vehicle weight ($W$) will lower grade and rolling resistances, thus reducing fuel consumption (all other factors held constant). Similarly, aerodynamic improvements such as a lower drag coefficient ($C_D$) and a reduced frontal area ($A_f$) can produce significant fuel savings. Finally, improved tire designs with lower rolling resistance can improve overall fuel efficiency.

## 2.9   PRINCIPLES OF BRAKING

In highway design and traffic analysis, the braking characteristics of road vehicles are arguably the single most important aspect of vehicle performance. The braking behavior of road vehicles is critical in the determination of stopping sight distance, roadway surface design, and accident avoidance systems. Moreover, ongoing advances in braking technology make it essential that transportation engineers have a basic comprehension of the underlying principles involved.

### 2.9.1   Braking Forces

To begin the discussion of braking principles, consider the force and moment-generating diagram in Fig. 2.6. During vehicle braking there is a load transfer from the rear to the front axle. To illustrate this, expressions for the normal loads on the front and rear axles can be written by summing the moments about roadway surface–tire contact points $A$ and $B$ (as was done in deriving Eqs. 2.14 and 2.15, with cos $\theta_g$ assumed equal to 1 because of the small grades encountered in highway applications):

$$W_f = \frac{1}{L}\left[Wl_r + h\left(ma - R_a \pm W\sin\theta_g\right)\right] \tag{2.23}$$

and

$$W_r = \frac{1}{L}\left[Wl_f - h\left(ma - R_a \pm W\sin\theta_g\right)\right] \tag{2.24}$$

where, in this case, the contribution of grade resistance ($W \sin \theta_g$) is negative for uphill grades and positive for downhill grades.

From the summation of forces along the vehicle's longitudinal axis,

$$F_b + f_{rl}W = ma - R_a \pm W\sin\theta_g \tag{2.25}$$

with the rolling resistance equal to the coefficient of rolling resistance multiplied by the vehicle weight (from Eq. 2.6, $R_{rl} = f_{rl}W$) and $F_b = F_{bf} + F_{br}$. Substituting Eq. 2.25 into Eqs. 2.23 and 2.24 gives

$$W_f = \frac{1}{L}\left[Wl_r + h\left(F_b + f_{rl}W\right)\right] \tag{2.26}$$

and

$$W_r = \frac{1}{L}\left[Wl_f - h\left(F_b + f_{rl}W\right)\right] \tag{2.27}$$

Because the maximum vehicle braking force ($F_{b\ max}$) is equal to the coefficient of road adhesion ($\mu$), multiplied by the vehicle weights normal to the roadway surface,

$$F_{bf\ max} = \mu W_f$$
$$= \frac{\mu W}{L}\left[l_r + h\left(\mu + f_{rl}\right)\right] \tag{2.28}$$

and

$$F_{br\ max} = \mu W_r$$
$$= \frac{\mu W}{L}\left[l_f - h\left(\mu + f_{rl}\right)\right] \tag{2.29}$$

**Figure 2.6** Forces acting on a vehicle during braking, with drivetrain resistance ignored.

$R_a$ = aerodynamic resistance in N,

$R_{rlf}$ = rolling resistance of the front tires in N,

$R_{rlr}$ = rolling resistance of the rear tires in N,

$F_{bf}$ = braking force on the front tires in N,

$F_{br}$ = braking force on the rear tires in N,

$W$ = total vehicle weight in N,

$W_f$ = weight of the vehicle on the front axle in N,

$W_r$ = weight of the vehicle on the rear axle in N,

$\theta_g$ = angle of the grade in degrees,

$m$ = vehicle mass in kg,

$a$ = acceleration in m/s$^2$,

$L$ = length of wheelbase,

$h$ = height of the center of gravity above the roadway surface,

$l_f$ = distance from the front axle to the center of gravity, and

$l_r$ = distance from the rear axle to the center of gravity.

To develop maximum braking forces, the tires should be at the point of an impending slide. If the tires begin to slide (the brakes lock), a significant reduction in road adhesion results. An indication of the extent of the reduction in road adhesion as the result of tire slide, under various pavement and weather conditions, is presented in Table 2.4. It is clear from this table that the braking forces decline dramatically when the wheels are locked (resulting in tire slide). Avoiding this locked condition is the function of antilock braking systems in cars. Such systems are discussed later in this chapter.

**Table 2.4** Typical Values of Coefficients of Road Adhesion

| Pavement | Coefficient of road adhesion | |
| --- | --- | --- |
| | Maximum | Slide |
| Good, dry | 1.00* | 0.80 |
| Good, wet | 0.90 | 0.60 |
| Poor, dry | 0.80 | 0.55 |
| Poor, wet | 0.60 | 0.30 |
| Packed snow or ice | 0.25 | 0.10 |

*In some instances, the coefficient of road adhesion values can exceed 1.0. See discussion at the end of Section 2.7.

*Source:* S. G. Shadle, L. H. Emery, and H. K. Brewer, "Vehicle Braking, Stability, and Control," *SAE Transactions,* vol. 92, paper 830562, 1983.

## 2.9.2 Braking Force Ratio and Efficiency

On a given roadway surface, the maximum attainable vehicle deceleration (using the vehicle's braking system) is equal to $\mu g$, where $\mu$ is the coefficient of road adhesion and $g$ is the gravitational constant (9.807 m/s$^2$). To approach this maximum vehicle deceleration, vehicle braking systems must correctly distribute braking forces between the vehicle's front and rear brakes. This is typically done by allocation of hydraulic pressures within the braking system. This front-rear proportioning of braking forces (within the vehicle's braking system) will be optimal (achieving a deceleration rate equal to $\mu g$) when it is in exactly the same proportion as the ratio of the maximum braking forces on the front and rear axles ($F_{bf\ max} / F_{br\ max}$). Thus maximum braking forces (with the tires at the point of impending slide) will be developed when the brake force ratio (front force over rear force) is

$$BFR_{f/r\ max} = \frac{l_r + h(\mu + f_{rl})}{l_f - h(\mu + f_{rl})} \tag{2.30}$$

*where*

$BFR_{f/r\ max}$ = the brake force ratio, allocated by the vehicle's braking system, that results in maximum (optimal) braking forces, and

Other terms are as defined previously.

It follows that the percentage of braking force that the braking system should allocate to the front axle ($PBF_f$) for maximum braking is

$$PBF_f = 100 - \frac{100}{1 + BFR_{f/r\ max}} \tag{2.31}$$

and the percentage of braking force that the braking system should allocate to the rear axle ($PBF_r$) for maximum braking is

$$PBF_r = \frac{100}{1 + BFR_{f/r\ max}} \tag{2.32}$$

**EXAMPLE 2.7** **BRAKE-FORCE PROPORTIONING**

A car has a wheelbase of 2.5 m and a center of gravity that is 1.0 m behind the front axle at a height of 600 mm. If the car is traveling at 130 km/h on a road with poor pavement that is wet, determine the percentage of braking forces that should be allocated to the front and rear brakes (by the vehicle's braking system) to ensure that maximum braking forces are developed.

**SOLUTION**

The coefficient of rolling resistance is

$$f_{rl} = 0.01\left(1 + \frac{130 \times 0.2778}{44.73}\right) = 0.0181$$

and $\mu = 0.6$ from Table 2.4 (maximum because we want the tires to be at the point of impending slide). Applying Eq. 2.30 gives

$$BFR_{f/r\,max} = \frac{l_r + h(\mu + f_{rl})}{l_f - h(\mu + f_{rl})}$$

$$= \frac{1.500 + 0.600(0.6 + 0.0181)}{1.000 - 0.600(0.6 + 0.0181)}$$

$$= 2.97$$

Using Eq. 2.31, the percentage of the force allocated to the front brakes should be

$$PBF_f = 100 - \frac{100}{1 + BFR_{f/r\,max}}$$

$$= 100 - \frac{100}{1 + 2.97}$$

$$= \underline{\underline{74.81\ \%}}$$

and using Eq. 2.32 (or simply $100 - PBF_f$), the percentage of the force allocated to the rear brakes should be

$$PBF_r = \frac{100}{1 + BFR_{f/r\,max}}$$

$$= \frac{100}{1 + 2.97}$$

$$= \underline{\underline{25.19\ \%}}$$

It is clear from Eq. 2.30 that the design of a vehicle's braking system is not an easy task because the optimal brake-force proportioning changes with both vehicle and road conditions. For example, the addition of vehicle cargo and/or passengers will change not only the weight of the vehicle (which affects $f_{rl}$ in Eq. 2.30), but also the distribution of the weight, shifting the height of the center of gravity and its location along the vehicle's longitudinal axis, and this will change the optimal brake force proportioning ($BFR_{f/r\,max}$). This is particularly problematic for trucks because of the large weight and center of gravity differences between loaded and unloaded conditions. Similarly, changes in road conditions produce different coefficients of adhesion, again changing optimal brake force proportioning. As a result of the uncertainties in weight and road conditions, vehicle designers often choose a compromise value of brake force proportioning that, on average, provides good braking but is rarely, if ever, optimal.

It is important to note that studies have indicated that if wheel lockup is to occur, it is preferable to have the front wheels lock first because having the rear wheels lock first can result in uncontrollable vehicle spin. Front-wheel lockup results in loss of steering control, but the vehicle will at least continue to brake in a straight line. Technological advancements in braking systems since the late 1970s have resulted in vehicles that are increasingly capable of proportioning brake forces in a manner that is closer to optimal and avoids the dangerous rear-wheel–first lockup due to front-wheel underbraking.

Because true optimal brake force proportioning is seldom achieved in standard non-antilock braking systems, it is useful to define a braking-efficiency term that reflects the degree to which the braking system is operating below optimal. Simply stated, braking efficiency is defined as the ratio of the maximum rate of deceleration, expressed in $g$'s ($g_{max}$), achievable prior to any wheel lockup to the coefficient of road adhesion:

$$\eta_b = \frac{g_{max}}{\mu} \tag{2.33}$$

*where*

$$\eta_b = \text{braking efficiency,}$$
$$g_{max} = \text{maximum deceleration in } g \text{ units (with the absolute maximum} = \mu), \text{ and}$$
$$\mu = \text{coefficient of road adhesion.}$$

### 2.9.3 Antilock Braking Systems

Many modern cars have braking systems designed to prevent the wheels from locking during braking applications (antilock braking systems). In theory, antilock braking systems serve two purposes. First, they prevent the coefficient of road adhesion from dropping to slide values (see Table 2.4). Second, they have the potential to raise the braking efficiency to 100%. In practice, designing an antilock braking system that avoids slide coefficients of adhesion and achieves 100% braking efficiency ($\eta_b = 1.0$) is a difficult task. This is because most antilock braking system technologies detect which wheels have locked and release them momentarily before reapplying the brake on the locking wheel. The wheel lock detection speed, speed of brake force reallocation, and braking system design (the amount of braking forces that can be accommodated by the vehicle's front and rear brake discs and calipers) all impact the overall effectiveness of the antilock braking system. Early antilock braking systems often fell short of achieving 100% braking efficiency, and in many cases, an expert driver operating a non-antilock braking car could modulate the brakes to achieve shorter stopping distances than cars equipped with antilock brakes. However, advances in antilock braking system technology continue to bring us closer to 100% braking efficiency.

### 2.9.4 Theoretical Stopping Distance

With a basic understanding of brake force proportioning and the resulting braking efficiency, attention can now be directed toward developing expressions for minimum stopping distances. By inspection of Fig. 2.6, it can be seen that the relationship among stopping distance, braking force, vehicle mass, and vehicle speed is

$$a \, ds = \left[ \frac{F_b + \sum R}{\gamma_b m} \right] ds$$

$$= V \, dV$$

(2.34)

*where*

$\gamma_b$  =  mass factor accounting for moments of inertia during braking, which is given the value of 1.04 for automobiles [Wong 2008], and

Other terms are as defined in Fig. 2.6.

Integrating to determine stopping distance ($S$) gives

$$S = \int_{V_2}^{V_1} \gamma_b m \frac{V \, dV}{F_b + \sum R}$$

(2.35)

Substituting in the resistances (see Fig. 2.6), we obtain

$$S = \gamma_b m \int_{V_2}^{V_1} \frac{V \, dV}{F_b + R_a + f_{rl} W \pm W \sin \theta_g}$$

(2.36)

*where*

$V_1$  =  initial vehicle speed in m/s,

$V_2$  =  final vehicle speed in m/s,

$f_{rl} W$  =  rolling resistance,

$W \sin \theta_g$  =  grade resistance (positive for uphill slopes and negative for downhill slopes), and

Other terms are as defined previously.

To simplify notation, let

$$K_a = \frac{\rho}{2} C_D A_f$$

(2.37)

so that Eq. 2.3 is

$$R_a = K_a V^2$$

(2.38)

Continuing, assume that the effect of speed on the coefficient of rolling resistance, $f_{rl}$, is constant and can be approximated by using the average of initial ($V_1$) and final ($V_2$) speeds in Eq. 2.5 [$V = (V_1 + V_2) / 2$]. With this assumption (which introduces only a very small amount of error), and letting $m = W/g$ and $F_b = \mu W$, integration of Eq. 2.36 gives

$$S = \frac{\gamma_b W}{2gK_a} \ln \left[ \frac{\mu W + K_a V_1^2 + f_{rl} W \pm W \sin\theta_g}{\mu W + K_a V_2^2 + f_{rl} W \pm W \sin\theta_g} \right] \qquad (2.39)$$

If the vehicle is assumed to stop ($V_2 = 0$),

$$S = \frac{\gamma_b W}{2gK_a} \ln \left[ 1 + \frac{K_a V_1^2}{\mu W + f_{rl} W \pm W \sin\theta_g} \right] \qquad (2.40)$$

With braking efficiency considered, the actual braking force is

$$F_b = \eta_b \mu W \qquad (2.41)$$

Therefore, by substitution into Eq. 2.40, the theoretical stopping distance is

$$S = \frac{\gamma_b W}{2gK_a} \ln \left[ 1 + \frac{K_a V_1^2}{\eta_b \mu W + f_{rl} W \pm W \sin\theta_g} \right] \qquad (2.42)$$

Similarly, Eq. 2.39 can be written to include braking efficiency. Finally, if aerodynamic resistance is ignored (due to its comparatively small contribution to braking), integration of Eq. 2.35 gives the theoretical stopping distance as

$$S = \frac{\gamma_b \left( V_1^2 - V_2^2 \right)}{2g \left( \eta_b \mu + f_{rl} \pm \sin\theta_g \right)} \qquad (2.43)$$

**EXAMPLE 2.8   THEORETICAL MINIMUM STOPPING DISTANCE**

A new experimental 11-kN car, with $C_D = 0.25$ and $A_f = 2.0$ m$^2$, is traveling at 145 km/h down a 10% grade. The coefficient of road adhesion is 0.7 and the air density is 1.2256 kg/m$^3$. The car has an advanced antilock braking system that gives it a braking efficiency of 100%. Determine the theoretical minimum stopping distance for the case where aerodynamic resistance is considered and the case where aerodynamic resistance is ignored.

*SOLUTION*

With aerodynamic resistance considered, Eq. 2.42 can be applied with $\gamma_b = 1.04$, $\theta_g = 5.71°$, and

$$f_{rl} = 0.01 \left( 1 + \frac{\left( \dfrac{145 \times 1000 / 3600 + 0}{2} \right)}{44.73} \right) = 0.0145$$

$$K_a = \frac{1.2256}{2}(0.25)(2) = 0.3064$$

Then

$$S = \frac{1.04(11000)}{2(9.807)(0.3064)} \ln \left[ 1 + \frac{0.3064(145 \times 1000 / 3600)^2}{(1.0)(0.7)(11000) + (0.0145)(11000) - 11000\sin(5.71°)} \right]$$

$$= \underline{\underline{134.97 \text{ m}}}$$

With aerodynamic resistance excluded, Eq. 2.43 is used:

$$S = \frac{1.04(145 \times 1000 / 3600)^2}{2(9.807)(0.7 + 0.0145 - \sin(5.71°))} = \underline{\underline{139.87 \text{ m}}}$$

**EXAMPLE 2.9**   **EFFECTS OF GRADE ON THEORETICAL MINIMUM STOPPING DISTANCE**

A car is traveling at 130 km/h and has a braking efficiency of 80%. The brakes are applied to miss an object that is 45 m from the point of brake application, and the coefficient of road adhesion is 0.85. Ignoring aerodynamic resistance and assuming theoretical minimum stopping distance, estimate how fast the car will be going when it strikes the object if (a) the surface is level and (b) the surface is on a 5% upgrade.

*SOLUTION*

In both cases, rolling resistance is approximated as

$$f_{rl} = 0.01 \left( 1 + \frac{\left( \dfrac{130 \times 1000 / 3600 + V_2}{2} \right)}{44.73} \right) = 0.014 + 0.0001118 V_2$$

Applying Eq. 2.43 for the level grade with $\gamma_b = 1.04$, $\theta_g = 0°$,

$$S = \frac{\gamma_b \left( V_1^2 - V_2^2 \right)}{2g \left( \eta_b \mu + f_{rl} \pm \sin \theta_g \right)}$$

$$45 = \frac{1.04 \left[ (130 \times 1000 / 3600)^2 - V_2^2 \right]}{2(9.807) \left[ 0.8(0.85) + (0.014 + 0.0001118 V_2) \pm 0 \right]}$$

$$V_2 = \underline{\underline{26.70 \text{ m/s}}} \quad \text{or} \quad \underline{\underline{96.11 \text{ km/h}}}$$

On a 5% grade with $\theta_g = 2.86°$,

$$45 = \frac{1.04\left[(130\times1000/3600)^2 - V_2^2\right]}{2(9.807)\left[0.8(0.85) + (0.014 + 0.0001118V_2) + 0.05\right]}$$

$$V_2 = \underline{25.89 \text{ m/s}} \quad \text{or} \quad \underline{93.20 \text{ km/h}}$$

**EXAMPLE 2.10    THEORETICAL MINIMUM STOPPING DISTANCE WITH AND WITHOUT ANTILOCK BRAKES**

A car is traveling up a 3% grade on a road that has good, wet pavement. The engine is running at 2500 revolutions per minute. The radius of the wheels is 380mm, the driveline slippage is 3%, and the overall gear reduction ratio is 2.5 to 1. A deer jumps out onto the road and the driver applies the brakes 89 m from it. The driver hits the deer at a speed of 30 km/h. If the driver did not have antilock brakes, and the wheels were locked the entire distance, would a deer-impact speed of 30 km/h be possible?

**SOLUTION**

The speed of the car must first be determined by applying Eq. 2.18 with $i = 0.03$, $\varepsilon_0 = 2.5$, $r = 0.38$ m, and $n_e = 2500/60$ revolutions per second:

$$V = \frac{2\pi r n_e (1-i)}{\varepsilon_0}$$

$$= \frac{2(3.141)(0.380)(2500/60)(1-0.03)}{2.5}$$

$$= 38.59 \text{ m/s } (138.93 \text{ km/h})$$

Next, using Eq. 2.5 the coefficient of rolling resistance is computed using the average speed, $(V_1 + V_2)/2$ as an approximation of $V$ (with a $V_2$ of 30 km/h):

$$f_{rl} = 0.01\left(1 + \frac{V}{44.73}\right) = 0.01\left(1 + \frac{\dfrac{38.59 + 30(1000/3600)}{2}}{44.73}\right) = 0.01525$$

If the car did not have an antilock braking system and the brakes were locked, the slide value on good, wet pavement is a coefficient of road adhesion of 0.6 ($\mu = 0.6$) from Table 2.4. Applying Eq. 2.43 with $\mu = 0.6$, a 3% grade (so $\sin\theta_g \approx 0.03$), and $\gamma_b = 1.04$,

$$S = \frac{\gamma_b\left(V_1^2 - V_2^2\right)}{2g\left(\eta_b\mu + f_{rl} \pm \sin\theta_g\right)}$$

$$89 = \frac{1.04\left(38.59^2 - (30(1000/3600))^2\right)}{2(9.807)\left[\eta_b(0.6) + 0.01525 + 0.03\right]}$$

or $\eta_b = \underline{\underline{1.33}}$

Because a braking efficiency of 1.33 is not possible, the driver would hit the deer at a higher speed if the wheels were locked the entire distance. Note that to achieve a deer-impact speed of 30 km/h or less, $\eta_b\mu$ in Eq. 2.43 must be 0.8 ($1.33 \times 0.6$) or greater. The maximum coefficient of road adhesion for good, wet pavements is 0.9 (from Table 2.4). So, to achieve a deer-impact speed of 30 km/h or less (with $\mu = 0.9$),

$$\eta_b\mu = 0.8 \text{ so,}$$
$$\eta_b = 0.8/\mu$$
$$= 0.8/0.9$$
$$= 0.89$$

Thus a braking efficiency of at least 89% is needed (if the vehicle has a functional antilock braking system) in order to achieve a deer-impact speed of 30 km/h or less.

### 2.9.5  Practical Stopping Distance

As mentioned earlier, one of the most critical concerns in the design of a highway is the provision of adequate driver sight distance to permit a safe stop. The theoretical assessment of vehicle stopping distance presented in the previous section provided the principles of braking for an individual vehicle under specified roadway surface conditions. However, highway engineers face a more complex problem because they must design for a variety of driver skill levels (which can affect whether or not the brakes lock and reduce the coefficient of road adhesion to slide values), vehicle types (with varying aerodynamics, weight distributions, and brake efficiencies), and weather conditions (which change the roadway's coefficient of adhesion). As a result of the wide variability inherent in the determination of braking distance, an equation is required that provides an estimate of typical observed braking distances and is more simplistic and usable than Eq. 2.42.

The basic physics equation on rectilinear motion, assuming constant deceleration, is chosen as the basis of a practical equation for stopping distance:

$$V_2^2 = V_1^2 + 2ad \tag{2.44}$$

*where*

$V_2$ = final vehicle speed in m/s,
$V_1$ = initial vehicle speed in m/s,
$a$ = acceleration (negative for deceleration) in m/s$^2$, and
$d$ = deceleration distance (practical stopping distance) in m.

Rearranging Eq. 2.44 and assuming $a$ is negative for deceleration gives

$$d = \frac{V_1^2 - V_2^2}{2a} \tag{2.45}$$

If $V_2 = 0$ (the vehicle comes to a complete stop), the practical-stopping-distance equation is

$$d = \frac{V_1^2}{2a} \tag{2.46}$$

To make this equation generally applicable for design purposes, a deceleration rate, $a$, must be chosen that is representative of appropriately conservative braking behavior. AASHTO [2011] recommends a deceleration rate of 3.4 m/s$^2$. Empirical studies [Fambro et al. 1997] have shown that approximately 90% of drivers decelerate at rates greater than this, and that this deceleration rate is well within a driver's capability to maintain steering control during a braking maneuver on wet surfaces. Additionally, empirical studies [Fambro et al. 1997] have confirmed that most vehicle braking systems and tire-pavement friction levels are capable of supporting this deceleration rate, even under wet conditions.

To account for the effect of grade, Eq. 2.46 is modified as follows:

$$d = \frac{V_1^2}{2g\left(\left(\dfrac{a}{g}\right) \pm G\right)} \tag{2.47}$$

*where*

> $g$ = gravitational constant, 9.807 m/s$^2$,
>
> $G$ = roadway grade (+ for uphill, − for downhill) in percent/100, and
>
> Other terms are as defined previously.

It is important to note that Eq. 2.47 is consistent with Eq. 2.43 (the theoretical stopping distance ignoring aerodynamic resistance). Rewriting Eq. 2.43 with the assumption that the vehicle comes to a stop ($V_2 = 0$), that $\sin \theta_g = \tan \theta_g = G$ (for small grades), and that $\gamma_b$ and $f_{rl}$ can be ignored due to their small and essentially offsetting effects, we have

$$S = \frac{V_1^2}{2g\left(\eta_b \mu \pm G\right)} \tag{2.48}$$

Recall that $\eta_b \mu = g_{max}$ (Eq. 2.33). However, rather than determining the maximum deceleration rate (in $g$'s) for a specific vehicle braking efficiency and specific coefficient of road adhesion, the AASHTO-recommended maximum deceleration rate (again, an appropriately conservative value for the overall driver and vehicle population) is used. Thus, a maximum deceleration of 0.35 $g$'s (3.4/9.807) is used for Eq. 2.47.

The recommended deceleration rate as determined empirically already accounts for the effects of aerodynamic resistance, braking efficiency, coefficient of road adhesion, and inertia during braking (the braking mass factor). This value reflects current vehicle technologies and driving behavior. It is important to recognize that as vehicle braking technology and other vehicle characteristics change, as well as possibly driver behavior, the recommended value of $a$ should be reviewed to determine if it is still applicable for highway design purposes. The relationship between changing vehicle characteristics and changing highway design guidelines is one that must always be kept in the design engineer's mind.

**EXAMPLE 2.11   BRAKING EFFICIENCY AND STOPPING DISTANCE**

A car ($W = 9.8$ kN, $C_D = 0.25$, $A_f = 2.0$ m$^2$) has an antilock braking system that gives it a braking efficiency of 100%. The car's stopping distance is tested on a level roadway with poor, wet pavement (with tires at the point of impending skid), and $\rho = 1.227$ kg/m$^3$. How inaccurate will the stopping distance predicted by the practical-stopping-distance equation be compared with the theoretical stopping distance, assuming the car is initially traveling 100 km/h? How inaccurate will the practical-stopping-distance equation be if the same car has a braking efficiency of 85%?

**SOLUTION**

First, to calculate the theoretical minimum stopping distance, Eq. 2.42 is applied with $\gamma_b = 1.04$, $\theta_g = 0°$, $\mu = 0.60$ (maximum for poor, wet pavement, from Table 2.4), and

$$f_{rl} = 0.01 \left( 1 + \frac{\left( \frac{100 \times 1000 / 3600 + 0}{2} \right)}{44.73} \right) = 0.0131$$

$$K_a = \frac{1.227}{2}(0.25)(2.0) = 0.3068$$

Thus, from Eq. 2.42,

$$S = \frac{1.04(9800)}{2(9.807)(0.3068)} \ln \left[ 1 + \frac{0.3068(100 \times 1000 / 3600)^2}{(1.0)(0.60)(9800) + (0.0131)(9800) \pm 0} \right] = 65.45 \text{ m}$$

For the same conditions but with a vehicle braking efficiency of 85%, Eq. 2.42 gives

$$S = \frac{1.04(9800)}{2(9.807)(0.3068)} \ln \left[ 1 + \frac{0.3068(100 \times 1000 / 3600)^2}{(0.85)(0.60)(9800) + (0.0131)(9800) \pm 0} \right]$$

$$= 76.46 \text{ m}$$

Now applying Eq. 2.46 (since $G = 0$) for the practical stopping distance, we find

$$d = \frac{V_1^2}{2a} = \frac{(100 \times 1000 / 3600)^2}{2(3.4)} = 113.47 \text{ m}$$

In the first case, the error is <u>48.02 m</u>. In the case of 85% braking efficiency, the error is <u>37.01 m</u>. Rearranging Eq. 2.46 to solve for $a$, we find that the stopping distances of 65.45 m and 76.46 m correspond to deceleration rates of 5.90 m/s$^2$ and 5.046 m/s$^2$, respectively. Studies [Fambro et al. 1997] have shown that most drivers decelerate at rates of 5.6 m/s$^2$ or greater in emergency stopping situations. Thus, this range of theoretical values is consistent with observed distances for situations in which minimum stopping distances are trying to be achieved. Comparing these theoretical values with the AASHTO recommended deceleration rate of 3.4 m/s$^2$, it is readily apparent that a considerable level of conservatism is built into the deceleration rate for practical stopping sight distance.

### 2.9.6   Distance Traveled During Driver Perception/Reaction

Until now the focus has been directed toward the distance required to stop the vehicle from the point of brake application. However, in providing sufficient sight distance for a driver to stop safely, it is also necessary to consider the distance traveled during the time the driver is perceiving and reacting to the need to stop. The distance traveled during perception/reaction ($d_r$) is given by

$$d_r = V_1 \times t_r \qquad (2.49)$$

*where*

$\quad V_1$ = initial vehicle speed in m/s, and

$\quad t_r$ = time required to perceive and react to the need to stop, in s.

The perception/reaction time of a driver is a function of a number of factors, including the driver's age, physical condition, and emotional state, as well as the complexity of the situation and the strength of the stimuli requiring a stopping action. For highway design, a conservative perception/reaction time has been determined to be 2.5 seconds [AASHTO 2011]. For comparison, average drivers have perception/reaction times of approximately 1.0 to 1.5 seconds.

Thus, the total required stopping distance is a combination of the braking distance, either theoretical (Eq. 2.42 or 2.43) or practical (Eq. 2.47), and the distance traveled during perception/reaction (Eq. 2.49), as shown in Eq. 2.50:

$$d_s = d + d_r \qquad (2.50)$$

*where*

$\quad d_s$ = total stopping distance (including perception/reaction) in m,

$\quad d$ = distance traveled during braking in m, and

$\quad d_r$ = distance traveled during perception/reaction in m.

The combination of practical stopping distance and the distance traveled during perception/reaction is a primary consideration in highway design, as will be discussed in detail in Chapter 3.

---

**EXAMPLE 2.12   PRACTICAL STOPPING DISTANCE AND PERCEPTION/REACTION TIMES**

Two drivers each have a reaction time of 2.5 seconds. One is obeying a 90-km/h speed limit and the other is traveling illegally at 120 km/h. How much distance will each of the drivers cover while perceiving/reacting to the need to stop, and what will the total stopping distance be for each driver (using practical stopping distance and assuming $G = -2.5\%$)?

*SOLUTION*

The distances traveled by each driver during perception/reaction will be calculated first, using Eq. 2.49. For the driver traveling at 90 km/h,

$$d_r = V_1 \times t_r = (90 \times 1000 / 3600)(2.5) = \underline{\underline{62.5 \text{ m}}}$$

For the driver traveling at 120 km/h,

$$d_r = V_1 \times t_r = (120 \times 1000 / 3600)(2.5) = \underline{\underline{83.33 \text{ m}}}$$

Therefore, driving at 120 km/h increases the distance traveled during perception/reaction by 20.83 m.

Next, the distance traveled during braking will be calculated for each driver, using the equation for practical stopping distance (Eq. 2.47). For the driver traveling at 90 km/h,

$$d = \frac{(90 \times 1000/3600)^2}{2(9.807)\left(\left(\frac{3.4}{9.807}\right) - 0.025\right)} = \frac{625}{6.31} = 99.05 \text{ m}$$

For the driver traveling at 120 km/h,

$$d = \frac{(120 \times 1000/3600)^2}{2(9.807)\left(\left(\frac{3.4}{9.807}\right) - 0.025\right)} = \frac{111.11}{6.31} = 176.09 \text{ m}$$

The total stopping distance for each driver is now calculated with Eq. 2.50. For the driver traveling at 90 km/h,

$$d_s = d + d_r = 62.5 + 99.05 = \underline{\underline{161.55 \text{ m}}}$$

For the driver traveling at 120 km/h,

$$d_s = d + d_r = 83.33 + 176.09 = \underline{\underline{259.42 \text{ m}}}$$

Therefore, driving at 120 km/h increases the total stopping distance by a very substantial 97.87 m.

## NOMENCLATURE FOR CHAPTER 2

| | | | |
|---|---|---|---|
| $a$ | acceleration (deceleration if negative) | $F_{br}$ | rear-axle braking force |
| $A_f$ | frontal area of vehicle | $F_{br\,max}$ | maximum rear-axle braking force |
| $BFR_{f/r\,max}$ | brake force ratio (front over rear) for maximum braking force | $F_e$ | engine-generated tractive effort |
| | | $F_f$ | available tractive effort at the front axle |
| $C_D$ | coefficient of aerodynamic drag | $F_{max}$ | maximum tractive effort |
| $d$ | practical stopping distance | $F_r$ | available tractive effort at the rear axle |
| $d_a$ | distance to accelerate | $f_{rl}$ | coefficient of rolling resistance |
| $d_r$ | distance traveled during driver perception/reaction | $G$ | roadway grade in m/m (percent grade divided by 100; $G = 0.05$ is a 5% grade) |
| $d_s$ | total stopping distance (vehicle braking distance plus perception/reaction distance) | $g$ | gravitational constant (9.807 m/s$^2$) |
| | | $g_{max}$ | maximum deceleration achieved before wheel lockup |
| $F$ | total available tractive effort | | |
| $F_b$ | total braking force | $h$ | height of vehicle's center of gravity above the roadway surface |
| $F_{bf}$ | front-axle braking force | | |
| $F_{bf\,max}$ | maximum front-axle braking force | $i$ | drive axle slippage |

| | | | |
|---|---|---|---|
| $K_a$ | elements of aerodynamic resistance that are not a function of speed | $r$ | radius of vehicle drive wheels |
| $L$ | vehicle wheelbase | $S$ | minimum theoretical stopping distance |
| $l_f$ | distance from vehicle's center of gravity to front axle | $t_r$ | driver perception/reaction time |
| | | $V$ | vehicle speed |
| $l_r$ | distance from vehicle's center of gravity to rear axle | $W$ | total vehicle weight |
| | | $W_f$ | vehicle weight acting normal to the roadway surface on the front axle |
| $M_e$ | engine torque | | |
| $m$ | mass | $W_r$ | vehicle weight acting normal to the roadway surface on the rear axle |
| $n_e$ | engine speed in crankshaft revolutions per second | $\gamma_b$ | braking mass factor |
| | | $\gamma_m$ | acceleration mass factor |
| $P_e$ | engine-generated power, measured in kW | $\varepsilon_0$ | gear reduction ratio |
| $PBF_f$ | optimal percent of braking force on the front axle | $\eta_b$ | braking efficiency |
| | | $\eta_d$ | drivetrain efficiency |
| $PBF_r$ | optimal percent of braking force on the rear axle | $\theta_g$ | angle of grade |
| | | $\mu$ | coefficient of road adhesion |
| $R_a$ | aerodynamic resistance | $\rho$ | air density |
| $R_g$ | grade resistance | | |
| $R_{rl}$ | rolling resistance | | |

# REFERENCES

AASHTO (American Association of State Highway and Transportation Officials). *A Policy on Geometric Design of Highways and Streets*, 6th ed. Washington, DC: AASHTO, 2011.

Brewer, H. K., and R. S. Rice. "Tires: Stability and Control." SAE Transactions, Vol. 92, paper 830561, 1983.

Campbell, C. *The Sports Car: Its Design and Performance*. Cambridge, MA: Robert Bently, 1978.

Fambro, D. B., K. Fitzpatrick, and R. J. Koppa. *Determination of Stopping Sight Distances*. NCHRP Report 400. Transportation Research Board, Washington, DC, 1997.

Taborek, J. J. "Mechanics of Vehicles." Machine Design, 1957.

Wong, J. Y. *Theory of Ground Vehicles*. Fourth Edition. New York: John Wiley & Sons, 2008.

# PROBLEMS

**2.1**   A new sports car has a drag coefficient of 0.29 and a frontal area of 1.9 m², and is traveling at 160 km/h. How much power is required to overcome aerodynamic drag if $\rho = 1.2256$ kg/m³?

**2.2**   A vehicle manufacturer is considering an engine for a new sedan ($C_D = 0.30$, $A_f = 2$ m²). The car is being designed to achieve a top speed of 160 km/h on a paved surface at sea level ($\rho = 1.2256$ kg/m³). The car currently weighs 9.3 kN, but the designer initially selected an underpowered engine because he did not account for aerodynamic and rolling resistances. If 9 N of additional vehicle weight is added for each kilowatt needed to overcome the neglected resistance, what will be the final weight of the car if it is to achieve the 160-km/h top speed?

**2.3**   For Example 2.3, how far back from the front axle would the center of gravity have to be to ensure that the maximum tractive effort developed for front- and rear-wheel–drive options is equal (assume that all other variables are unchanged)?

**2.4**   A rear-wheel–drive 13.3-kN drag race car has a 5.10-m wheelbase and a center of gravity 500 mm above the pavement and 3.55 m behind the front axle. The owners wish to achieve an initial acceleration from rest of 4.6 m/s² on a level paved surface. What is the minimum coefficient of road adhesion needed to achieve this acceleration? (Assume $\gamma_m = 1.00$.)

**2.5**   If the race car in Problem 2.4 has a center of gravity 0.9 m above the roadway and is run on a pavement with a coefficient of road adhesion of 1.0, how far back from the front axle would the center of gravity have to be to develop a maximum acceleration from rest of 1.0 $g$ (9.807 m/s$^2$)? (Assume $\gamma_m = 1.00$.)

**2.6** A rear-wheel–drive car weighs 12 kN, has 355-mm–radius wheels, a driveline efficiency of 95%, and an engine that develops 400 N-m of torque. Its wheelbase is 2.5 m, and the center of gravity is 450 mm above the road surface and 1 m behind the front axle. What is the lowest gear reduction ratio that would allow this car to achieve the highest possible acceleration from rest on good, dry pavement?

**2.7** A newly designed car has a 2.8-m wheelbase, is rear-wheel drive, and has a center of gravity 450 mm above the road and 1.3 m behind the front axle. The car weighs 10.9 kN, the mechanical efficiency of the driveline is 90%, and the wheel radius is 355 mm. The base engine develops 250 N-m of torque, and a modified version of the engine develops 290 N-m of torque. If the overall gear reduction ratio is 9 to 1, what is the maximum acceleration from rest for the car with the 250-N-m engine and for the car with the 290–N-m engine? (It is on good, dry, and level pavement.)

**2.8** A 13.3-kN car with $C_D = 0.35$, $A_f = 2$ m$^2$, and $\rho = 1.2256$ kg/m$^3$ is traveling on a paved road. Its engine is running at 3000 rev/min and is producing 340 N-m of torque. The car's gear reduction ratio is 3.5 to 1, driveline efficiency is 90%, driveline slippage is 3.5%, and the road-wheel radius is 380 mm. What will the car's maximum acceleration rate be under these conditions on a level road? (Assume that the available tractive effort is the engine-generated tractive effort.)

**2.9** A 9.5-kN car is traveling at sea level at a constant speed. Its engine is running at 4500 rev/min and is producing 200 N-m of torque. It has a driveline efficiency of 90%, a driveline slippage of 2%, 380-mm–radius wheels, and an overall gear reduction ratio of 3 to 1. If the car's frontal area is 1.8 m$^2$, what is the drag coefficient?

**2.10** An 11-kN car has a maximum speed (at sea level and on a level, paved surface) of 240 km/h with 355-mm–radius wheels, a gear reduction of 3 to 1, and a driveline efficiency of 90%. It is known that at the car's top speed the engine is producing 270 N-m of torque. If the car's frontal area is 2.3 m$^2$, what is its drag coefficient?

**2.11** An 11-kN car ($C_D = 0.35$, $A_f = 2.3$ m$^2$, $\rho = 1.2256$ kg/m$^3$) has 355-mm–radius wheels, a driveline efficiency of 90%, an overall gear reduction ratio of 3.2 to 1, and driveline slippage of 3.5%. The engine develops a maximum torque of 270 N-m at 3500 rev/min. What is the maximum grade this vehicle could ascend, on a paved surface while the engine is developing maximum torque? (Assume that the available tractive effort is the engine-generated tractive effort.)

**2.12** A 12-kN car is traveling in third gear (overall gear reduction ratio of 2.5 to 1) on a level road at its top speed of 200 km/h. The air density is 1.06 kg/m$^3$. The car has a frontal area of 1.8 m$^2$, a drag coefficient of 0.28, a wheel radius of 320 mm, a driveline slippage of 0.03, and a driveline efficiency of 90%. At this vehicle speed, what torque is the engine producing and what is the engine speed (in revolutions per minute)?

**2.13** A rear-wheel–drive car weighs 11 kN, has a 2.03-m wheelbase, a center of gravity 500 mm above the roadway surface and 760 mm behind the front axle, a driveline efficiency of 75%, 355-mm–radius wheels, and an overall gear reduction of 11 to 1. The car's torque/engine speed curve is given by

$$M_e = 6n_e - 0.045n_e^2$$

If the car is on a paved, level roadway surface with a coefficient of adhesion of 0.75, determine its maximum acceleration from rest.

**2.14** Consider the car in Problem 2.13. If it is known that the car achieves maximum speed at an overall gear reduction ratio of 2 to 1 with a driveline slippage of 3.5%, how fast would the car be going if it could achieve its maximum speed when its engine is producing maximum power?

**2.15** Consider the situation described in Example 2.5. If the vehicle is redesigned with 330-mm–radius wheels (assume that the mass factor is unchanged) and a center of gravity located at the same height but at the midpoint of the wheelbase, determine the acceleration for front- and rear-wheel–drive options.

**2.16** An engineer designs a rear-wheel–drive car (without an engine) that weighs 8.9 kN and has a 2.55-m wheelbase, a driveline efficiency of 80%, 355-mm–radius wheels, an overall gear reduction ratio of 10 to 1, and a center of gravity (without engine) that is 550 mm above the roadway surface and 1.40 m behind the front axle. An engine that weighs 13 N for each N-m of developed torque is to be placed in the front portion of the car. Calculations show that for every 89 N of engine weight added, the car's center of gravity moves 25 mm closer to the front axle (but stays at the same height above the roadway surface). If the car is starting from rest on a level paved roadway with a coefficient of adhesion of 0.8, select an engine size (weight and associated torque) that will result in the highest possible available tractive effort.

**2.17** If the car in Example 2.8 had $C_D = 0.45$ and $A_f = 2.3$ m$^2$, what would have been the difference in minimum theoretical stopping distances with and without aerodynamic resistance considered (all other factors the same as in Example 2.8)?

**2.18** A 15.6-kN vehicle ($C_D = 0.40$, $A_f = 2.4$ m$^2$, $\rho = 1.2256$ kg/m$^3$) is driven on a surface with a coefficient of adhesion of 0.5, and the coefficient of rolling friction is approximated as 0.015 for all speeds. Assuming minimum theoretical stopping distances, if the vehicle comes to a stop 76 m after brake application on a level surface and has a braking efficiency of 0.78, what was its initial speed (a) if aerodynamic resistance is considered and (b) if aerodynamic resistance is ignored?

**2.19** A level test track has a coefficient of road adhesion of 0.75, and a car being tested has a coefficient of rolling friction that is approximated as 0.018 for all speeds. The vehicle is tested unloaded and achieves the theoretical minimum stop in 61 m (from brake application). The initial speed was 100 km/h. Ignoring aerodynamic resistance, what is the unloaded braking efficiency?

**2.20** A driver is traveling at 180 km/h down a 3% grade on good, wet pavement. An accident investigation team noted that braking skid marks started 180 m before a parked car was hit at an estimated 90 km/h. Ignoring air resistance, and using theoretical stopping distance, what was the braking efficiency of the car?

**2.21** A small truck is to be driven down a 4% grade at 120 km/h. The coefficient of road adhesion is 0.95, and it is known that the braking efficiency is 80% when the truck is empty and decreases by one percentage point for every 445 N of cargo added. Ignoring aerodynamic resistance, if the driver wants the truck to be able to achieve a minimum theoretical stopping distance of 90 m from the point of brake application, what is the maximum amount of cargo (in newtons) that can be carried?

**2.22** Consider the conditions in Example 2.8. The car has $C_D = 0.5$, $A_f = 2.3$ m$^2$, $W = 15.6$ kN, $\rho = 1.2256$ kg/m$^3$, and a coefficient of rolling friction that is approximated as 0.018 for all speed conditions. If aerodynamic resistance is considered in stopping, estimate how fast the car will be going when it strikes

the object on a level and a +5% grade [all other conditions (speed, etc.) as described in Example 2.8].

**2.23** A car is traveling at 120 km/h down a 3% grade on poor, wet pavement. The car's braking efficiency is 90%. The brakes were applied 90 m before impacting an object. The car had an antilock braking system, but the system failed 60 m after the brakes had been applied (wheels locked). What speed was the car traveling at just before it impacted the object? (Assume theoretical stopping distance, ignore air resistance, and let $f_{rl} = 0.015$.)

**2.24** A driver traveling down a 4% grade collides with a roadside object in rainy conditions, and is issued a ticket for driving too fast for conditions. The posted speed limit is 105 km/h. The accident investigation team determined the following: The vehicle was traveling 65 km/h when it struck the object, braking skid marks started 61 m before the struck object, the pavement is in good condition, and the braking efficiency of the vehicle was 0.95. Using theoretical stopping distance, assuming aerodynamic resistance is negligible, and with the coefficient rolling resistance approximated as 0.015, should the driver appeal the ticket? Why or why not?

**2.25** A driver is traveling 110 km/h on a road with a negative 3% grade. There is a stalled car on the road 305 m ahead of the driver. The driver's vehicle has a braking efficiency of 90%, and it has antilock brakes. The road is in good condition and is initially dry, but it becomes wet 45 m before the stalled car (and stays wet until the car is reached). What is the minimum distance from the stalled car at which the driver could apply the brakes and still stop before hitting it? (Assume theoretical stopping distance, ignore air resistance, and let $f_{rl} = 0.013$.)

**2.26** Two cars are traveling on level terrain at 100 km/h on a road with a coefficient of adhesion of 0.8. The driver of car 1 has a 2.5-s perception/reaction time and the driver of car 2 has a 2.0-s perception/reaction time. Both cars are traveling side by side, and the drivers are able to stop their respective cars in the same amount of distance after first seeing a roadway obstacle (perception/reaction plus vehicle stopping distance). If the braking efficiency of car 2 is 0.75, determine the braking efficiency of car 1. (Assume minimum theoretical stopping distance and ignore aerodynamic resistance.)

**2.27** An engineering student is driving on a level roadway and sees a construction sign 185 m ahead in the middle of the roadway. The student strikes the sign at a speed of 55 km/h. If the student was traveling at 90 km/h when the sign was first spotted, what was the student's associated perception/reaction time (use practical stopping distance)?

**2.28** An engineering student claims that a country road can be safely negotiated at 110 km/h in rainy weather. Because of the winding nature of the road, one stretch of level pavement has a sight distance of only 180 m. Assuming practical stopping distance, comment on the student's claim.

**2.29** A driver is traveling at 90 km/h on a wet road. An object is spotted on the road 140 m ahead and the driver is able to come to a stop just before hitting the object. Assuming standard perception/reaction time and practical stopping distance, determine the grade of the road.

**2.30** A test of a driver's perception/reaction time is being conducted on a special testing track with wet pavement and a driving speed of 90 km/h. When the driver is sober, a stop can be made just in time to avoid hitting an object that is first visible 160 m ahead. After a few drinks under exactly the same conditions, the driver fails to stop in time and strikes the object at a speed of 55 km/h. Determine the driver's perception/reaction time before and after drinking. (Assume practical stopping distance.)

# Chapter 3

# Geometric Design of Highways

## 3.1 INTRODUCTION

With the understanding of vehicle performance provided in Chapter 2, attention can now be directed toward highway design. The design of highways necessitates the determination of specific design elements, which include the number of lanes, lane width, median type (if any) and width, length of acceleration and deceleration lanes for on- and off-ramps, need for truck climbing lanes for steep grades, curve radii required for vehicle turning, and the alignment required to provide adequate stopping and passing sight distances. Many of these design elements are influenced by the performance characteristics of vehicles. For example, vehicle acceleration and deceleration characteristics have a direct impact on the design of acceleration and deceleration lanes (the length needed to provide a safe and orderly flow of traffic) and the highway alignment needed to provide adequate passing and stopping sight distances. Furthermore, vehicle performance characteristics determine the need for truck climbing lanes on steep grades (where the poor performance of large trucks necessitates a separate lane) as well as the number of lanes required because the observed spacing between vehicles in traffic is directly related to vehicle performance characteristics (this will be discussed further in Chapter 5). In addition, the physical dimensions of vehicles affect a number of design elements, such as the radii required for low-speed turning, height of highway overpasses, and lane widths.

When one considers the diversity of vehicles' performance and physical dimensions, and the interaction of these characteristics with the many elements constituting highway design, it is clear that proper design is a complex procedure that requires numerous compromises. Moreover, it is important that design guidelines evolve over time in response to changes in vehicle performance and dimensions, and in response to evidence collected on the effectiveness of existing highway design practices, such as the relationship between crash rates and various roadway design characteristics. Current guidelines of highway design are presented in detail in *A Policy on Geometric Design of Highways and Streets*, 6th Edition, published by the American Association of State Highway and Transportation Officials [AASHTO 2011].

Because of the sheer number of geometric elements involved in highway design, a detailed discussion of each design element is beyond the scope of this book, and the reader is referred to [AASHTO 2011] for a complete discussion of current design practices. Instead, this book focuses exclusively on the key elements of highway alignment, which are arguably the most important components of geometric design. As will be shown, the alignment topic is particularly well suited for demonstrating the effect of vehicle performance (specifically braking performance) and vehicle dimensions (such as driver's eye height, headlight height, and taillight height) on the design of highways. By concentrating on the specifics of the highway alignment problem, the reader will develop an understanding of the procedures and compromises inherent in the design of all highway-related geometric elements.

## 3.2   PRINCIPLES OF HIGHWAY ALIGNMENT

The alignment of a highway is a three-dimensional problem measured in $x$, $y$, and $z$ coordinates. This is illustrated, from a driver's perspective, in Fig. 3.1. However, in highway design practice, three-dimensional design computations are cumbersome, and, what is perhaps more important, the actual implementation and construction of a design based on three-dimensional coordinates has historically been prohibitively difficult. As a consequence, the three-dimensional highway alignment problem is reduced to two two-dimensional alignment problems, as illustrated in Fig. 3.2. One of the alignment problems in this figure corresponds roughly to $x$ and $z$ coordinates and is referred to as horizontal alignment. The other corresponds to highway length (measured along some constant elevation) and $y$ coordinates (elevation) and is referred to as vertical alignment. Referring to Fig. 3.2, note that the horizontal alignment of a highway is referred to as the plan view, which is roughly equivalent to the perspective of an aerial photo of the highway. The vertical alignment is represented in a profile view, which gives the elevation of all points measured along the length of the highway (again, with length measured along a constant elevation reference).

Aside from considering the alignment problem as two two-dimensional problems, one further simplification is made: instead of using $x$ and $z$ coordinates, highway positioning and length are defined as the distance along the highway (usually measured along the centerline of the highway, on a horizontal, constant-elevation plane) from a specified point. This distance is measured in terms of stations, with each station consisting of 1000 m of highway alignment distance.

The notation for stationing distance is such that a point on a highway 1295.3 m from a specified point of origin is said to be at station 1 + 295.300, that is, 1 station and 295.300 m, with the point of origin being at station 0 + 000. This stationing concept,

**Figure 3.1** Highway alignment in three dimensions.

Plan view (horizontal alignment)

Profile view (vertical alignment)

**Figure 3.2**  Highway alignment in two-dimensional views.

combined with the highway's alignment direction given in the plan view (horizontal alignment) and the elevation corresponding to stations given in the profile view (vertical alignment), gives a unique identification of all highway points in a manner that is virtually equivalent to using true $x$, $y$, and $z$ coordinates.

## 3.3   VERTICAL ALIGNMENT

Vertical alignment specifies the elevation of points along a roadway. The elevation of these roadway points is usually determined by the need to provide an acceptable level of driver safety, driver comfort, and proper drainage (from rainfall runoff). A primary concern in vertical alignment is establishing the transition of roadway elevations between two grades. This transition is achieved by means of a vertical curve.

Vertical curves can be broadly classified into crest vertical curves and sag vertical curves, as illustrated in Fig. 3.3. Note that in Fig. 3.3, the distance from the *PVC* to the *PVI* is $L/2$. This is used in this figure because in practice the vast majority of vertical curves are arranged such that half of the curve length is positioned before the *PVI* and half after. Curves that satisfy this criterion are called equal-tangent vertical curves.

For referencing points on a vertical curve, it is important to note that the profile views presented in Fig. 3.3 correspond to all highway points even if a horizontal curve occurs concurrently with a vertical curve (as in Figs. 3.1 and 3.2). Thus, each roadway point is uniquely defined by stationing (which is measured along a horizontal plane) and elevation. This will be made clearer through forthcoming examples.

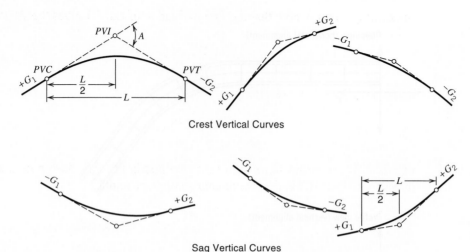

Crest Vertical Curves

Sag Vertical Curves

**Figure 3.3** Types of vertical curves.

Used by permission from American Association of State Highway and Transportation Officials, *A Policy on Geometric Design of Highways and Streets*, 6th Edition, Washington, DC, 2011.

$G_1$ = initial roadway grade in percent or m/m (this grade is also referred to as the initial tangent grade, viewing Fig. 3.3 from left to right),

$G_2$ = final roadway (tangent) grade in percent or m/m,

$A$ = absolute value of the difference in grades (initial minus final, usually expressed in percent),

$L$ = length of the curve in stations or m measured in a constant-elevation horizontal plane.

$PVC$ = point of the vertical curve (the initial point of the curve),

$PVI$ = point of vertical intersection (intersection of initial and final grades), and

$PVT$ = point of vertical tangent, which is the final point of the vertical curve (the point where the curve returns to the final grade or, equivalently, the final tangent).

### 3.3.1    Vertical Curve Fundamentals

In connecting roadway grades (tangents) with an appropriate vertical curve, a mathematical relationship defining elevations at all points (or equivalently, stations) along the vertical curve is needed. A parabolic function has been found suitable in this regard because, among other things, it provides a constant rate of change of slope and implies equal curve tangents. The general form of the parabolic equation, as applied to vertical curves, is

$$y = ax^2 + bx + c \tag{3.1}$$

*where*

$y$ = roadway elevation at distance $x$ from the beginning of the vertical curve (the $PVC$) in stations or m,

$x$ = distance from the beginning of the vertical curve in stations or m,

$a, b$ = coefficients defined below, and

$c$ = elevation of the $PVC$ (because $x = 0$ corresponds to the $PVC$) in m.

In defining $a$ and $b$, note that the first derivative of Eq. 3.1 gives the slope and is

$$\frac{dy}{dx} = 2ax + b \qquad \text{(3.2)}$$

At the *PVC*, $x = 0$, so, using Eq. 3.2,

$$b = \frac{dy}{dx} = G_1 \qquad \text{(3.3)}$$

where $G_1$ is the initial slope in m/m, as defined in Fig. 3.3. Also note that the second derivative of Eq. 3.1 is the rate of change of slope and is

$$\frac{d^2y}{dx^2} = 2a \qquad \text{(3.4)}$$

However, the average rate of change of slope, by observation of Fig. 3.3, can also be written as

$$\frac{d^2y}{dx^2} = \frac{G_2 - G_1}{L} \qquad \text{(3.5)}$$

Equating Eqs. 3.4 and 3.5 gives

$$a = \frac{G_2 - G_1}{2L} \qquad \text{(3.6)}$$

with all terms as defined previously (see Fig. 3.3). Please note that the units for coefficients $a$ and $b$ in Eqs. 3.3 and 3.6 must be such that they provide m when multiplied by $x^2$ and $x$, respectively. The preceding equations define all the terms in the parabolic vertical curve equation (Eq. 3.1). The following example gives a typical application of this equation.

**EXAMPLE 3.1   VERTICAL CURVE STATIONS AND ELEVATIONS**

A 200-m equal-tangent sag vertical curve has the *PVC* at station 3 + 700.000 and elevation 321 m. The initial grade is −3.5% and the final grade is +0.5%. Determine the stationing and elevation of the *PVI*, the *PVT*, and the lowest point on the curve.

*SOLUTION*

Since the curve is equal tangent, the *PVI* will be 100 m (measured in a horizontal plane) from the *PVC*, and the *PVT* will be 200 m from the *PVC*. Therefore, the stationing of the *PVI* and *PVT* is 3 + 800.000 and 3 + 900.000, respectively. For the elevations of the *PVI* and *PVT*, it is known that a −3.5% grade can be equivalently written as −0.035 m/m. Since the *PVI* is 100 m from the *PVC*, which is known to be at elevation 321 m, the elevation of the *PVI* is

$$321 - 0.035 \text{ m/m} \times (100 \text{ m}) = \underline{317.5 \text{ m}}$$

Similarly, with the *PVI* at elevation 317.5 m, the elevation of the *PVT* is

$$317.5 + 0.005 \text{ m/m} \times (100 \text{ m}) = \underline{318.0 \text{ m}}$$

It is clear from the values of the initial and final grades that the lowest point on the vertical curve will occur when the first derivative of the parabolic function (Eq. 3.1) is zero because the initial and final grades are opposite in sign. When initial and final grades are not opposite in sign, the low (or high) point on the curve will not be where the first derivative is zero because the slope along the curve will never be zero. For example, a sag curve with an initial grade of −2.0% and a final grade of −1.0% will have its lowest elevation at the *PVT*, and the first derivative of Eq. 3.1 will not be zero at any point along the curve. However, in our example problem the derivative will be equal to zero at some point, so the low point will occur when

$$\frac{dy}{dx} = 2ax + b = 0$$

From Eq. 3.3 we have

$$b = G_1 = -0.035$$

and from Eq. 3.6,

$$a = \frac{0.005 - (-0.035)}{2(200)} = 0.0001$$

Substituting for *a* and *b* gives

$$\frac{dy}{dx} = 2(0.0001)x + (-0.035) = 0$$
$$x = 175 \text{ m}$$

This gives the stationing of the low point at $\underline{3 + 875.000}$ (175 m from the *PVC*). For the elevation of the lowest point on the vertical curve, the values of *a*, *b*, *c* (elevation of the *PVC*), and *x* are substituted into Eq. 3.1, giving

$$y = 0.0001(175)^2 + (-0.035)(175) + 321$$
$$= \underline{317.94 \text{ m}}$$

Another interesting vertical curve problem that is sometimes encountered is one in which the curve must be designed so that the elevation of a specific location is met. An example might be to have the roadway connect with another (at the same elevation) or to have the roadway at some specified elevation so as to pass under another roadway. This type of problem is referred to as a curve-through-a-point problem and is demonstrated by the following example.

**EXAMPLE 3.2     ELEMENTS OF VERTICAL CURVE DESIGN**

An equal-tangent vertical curve is to be constructed between grades of −2.0% (initial) and +1.0% (final). The *PVI* is at station 3 + 350.000 and at elevation 130 m. Due to a street crossing the roadway, the elevation of the roadway at station 3 + 415.000 must be at 131 m. Design the curve.

*SOLUTION*

The design problem is one of determining the length of the curve required to ensure that station 3 + 415.000 is at elevation 131 m. To begin, we use Eq. 3.1:

$$y = ax^2 + bx + c$$

From Eq. 3.3,

$$b = G_1 = -0.02$$

and from Eq. 3.6,

$$a = \frac{G_2 - G_1}{2L}$$

Substituting $G_1 = -0.02$ and $G_2 = 0.01$, we have

$$a = \frac{G_2 - G_1}{2L} = \frac{0.01 - (-0.02)}{2L} = \frac{0.015}{L}$$

Now note that $c$ (the elevation of the *PVC*) in Eq. 3.1 will be equal to the elevation of the *PVI* plus $G_1 \times 0.5L$ (this is simply using the slope of the initial grade to determine the elevation difference between the *PVI* and *PVC*). With $G_1$ in m/m and the curve length $L$ in meters, we have

$$c = 130 + 0.02(0.5L) = 130 + 0.01L$$

Finally, the value of $x$ to be used in Eq. 3.1 will be $0.5L + 65$ because the point of interest (station 3 + 415.000) is 65 meters from the *PVI* (which is at station 3 + 350.000). Substituting $b = -0.02$, the expressions for $a$, $c$, and $x$, and $y = 131$ m (the given elevation) into Eq. 3.1 gives

$$131 = (0.015/L)(0.5L + 65)^2 + (-0.02)(0.5L + 65) + (130 + 0.01L)$$

$$0 = 0.00375\, L^2 - 1.325L + 63.375$$

Solving this quadratic equation gives $L = 57.04$ m (which is not feasible because we know that the point of interest is 65 m beyond the *PVI*, so the curve length must be longer than 57.04 m) or $L = 296.30$ m (which is the only feasible solution). This means that the curve must be 296.30 m long. Using this value of $L$,

$$\text{elevation of } PVC = c = 130 + 0.01L = 130 + 2.963 = 132.96 \text{ m}$$

$$\text{station of } PVC = 3 + 350.000 - (296.30/2) = 3 + 201.850$$

$$\text{elevation of } PVT = \text{elevation of } PVI + (0.5L)G_2 = 130 + \left[0.5(296.30)\right](0.01) = 131.48 \text{ m}$$

$$\text{station of } PVT = 3 + 350.000 + (296.30/2) = 3 + 498.150$$

and

$$x = 0.5L + 65 = 148.15 + 65 = 213.15 \text{ m from the } PVC$$

To check the elevation of the curve at station $3 + 415.000$, we apply Eq. 3.1 with $x = 213.15$:

$$y = ax^2 + bx + c$$

$$= \left(\frac{0.03}{2(296.30)}\right)(213.15)^2 + (-0.02)(213.15) + 132.96$$

$$= 131.00 \text{ m}$$

Therefore, all calculations are correct.

Some additional properties of vertical curves can now be formalized. For example, offsets, which are vertical distances from the initial tangent to the curve, as illustrated in Fig. 3.4, are extremely important in vertical curve design and construction.

**Figure 3.4** Offsets for equal-tangent vertical curves.

$G_1$ = initial roadway grade in percent or m/m (this grade is also referred to as the initial tangent grade, viewing Fig. 3.4 from left to right),

$G_2$ = final roadway (tangent) grade in percent or m/m,

$Y$ = offset at any distance $x$ from the $PVC$ in m,

$Y_m$ = midcurve offset in m,

$Y_f$ = offset at the end of the vertical curve in m,

$x$ = distance from the $PVC$ in m,

$L$ = length of the curve in stations or m measured in a constant-elevation horizontal plane,

$PVC$ = point of the vertical curve (the initial point of the curve),

$PVI$ = point of vertical intersection (intersection of initial and final grades), and

$PVT$ = point of vertical tangent, which is the final point of the vertical curve (the point where the curve returns to the final grade or, equivalently, the final tangent).

Referring to the elements shown in Fig. 3.4, the properties of an equal-tangent parabola can be used to give

$$Y = \frac{A}{200L} x^2 \qquad (3.7)$$

*where*

A = absolute value of the difference in grades ($|G_1 - G_2|$) expressed in percent, and
Other terms are as defined in Fig. 3.4.

Note that in this equation, 200 is used in the denominator instead of 2 because $A$ is expressed in percent instead of m/m (this division by 100 also applies to Eqs. 3.8 and 3.9 below). It follows from Fig. 3.4 that

$$Y_m = \frac{AL}{800} \qquad (3.8)$$

and

$$Y_f = \frac{AL}{200} \qquad (3.9)$$

Another useful vertical curve property is one that gives the length of curve required to effect a 1% change in slope. Because the parabolic equation used for roadway elevations (Eq. 3.1) gives a constant rate of change of slope, it can be shown that the horizontal distance required to change the slope by 1% is

$$K = \frac{L}{A} \qquad (3.10)$$

*where*

K = value that is the horizontal distance, in m, required to affect a 1% change in the slope of the vertical curve,
L = length of curve in m, and
A = absolute value of the difference in grades ($|G_1 - G_2|$) expressed as a percentage.

This $K$-value can also be used to compute the high and low point locations of crest and sag vertical curves, respectively (provided the high or low point does not occur at the *PVC* or *PVT*). As shown in Example 3.1, setting $dy/dx = 0$ in Eq. 3.2 and solving for $x$ gives the distance from the *PVC* to the high/low point. If Eq. 3.6 is used to substitute for $a$ in Eq. 3.2 (with $L = KA$), it can be shown that setting $dy/dx = 0$ in Eq. 3.2 gives

$$x_{hl} = K \times |G_1| \qquad (3.11)$$

*where*

$x_{hl}$ =   distance from the *PVC* to the high/low point in m, and

Other terms are as defined previously.

In addition to high/low point computations, *K*-values have an important application in the design of vertical curves, as will be demonstrated in Sections 3.3.3 and 3.3.4.

**EXAMPLE 3.3**   **VERTICAL CURVE DESIGN WITH *K*-VALUES**

A curve has initial and final grades of +3% and −4%, respectively, and is 210 m long. The *PVC* is at elevation 100 m. Graph the vertical curve elevations and the slope of the curve against the length of the curve. Compute the *K*-value and use it to locate the high point of the curve (distance from the *PVC*).

**SOLUTION**

Recall that to find the slope at any point on the curve, we take the derivative of Eq. 3.1, which gives Eq. 3.2. To apply this equation, *a* and *b* need to be determined. From Eq. 3.6,

$$a = \frac{-0.04 - (0.03)}{2(210)} = -0.000166\overline{6}$$

and from Eq. 3.3,

$$b = G_1 = 0.03$$

The results of applying Eq. 3.2 and solving for the slope at all points along the curve, as well as a profile view of the curve itself (by application of Eq. 3.1), are shown graphically in Fig. 3.5 (exaggerating the vertical scale). Figure 3.5 shows the constant rate of change of the slope along the length of the curve. The circular points on the slope-of-curve line correspond to changes in grade of 1%, and these points occur at equal intervals of 30 m.

To show that this is consistent with the *K*-value, Eq. 3.10 gives

$$K = \frac{L}{A} = \frac{210}{|3 - (-4)|} = 30$$

**Figure 3.5** Profile view of vertical curve for Example 3.3 with the graph of the slope at all points along the curve overlaid.

This indicates that there should be a change in grade of 1% for every 30 m of curve length (measured in the horizontal plane), and this is consistent with Fig. 3.5. Applying Eq. 3.11 with the $K$-value of 30 m gives the high point at 90 m from the beginning of the curve ($x_{hl} = 30 \times 3 = 90$ m). This is shown in Fig. 3.5, where the slope of the curve at 90 m is zero (the same result obtained by setting the derivative of Eq. 3.2 equal to zero and solving for $x$). This result can also be explained conceptually based on the definition of the $K$-value. The $K$-value gives the horizontal distance required to effect a 1% change in the slope of the curve, and for this curve, that value is 30 m. Thus, to go from an initial grade ($G_1$) of 3% to a grade of 0% (the high point), it takes a horizontal distance equal to $K \times 3$, or 90 m.

## EXAMPLE 3.4    VERTICAL CURVE DESIGN USING OFFSETS

A vertical curve crosses a 1-m–diameter pipe at right angles. The pipe is located at station 3 + 420 and its centerline is at elevation 333 m. The *PVI* of the vertical curve is at station 3 + 400 and elevation 335 m. The vertical curve is equal tangent, 180 m long, and connects an initial grade of +1.20% and a final grade of −1.08%. Using offsets, determine the depth, below the surface of the curve, of the top of the pipe and determine the station of the highest point on the curve.

## SOLUTION

The *PVC* is at station 3 + 310 (3 + 400 minus 0 + 090, which is half of the curve length), so the pipe is 110 m (3 + 420 minus 3 + 310) from the beginning of the curve (*PVC*). The elevation of the *PVC* will be the elevation of the *PVI* minus the drop in grade over one-half the curve length,

$$335 - (90 \text{ m} \times 0.012 \text{ m/m}) = 333.92 \text{ m}$$

Using this, the elevation of the initial tangent above the pipe is

$$333.92 + (110 \text{ m} \times 0.012 \text{ m/m}) = 335.24 \text{ m}$$

Using Eq. 3.7 to determine the offset above the pipe at $x = 110$ m (the distance of the pipe from the *PVC*), we have

$$Y = \frac{A}{200L} \, x^2$$

$$Y = \frac{|1.2 - (-1.08)|}{200(180)} (110)^2 = 0.77 \text{ m}$$

Thus the elevation of the curve above the pipe is 334.47 m (335.24 − 0.77). The elevation of the top of the pipe is 333.5 m (elevation of the centerline plus one-half of the pipe's diameter), so the pipe is 0.97 m below the surface of the curve (334.47 − 333.5).

To determine the location of the highest point on the curve, we find $K$ from Eq. 3.10 as

$$K = \frac{180}{|1.2 - (-1.08)|} = 78.95$$

and the distance from the *PVC* to the highest point is (from Eq. 3.11)

$$x_{hl} = K \times |G_1| = 78.95 \times 1.2 = 94.73 \text{ m}$$

This gives the station of the highest point at <u>3 + 404.73</u> (3 + 310 plus 0 + 094.73). Note that this example could also be solved by applying Eq. 3.1, setting Eq. 3.2 equal to zero (to determine the location of the highest point on the curve), and following the procedure used in Example 3.1.

### 3.3.2    Stopping Sight Distance

Construction of a vertical curve is generally a costly operation requiring the movement of significant amounts of earthen material. Thus one of the primary challenges facing highway designers is to minimize construction costs (usually by making the vertical curve as short as possible) while still providing an adequate level of safety. An appropriate level of safety is usually defined as that level of safety that gives drivers sufficient sight distance to allow them to safely stop their vehicles to avoid collisions with objects obstructing their forward motion. The provision of adequate roadway drainage is sometimes an important concern as well, but is not discussed in terms of vertical curves in this book (see [AASHTO 2011]). Referring back to the vehicle braking performance concepts discussed in Chapter 2, we can compute this necessary stopping sight distance (SSD) as the summation of vehicle practical stopping distance (Eq. 2.47) and the distance traveled during driver perception/reaction time (Eq. 2.49). That is,

$$\text{SSD} = \frac{V_1^2}{2g\left(\left(\dfrac{a}{g}\right) \pm G\right)} + V_1 \times t_r \tag{3.12}$$

*where*

$$
\begin{aligned}
\text{SSD} &= \text{stopping sight distance in m,} \\
V_1 &= \text{initial vehicle speed in m/s,} \\
g &= \text{gravitational constant, } 9.807 \text{ m/s}^2, \\
a &= \text{deceleration rate in m/s}^2, \\
G &= \text{roadway grade (+ for uphill and − for downhill) in percent/100, and} \\
t_r &= \text{perception/reaction time in s.}
\end{aligned}
$$

Recall from Sections 2.9.5 and 2.9.6 that a value of 3.4 m/s$^2$ for $a$ and a value of 2.5 s for $t_r$ were recommended for roadway design purposes. The design speed of the highway is defined as the maximum safe speed at which a highway can be negotiated assuming near-worst-case conditions (wet-weather conditions). The application of Eq. 3.12 (assuming $G = 0$) produces the stopping sight distances presented in Table 3.1.

**Table 3.1** Stopping Sight Distance

| Design speed (km/h) | Brake reaction distance (m) | Braking distance on level (m) | Stopping sight distance | |
|---|---|---|---|---|
| | | | Calculated (m) | Design (m) |
| 30 | 20.9 | 10.3 | 31.2 | 35 |
| 40 | 27.8 | 18.4 | 46.2 | 50 |
| 50 | 34.8 | 28.7 | 63.5 | 65 |
| 60 | 41.7 | 41.3 | 83.0 | 85 |
| 70 | 48.7 | 56.2 | 104.9 | 105 |
| 80 | 55.6 | 73.4 | 129.0 | 130 |
| 90 | 62.6 | 92.9 | 155.5 | 160 |
| 100 | 69.5 | 114.7 | 184.2 | 185 |
| 110 | 76.5 | 138.8 | 215.3 | 220 |
| 120 | 83.4 | 165.2 | 248.6 | 250 |
| 130 | 90.4 | 193.8 | 284.2 | 285 |

*Note:* Brake reaction distance is based on a time of 2.5 s; a deceleration rate of 3.4 m/s$^2$ is used to determine calculated stopping sight distance.

*Source:* American Association of State Highway and Transportation Officials, *A Policy on Geometric Design of Highways and Streets*, 6th Edition, Washington, DC, 2011. Used by permission.

### 3.3.3 Stopping Sight Distance and Crest Vertical Curve Design

The length of curve ($L$ in Fig. 3.3) is the critical element in providing sufficient SSD on a vertical curve. Longer curve lengths provide more SSD, all else being equal, but are more costly to construct. Shorter curve lengths are less expensive to construct but may not provide adequate SSD due to more rapid changes in slope. What is needed, then, is an expression for minimum curve length given a required SSD. In developing such an expression, crest and sag vertical curves are considered separately.

The case of designing a crest vertical curve for adequate stopping sight distance is illustrated in Fig. 3.6. To determine the minimum length of curve for a required sight distance, the properties of a parabola for an equal tangent curve can be used to show that

For $S < L$

$$L_m = \frac{AS^2}{200\left(\sqrt{H_1} + \sqrt{H_2}\right)^2} \qquad (3.13)$$

For $S > L$

$$L_m = 2S - \frac{200\left(\sqrt{H_1} + \sqrt{H_2}\right)^2}{A} \qquad (3.14)$$

*where*

$L_m$ = minimum length of vertical curve in m,

$A$ = absolute value of the difference in grades ($|G_1-G_2|$), expressed as a percentage, and

Other terms are as defined in Fig. 3.6.

For the sight distance required to provide adequate SSD, current AASHTO design guidelines [2011] use a driver eye height, $H_1$, of 1080 mm and a roadway object height, $H_2$, of 600 mm (the height of an object to be avoided by stopping before a collision). In applying Eqs. 3.13 and 3.14 to determine the minimum length of curve required to provide adequate SSD, we set the sight distance, $S$, equal to the stopping sight distance, SSD (note that the relatively small distance from the driver's eye position to the front of the vehicle is ignored). Substituting AASHTO guidelines for $H_1$ and $H_2$ and letting $S$ = SSD in Eqs. 3.13 and 3.14 gives

For SSD < $L$

$$L_m = \frac{A \times \text{SSD}^2}{658}$$ (3.15)

For SSD > $L$

$$L_m = 2 \times \text{SSD} - \frac{658}{A}$$ (3.16)

**Figure 3.6** Stopping sight distance considerations for crest vertical curves.

$S$ = sight distance in m,

$H_1$ = height of driver's eye above roadway surface in m,

$H_2$ = height of object above roadway surface in m,

$L$ = length of the curve in m,

$PVC$ = point of vertical curve (the initial point of the curve),

$PVI$ = point of vertical intersection (intersection of initial and final grades), and

$PVT$ = point of vertical tangent, which is the final point of the vertical curve (the point where the curve returns to the final grade or, equivalently, the final tangent).

**EXAMPLE 3.5    DESIGN SPEED AND CREST VERTICAL CURVE DESIGN**

A highway is being designed to AASHTO guidelines with a 120-km/h design speed, and at one section, an equal-tangent vertical curve must be designed to connect grades of +1.0% and −2.0%. Determine the minimum length of curve necessary to meet SSD requirements.

*SOLUTION*

If we ignore the effect of grades ($Gs = 0$), the SSD can be read directly from Table 3.1. In this case, the SSD corresponding to a speed of 120 km/h is 250 m. If we assume that $L >$ SSD (an assumption that is typically made), Eq. 3.15 gives

$$L_m = \frac{A \times \text{SSD}^2}{658} = \frac{3 \times 250^2}{658} = \underline{\underline{284.95 \text{ m}}}$$

Since 284.95 > 250, the assumption that $L >$ SSD was correct.

The assumption that $G = 0$, made at the beginning of Example 3.5, is not really correct. If $G \neq 0$, we cannot use the SSD values in Table 3.1 and instead must apply Eq. 3.12 with the appropriate $G$ value. In this problem, if we use the initial grade in Eq. 3.12 (+1.0%), we will underestimate the stopping sight distance because the vertical curve has a slope as steeply positive as this only at the *PVC*. If we use the final grade in Eq. 3.12 (−2.0%), we will overestimate the stopping sight distance because the vertical curve has a slope as steeply negative as this only at the *PVT*. If we knew where the vehicle began to brake, we could use the first derivative of the parabolic curve function (from Eq. 3.2) to give $G$ in Eq. 3.12 and set up the equation to solve for SSD exactly. In practice, policies vary as to how this grade issue is handled. Fortunately, because sight distance tends to be greater on downgrades (which require longer stopping distances) than on upgrades, a self-correction for the effect of grades is generally provided. As a consequence, some design agencies ignore the effect of grades completely, while others assume $G$ is equal to zero for grades less than 3% and use simple adjustments to the SSD, depending on the initial and final grades, for grades of 3% or more. For the remainder of this chapter, we will ignore the effect of grades ($G = 0$ will be used in Eq. 3.12). However, it must be pointed out that the use of SSD grade corrections is very easy and straightforward, and all of the equations presented herein still apply.

The use of Eqs. 3.15 and 3.16 can be simplified if the initial assumption that $L >$ SSD is made, in which case Eq. 3.15 is always used. The advantage of this assumption is that the relationship between $A$ and $L_m$ is linear, and Eq. 3.10 can be used to give

$$L_m = KA \tag{3.17}$$

where $K =$ horizontal distance, in m, required to effect a 1% change in the slope (as in Eq. 3.10), defined as

$$K = \frac{\text{SSD}^2}{658}$$      **(3.18)**

With known SSD for a given design speed (assuming $G = 0$), $K$-values can be computed for crest vertical curves as shown in Table 3.2. Thus the minimum curve length can be obtained (as shown in Eq. 3.17) simply by multiplying $A$ by the $K$-value read from Table 3.2.

Some discussion about the assumption that $L > $ SSD is warranted. This assumption is made because there are two complications that could arise when SSD $> L$. First, if SSD $> L$ the relationship between $A$ and $L_m$ is not linear, so $K$-values cannot be used in the $L = KA$ formula (Eq. 3.10). Second, at low values of $A$, it is possible to get negative minimum curve lengths (see Eq. 3.16). As a result of these complications, the assumption that $L > $ SSD is almost always made in practice, and Eqs. 3.17 and 3.18 and the $K$-values presented in Table 3.2 are used. It is important to note that the assumption that $L > $ SSD (upon which Eqs. 3.17 and 3.18 are based) is a good one because in many cases, $L$ is greater than SSD, and when it is not (SSD $> L$), use of the $L > $ SSD formula (Eq. 3.15 instead of Eq. 3.16) gives longer curve lengths and thus the error is on the conservative, safe side.

A final point relates to the smallest allowable length of curve. Very short vertical curves can be difficult to construct and may not be warranted for safety purposes. As a result, it is common practice to set minimum curve length limits that range from 30 to 100 m depending on individual jurisdictional guidelines. A common alternative to these limits is to set the minimum curve length limit at three times the design speed (with speed in km/h and length in m) [AASHTO 2011].

**Table 3.2** Design Controls for Crest Vertical Curves Based on Stopping Sight Distance

| Design speed (km/h) | Stopping sight distance (m) | Rate of vertical curvature, $K^*$ | |
|---|---|---|---|
| | | Calculated | Design |
| 30 | 35 | 1.9 | 2 |
| 40 | 50 | 3.8 | 4 |
| 50 | 65 | 6.4 | 7 |
| 60 | 85 | 11.0 | 11 |
| 70 | 105 | 16.8 | 17 |
| 80 | 130 | 25.7 | 26 |
| 90 | 160 | 38.9 | 39 |
| 100 | 185 | 52.0 | 52 |
| 110 | 220 | 73.6 | 74 |
| 120 | 250 | 95.0 | 95 |
| 130 | 285 | 123.4 | 124 |

*Rate of vertical curvature, $K$, is the length of curve per percent algebraic difference in intersecting grades ($A$): $K = L/A$.

*Source:* American Association of State Highway and Transportation Officials, *A Policy on Geometric Design of Highways and Streets*, 6th Edition, Washington, DC, 2011. Used by permission.

**EXAMPLE 3.6    DESIGN SPEED AND CREST VERTICAL CURVE DESIGN WITH K-VALUES**

Solve Example 3.5 using the K-values in Table 3.2.

*SOLUTION*

From Example 3.5, $A = 3$. For a 120-km/h design speed, $K = 95.0$ (from Table 3.2). Therefore, application of Eq. 3.17 gives

$$L_m = KA = 95.0(3) = \underline{285 \text{ m}}$$

which is almost identical to the 284.95 m obtained in Example 3.5. This difference is due to rounding. In this example the rounded K of 95 was used as opposed to the calculated K of 94.98. The rounded values are typically used in design for computational convenience. Note, however, that fractional calculated values are always rounded up to the nearest integer value, to be conservative.

**EXAMPLE 3.7    STOPPING-SIGHT DISTANCE AND CREST VERTICAL CURVE DESIGN**

If the grades in Example 3.5 intersect at station $3 + 000$, determine the stationing of the *PVC*, *PVT*, and curve high point for the minimum curve length based on SSD requirements.

*SOLUTION*

Using the curve length from Example 3.6, $L = 285$ m. Since the curve is equal tangent (as are virtually all curves used in practice), one-half of the curve will occur before the *PVI* and one-half after, so that

$$PVC \text{ is at } 3 + 000 - L/2 = 3 + 000 \text{ minus } 0 + 142.5 = \underline{2 + 857.5}$$

$$PVT \text{ is at } 3 + 000 + L/2 = 3 + 000 \text{ plus } 0 + 142.5 = \underline{3 + 142.5}$$

For the stationing of the high point, Eq. 3.11 is used:

$$x_{hl} = K \times |G_1| = 95(1) = 95 \text{ m}$$

or

$$\text{station } 2 + 857.5 \text{ plus } 0 + 095 = \underline{2 + 952.5}$$

## 3.3.4    Stopping Sight Distance and Sag Vertical Curve Design

Sag vertical curve design differs from crest vertical curve design in the sense that sight distance is governed by nighttime conditions because in daylight, sight distance on a sag vertical curve is unrestricted. Thus the critical concern for sag vertical curve design is the length of roadway illuminated by the vehicle headlights, which is a function of the height of the headlight above the roadway and the inclined angle of the headlight beam, relative to the horizontal plane of the car. The sag vertical curve sight distance design problem is illustrated in Fig. 3.7.

**Figure 3.7** Stopping sight distance considerations for sag vertical curves.

$S$ = sight distance in m,

$H$ = height of headlight in m,

$\beta$ = inclined angle of headlight beam in degrees,

$L$ = length of the curve in m

$PVC$ = point of the vertical curve (the initial point of the curve)

$PVI$ = point of vertical intersection (intersection of initial and final grades), and

$PVT$ = point of vertical tangent, which is the final point of the vertical curve (the point where the curve returns to the final grade or, equivalently, the final tangent).

To determine the minimum length of curve for a required sight distance, the properties of a parabola for an equal-tangent curve can be used to show that

For $S < L$

$$L_m = \frac{AS^2}{200\left(H + S \tan \beta\right)}$$ (3.19)

For $S > L$

$$L_m = 2S - \frac{200\left(H + S \tan \beta\right)}{A}$$ (3.20)

*where*

$L_m$ = minimum length of vertical curve in m,

$A$ = absolute value of the difference in grades ($|G_1-G_2|$), expressed as a percentage, and

Other terms are as defined in Fig. 3.7.

For the sight distance required to provide adequate SSD, current AASHTO design guidelines [2011] use a headlight height of 600 mm and an upward angle of one degree. Substituting these design guidelines and $S$ = SSD (as was done in the crest vertical curve case, and again ignoring the relatively small distance from the driver's eye position to the front of the vehicle) into Eqs. 3.19 and 3.20 gives

For SSD < L

$$L_m = \frac{A \times \text{SSD}^2}{120 + 3.5 \times \text{SSD}} \qquad (3.21)$$

For SSD > L

$$L_m = 2 \times SSD - \frac{120 + 3.5 \times \text{SSD}}{A} \qquad (3.22)$$

*where*

SSD = stopping sight distance in m, and
Other terms are as defined previously.

As was the case for crest vertical curves, *K*-values can be computed by assuming *L* > SSD, which gives us the linear relationship between $L_m$ and *A* as shown in Eq. 3.21. Thus for sag vertical curves (with $L_m = KA$),

$$K = \frac{\text{SSD}^2}{120 + 3.5 \times \text{SSD}} \qquad (3.23)$$

*where*

*K* = horizontal distance, in m, required to effect a 1% change in the slope (as in Eq. 3.10), and
Other terms are as defined previously.

The *K*-values corresponding to design-speed-based SSDs are presented in Table 3.3. As was the case for crest vertical curves, some caution should be exercised in using this table because the assumption that *G* = 0 (for determining SSD) is used. Also, assume that *L* > SSD is a safe, conservative assumption (as was the case for crest vertical curves) and the smallest allowable curve lengths for sag curves are the same as those for crest curves (see discussion in Section 3.3.3).

---

**EXAMPLE 3.8    SAG VERTICAL CURVE FUNDAMENTALS WITH DESIGN SPEED**

An equal tangent sag vertical curve has an initial grade of –2.5%. It is known that the final grade is positive and that the low point is at elevation 82 m and station 1 + 410.000. The *PVT* of the curve is at elevation 83.5 m and the design speed of the curve is 60 km/h. Determine the station and elevation of the *PVC* and *PVI*.

**Table 3.3** Design Controls for Sag Vertical Curves Based on Stopping Sight Distance

| Design speed (km/h) | Stopping sight distance (m) | Rate of vertical curvature, $K$* | |
|---|---|---|---|
| | | Calculated | Design |
| 30 | 35 | 5.1 | 6 |
| 40 | 50 | 8.5 | 9 |
| 50 | 65 | 12.2 | 13 |
| 60 | 85 | 17.3 | 18 |
| 70 | 105 | 22.6 | 23 |
| 80 | 130 | 29.4 | 30 |
| 90 | 160 | 37.6 | 38 |
| 100 | 185 | 44.6 | 45 |
| 110 | 220 | 54.4 | 55 |
| 120 | 250 | 62.8 | 63 |
| 130 | 285 | 72.7 | 73 |

*Rate of vertical curvature, $K$, is the length of curve per percent algebraic difference in intersecting grades ($A$): $K = L/A$.

*Source*: American Association of State Highway and Transportation Officials, *A Policy on Geometric Design of Highways and Streets*, 6th Edition, Washington, DC, 2011. Used by permission.

*SOLUTION*

From Table 3.3 it can be seen the $K = 18$ for a design speed of 60 km/h. With this, equation 3.11 is used to find the distance of the low point from the *PVC*:

$$x_{hl} = K \times |G_1| = 18(2.5) = 45 \text{ m}$$

Knowing the elevation of the low point (82 m) and the distance of the low point from the *PVC* (45 m), Eq. 3.1 can be applied to determine the elevation of the *PVC* ($c$ in the Eq. 3.1):

$$y = ax^2 + bx + c$$

From Eq. 3.3,

$$b = G_1 = -0.025$$

and from Eq. 3.6,

$$a = \frac{G_2 - G_1}{2L}$$

Because the final grade is known to be positive (and with $G_2$ being negative),

$$G_2 - G_1 = |G_1 - G_2| = A,$$

Using $L = KA$ from Eq. 3.17, Eq. 3.6 becomes

$$a = \frac{G_2 - G_1}{2L} = \frac{A/100}{2KA} = \frac{0.01}{2K} = \frac{0.01}{2(18)} = 0.000278$$

Note that $A$ is divided by 100 to make certain that the units are consistent with the denominator since $L$ is in m from the $L = KA$ equation. At the low point, $y = 82$ m so solving for $c$ in Eq. 3.1 with $x = 45$, $a = 0.000278$ and $b = -0.025$ gives

$$82 = 0.000278(45)^2 + (-0.025)(45) + c$$
$$c = \underline{82.562 \text{ m}} = \text{elevation of the } PVC$$

Knowing that the station of the low point is $1 + 410.000$ and the distance from the $PVC$ to the low point is 82 m,

$$\text{station of } PVC = 1 + 410.000 \text{ minus } 0 + 082.000 = \underline{\underline{1 + 328.000}}$$

Next, the length of the curve is determined by applying Eq. 3.1. Because it is known that the elevation of the $PVT$ is 83.5 m, using $y = 83.5$ means that $x = L$ in Eq. 3.1, so (with $c = 82.562$)

$$y = aL^2 + bL + c$$
$$83.5 = 0.000278L^2 + (-0.025)L + 82.562$$
$$L = \underline{\underline{118.42 \text{ m}}}$$

The station of the $PVI$ is

$$\text{station of } PVI = \text{station of } PVC + L/2$$
$$= 1 + 328.000 \text{ plus } (118.42/2) = \underline{\underline{1 + 387.210}}$$

Finally, the elevation of the $PVI$ is determined as

$$\text{elevation of } PVI = \text{elevation of } PVC + G_1(L/2)$$
$$= 82.562 - 0.025(118.42/2) = \underline{\underline{81.082 \text{ m}}}$$

## EXAMPLE 3.9    COMBINED SAG AND CREST VERTICAL CURVES WITHOUT A CONSTANT GRADE CONNECTION

An existing tunnel needs to be connected to a newly constructed bridge with sag and crest vertical curves. The profile view of the tunnel and bridge is shown in Fig. 3.8. Develop a vertical alignment to connect the tunnel and bridge by determining the highest possible common design speed for the sag and crest (equal-tangent) vertical curves needed. Compute the stationing and elevations of $PVC$, $PVI$, and $PVT$ curve points.

**Figure 3.8** Profile view (vertical alignment diagram) for Example 3.9.

**SOLUTION**

From left to right (see Fig. 3.8), a sag vertical curve (with subscript $s$) and a crest vertical curve (with subscript $c$) are needed to connect the tunnel and bridge. From the given information, it is known that $G_{1s} = 0\%$ (the initial slope of the sag vertical curve) and $G_{2c} = 0\%$ (the final slope of the crest vertical curve). To obtain the highest possible design speed, we want to use all of the horizontal distance available. This means we want to connect the curve so that the *PVT* of the sag curve ($PVT_s$) will be the *PVC* of the crest curve ($PVC_c$). If this is the case, $G_{2s} = G_{1c}$ and since $G_{1s} = G_{2c} = 0$, $A_s = A_c = A$, the common algebraic difference in the grades.

Since 310 m separates the tunnel and bridge,

$$L_s + L_c = 310$$

Also, the summation of the end-of-curve offset for the sag curve and the beginning-of-curve offset (relative to the final grade) for the crest curve must equal 12 m. Using the equation for the final offset, Eq. 3.9, we have

$$\frac{AL_s}{200} + \frac{AL_c}{200} = 12$$

Rearranging,

$$\frac{A}{200}\left(L_s + L_c\right) = 12$$

and since $L_s + L_c = 310$,

$$\frac{A}{200}(310) = 12$$

Solving for $A$ gives 7.742%. The problem now becomes one of finding $K$-values that allow $L_s + L_c = 310$. Since $L = KA$ (Eq. 3.17), we can write

$$K_s A + K_c A = 310$$

Substituting $A = 7.742$,

$$K_s + K_c = 40$$

To find the highest possible design speed, Tables 3.2 and 3.3 are used to arrive at $K$-values to solve $K_s + K_c = 40$. From Tables 3.2 and 3.3 it is apparent that the highest possible design speed is 70 km/h, at which speed $K_c = 17$ and $K_s = 23$ (the summation of $K$'s is 40).

To arrive at the stationing of curve points, we first determine curve lengths as

$$L_s = K_s A = 23(7.742) = 178 \text{ m}$$
$$L_c = K_c A = 17(7.742) = 132 \text{ m}$$

Since the station of $PVC_s$ is $\underline{0 + 000}$ (given), it is clear that the $PVI_s = \underline{0 + 089}$, $PVT_s = PVC_c = \underline{0 + 178}$, $PVI_c = \underline{0 + 244}$, and $PVT_c = \underline{0 + 310}$. For elevations, $PVC_s = PVI_s = \underline{30 \text{ m}}$ and $PVI_c = PVT_c = \underline{42 \text{ m}}$. Finally, the elevation of $PVT_s$ and $PVC_c$ can be computed as

$$30 + \frac{AL_s}{200} = 30 + \frac{7.742(178)}{200} = \underline{\underline{36.9 \text{ m}}}$$

**EXAMPLE 3.10**  **COMBINED SAG AND CREST VERTICAL CURVES WITH A CONSTANT GRADE CONNECTION**

Consider the conditions described in Example 3.9. Suppose a design speed of only 60 km/h is needed. Determine the lengths of curves required to connect the bridge and tunnel while keeping the connecting grade as small as possible.

**SOLUTION**

It is known that the 310 m separating the tunnel and bridge are more than enough to connect a 60-km/h alignment because Example 3.9 showed that 70 km/h is possible. Therefore, to connect the tunnel and bridge and keep the connecting grade as small as possible, we will place a constant-grade section between the sag and crest curves (as shown in Fig. 3.9).

The elevation change will be the final offsets of the sag and crest curves plus the change in elevation resulting from the constant-grade section connecting the two curves. Let $G_{con}$ be the grade of the constant-grade section. This means that $G_{2s} = G_{1c} = G_{con}$, and since $G_{1s} = G_{2c} = 0$ (as in Example 3.8), $G_{con} = A_s = A_c = A$. The equation that will solve the vertical alignment for this problem is

$$\frac{AL_s}{200} + \frac{AL_c}{200} + \frac{A(310 - L_s - L_c)}{100} = 12$$

where the third term accounts for the elevation difference attributable to the constant-grade section connecting the sag and crest curves (the 100 in the denominator of this term converts $A$ from percent to m/m). Using $L = KA$, we have

$$\frac{A^2 K_s}{200} + \frac{A^2 K_c}{200} + \frac{A(310 - K_s A - K_c A)}{100} = 12$$

From Table 3.2, $K_c = 11$, and from Table 3.3, $K_s = 18$. Putting these values in the above equation gives

$$0.145 A^2 + 3.10 A - 0.29 A^2 = 12$$
$$-0.145 A^2 + 3.10 A - 12 = 0$$

Solving this gives $A = 5.076$ and $A = 16.303$; $A = 5.076$ is chosen because we want to minimize the grade. For this value of $A$, the curve lengths are

$$L_s = K_s A = 18(5.076) = \underline{91.368 \text{ m}}$$

$$L_c = K_c A = 11(5.076) = \underline{55.836 \text{ m}}$$

and the length of the constant-grade section will be 162.796 m. This means that about 8.264 m of the elevation difference will occur in the constant-grade section, with the remainder of the elevation difference attributable to the final curve offsets.

**Figure 3.9** Profile view (vertical alignment diagram) for Example 3.10.

Another variation of this type of problem is the case when the initial and final grades are not equal to zero, as in the following example.

## EXAMPLE 3.11 COMBINED SAG AND CREST VERTICAL CONNECTING HIGHWAY SEGMENTS WITH NON-ZERO GRADES

Two sections of highway are separated by 700 m, as shown in Fig. 3.10. Determine the curve lengths required for a 100-km/h vertical alignment to connect these two highway segments while keeping the connecting grade as small as possible.

**Figure 3.10** Profile view (vertical alignment diagram) for Example 3.11.

***SOLUTION***

Let $Y_{fc}$ and $Y_{fs}$ be the final offsets of the crest and sag curves, respectively. Let $G_{con}$ be the slope of a constant-grade section connecting the crest and sag curves (we will assume that the horizontal distance is sufficient to connect the highway with a 100-km/h alignment; if this assumption is incorrect, the following equations will produce an obviously erroneous answer and a lower design speed will have to be chosen). Finally, let $\Delta y_{con}$ be the change in elevation over the constant-grade section, and let $\Delta y_c$ and $\Delta y_s$ be the changes in elevation due to the extended curve tangents. The elevation equation is then (see Fig. 3.10)

$$Y_{fc} + Y_{fs} + \Delta y_{con} + \Delta y_s = 10 + \Delta y_c$$

Substituting offset equations and equations for elevation changes (with subscripts $c$ for crest and $s$ for sag),

$$\frac{A_c L_c}{200} + \frac{A_s L_s}{200} + \frac{G_{con}(700 - L_c - L_s)}{100} + \frac{1.0\, L_s}{100} = 10 + \frac{3.0\, L_c}{100}$$

Using $L = KA$, this equation becomes

$$\frac{A_c^2 K_c}{200} + \frac{A_s^2 K_s}{200} + \frac{G_{con}(700 - K_c A_c - K_s A_s)}{100} + \frac{1.0\, K_s A_s}{100} = 10 + \frac{3.0\, K_c A_c}{100}$$

From Tables 3.2 and 3.3, $K_c = 52$ and $K_s = 45$. Substituting and defining $A$'s (and arranging the equation so that $G_{con}$ will be positive, and assuming $G_{con}$ will be greater than 1%) gives

$$\frac{(3 + G_{con})^2\, 52}{200} + \frac{(G_{con} - 1)^2\, 45}{200} + \frac{G_{con}(700 - 52(3 + G_{con}) - 45(G_{con} - 1))}{100}$$
$$+ \frac{1.0(45(G_{con} - 1))}{100} = 10 + \frac{3.0(52(3 + G_{con}))}{100}$$

or

$$-0.485 G_{con}^2 + 5.890 G_{con} - 12.565 = 0$$

which gives $G_{con} = 2.761$ (the other possible solution is 9.383, which is rejected because we want to minimize the grade). Using $L = KA$ gives $L_c = \underline{299.572\ \text{m}}$ ($52 \times 5.761$) and $L_s = \underline{79.245\ \text{m}}$ ($45 \times 1.761$). Accordingly, the length of the constant-grade section is 321.183 m ($700 - L_c - L_s$). Elevations and the locations of curve points can be readily computed with this information.

### 3.3.5   Passing Sight Distance and Crest Vertical Curve Design

In addition to stopping sight distance, in some instances it may be desirable to provide adequate passing sight distance, which can be an important issue in two-lane highway design (one lane in each direction). Passing sight distance is a factor only in crest vertical curve design because, for sag curves, the sight distance is unobstructed looking up or down the grade, and at night, the headlights of oncoming or opposing vehicles will be seen. In determining the sight distance required to pass on a crest vertical curve, Eqs. 3.13 and 3.14 will apply; however, whereas the driver's eye height, $H_1$, will remain at 1080 mm, $H_2$ will now also be set to 1080 mm. This value for $H_2$ is the assumed value for the portion of a vehicle's height necessary to be visible such that it can be recognized as an opposing vehicle to a driver performing a passing maneuver. Using the same height for both $H_1$ and $H_2$ provides a reciprocal design relationship; that is, if the driver of the passing vehicle can see the opposing vehicle, then the opposing vehicle driver can see the passing vehicle. Substituting these $H$ values into Eqs. 3.13 and 3.14 and letting the sight distance $S$ equal the passing sight distance, PSD, gives

For PSD $< L$

$$L_m = \frac{A \times PSD^2}{864} \qquad (3.24)$$

For PSD $> L$

$$L_m = 2 \times PSD - \frac{864}{A} \qquad (3.25)$$

*where*

$\quad L_m$ = minimum length of vertical curve in m,

$\quad A$ = absolute value of the difference in grades ($|G_1 - G_2|$), expressed as a percentage, and

$\quad PSD$ = passing sight distance in m.

As was the case for stopping sight distance, it is typically assumed that the length of curve is greater than the required sight distance (in this case $L > PSD$), so

$$K = \frac{PSD^2}{864} \qquad (3.26)$$

*where*

$\quad K$ = horizontal distance, in m, required to effect a 1% change in the slope (as in Eq. 3.10), and

$\quad PSD$ = passing sight distance in m.

The passing sight distance (PSD) used for design is assumed to consist of four distances: (1) the initial maneuver distance (which includes the driver's perception/reaction time and the time it takes to bring the vehicle from its trailing speed to the point of encroachment on the left lane), (2) the distance that the passing vehicle traverses while occupying the left lane, (3) the clearance length between the passing and opposing vehicles at the end of the passing maneuver, and (4) the distance traversed by the opposing vehicle during two-thirds of the time the passing vehicle occupies the left lane. The determination of these distances is undertaken using assumptions regarding the time of the initial maneuver, average vehicle acceleration, and the speeds of passing, passed, and opposing vehicles. The sum of these four distances gives the required passing sight distance. The reader is referred to [AASHTO 2011] for a complete description of the assumptions made in determining required passing sight distances.

The minimum distances needed to pass (PSDs) at various design speeds, along with the corresponding $K$-values as computed from Eq. 3.26, are presented in Table 3.4. Notice that the $K$-values in this table are much higher than those required for stopping sight distance (as given in Table 3.2). As a result, designing a crest curve to provide adequate passing sight distance is often an expensive proposition (due to the length of curve required).

**Table 3.4** Design Controls for Crest Vertical Curves Based on Passing Sight Distance.

| Design speed (km/h) | Passing sight distance (m) | Rate of vertical curvature, $K*$ |
| --- | --- | --- |
| 30 | 120 | 17 |
| 40 | 140 | 23 |
| 50 | 160 | 30 |
| 60 | 180 | 38 |
| 70 | 210 | 51 |
| 80 | 245 | 69 |
| 90 | 280 | 91 |
| 100 | 320 | 119 |
| 110 | 355 | 146 |
| 120 | 395 | 181 |
| 130 | 440 | 224 |

*Rate of vertical curvature, $K$, is the length of curve per percent algebraic difference in intersecting grades ($A$): $K = L/A$.

*Source*: American Association of State Highway and Transportation Officials, *A Policy on Geometric Design of Highways and Streets*, Washington, DC, 2011. Used by permission.

---

**EXAMPLE 3.12   VERTICAL CURVE DESIGN WITH PASSING SIGHT DISTANCE**

An equal-tangent crest vertical curve is 332 m long and connects a +2.5% and a −1.5% grade. If the design speed of the roadway is 90 km/h, does this curve have adequate passing sight distance?

**SOLUTION**

To determine the length of curve required to provide adequate passing sight distance at a design speed of 90 km/h, we use $L = KA$ with $K = 91$ (as read from Table 3.4). This gives

$$L = 91(4.0) = 364 \text{ m}$$

Since the curve is only 330 m long, it is not long enough to provide adequate passing sight distance. Alternatively, the $K$-value for the existing design can be compared with that required for a PSD-based design. The $K$-value for the existing design is

$$K = \frac{332}{4} = 83$$

Since the $K$-value of 83 for the existing curve design is less than 91, this curve does not provide adequate PSD for a 90-km/h design speed.

### 3.3.6 Underpass Sight Distance and Sag Vertical Curve Design

As mentioned in Section 3.3.4, design for sag curves is based on nighttime conditions because during daytime conditions a driver can see the entire sag curve. However, in the case of a sag curve being built under an overhead structure (such as roadway or railroad crossing), a driver's line of sight may be restricted so that the entire curve length is not visible. An example of this situation is shown in Fig. 3.11.

In designing the sag curve, it is essential that the curve be long enough to provide a suitably gradual rate of curvature such that the overhead structure does not block the line of sight and allows the required stopping sight distance for the specified design speed to be maintained.

**Figure 3.11** Stopping sight distance considerations for underpass sag curves.

Used by permission from American Association of State Highway and Transportation Officials, *A Policy on Geometric Design of Highways and Streets*, 6$^{th}$ Edition, Washington, DC, 2011.

$S$ = sight distance in m,

$H_1$ = height of driver's eye in m,

$H_2$ = height of object in m,

$H_c$ = clearance height of overpass structure above roadway in m,

$L$ = length of the curve in m

$G_1$ = initial roadway grade in percent or m/m,

$G_2$ = final roadway grade in percent or m/m,

$PVC$ = point of the vertical curve (the initial point of the curve), and

$PVT$ = point of vertical tangent, which is the final point of the vertical curve (the point where the curve returns to the final grade or, equivalently, the final tangent).

Again, by using the properties of a parabola for an equal-tangent vertical curve, it can be shown that the minimum length of sag curve for a required sight distance and clearance height is

For $S < L$

$$L_m = \frac{AS^2}{800\left(H_c - \left(\frac{H_1 + H_2}{2}\right)\right)}$$ (3.27)

For $S > L$

$$L_m = 2S - \frac{800\left(H_c - \left(\frac{H_1 + H_2}{2}\right)\right)}{A} \qquad (3.28)$$

*where*

$L_m$ =   minimum length of vertical curve in m,

$A$ =   absolute value of the difference in grades ($|G_1 - G_2|$), expressed as a percentage, and

Other terms are as defined in Fig. 3.11.

Current AASHTO design guidelines [2011] use a driver eye height, $H_1$, of 2.4 m for a truck driver, and an object height, $H_2$, of 600 mm for the taillights of a vehicle. Substituting these values and $S$ = SSD into Eqs. 3.27 and 3.28 gives

For SSD $< L$

$$L = \frac{A \times \text{SSD}^2}{800(H_c - 1.5)} \qquad (3.29)$$

For SSD $> L$

$$L = 2 \times \text{SSD} - \frac{800(H_c - 1.5)}{A} \qquad (3.30)$$

*where*

SSD   =   stopping sight distance in m, and

Other terms are as defined previously.

In the case where there is an existing sag curve alignment and a new overpass structure is to be built over it, the above equations can be rearranged to solve for the necessary clearance height, $H_c$, of the overpass structure to provide for the required stopping sight distance. When the clearance height is determined in this manner, it is necessary to check this value against the minimum clearance heights based on maximum vehicle height regulations and AASHTO recommendations. Maximum vehicle heights as regulated by state laws range from 4.1 to 4.4 m. AASHTO [2011] recommends a minimum structure clearance height of 4.4 m and a desirable clearance height of 5.0 m. AASHTO [2011] also recommends that clearance heights be no less than 300 mm greater than the maximum allowable vehicle height. This provides a margin for snow or ice accumulation, some over-height vehicles, and future roadway resurfacings. Thus, in building a new overpass structure over an existing sag curve alignment, the clearance height must be determined for both required stopping sight distance and maximum allowable vehicle height for that roadway, and the greater of the two values should be used.

**EXAMPLE 3.13 UNDERPASS VERTICAL CURVE CLEARANCE**

An equal-tangent sag curve has an initial grade of −4.0%, a final grade of +3.0%, and a length of 385 m. An overhead guide sign is being placed directly over the *PVI* of this curve. At what height above the roadway should the bottom of this sign be placed?

***SOLUTION***

For this situation, Eq. 3.29 or 3.30 must be used to solve for the necessary clearance height based on stopping sight distance. Thus, the required SSD must be determined for the given sag curve specifications, based on the design speed. The design speed for the curve can be determined from the *K*-value by applying Eq. 3.10, as follows:

$$K = \frac{L}{A} = \frac{385}{|-4-3|} = 55$$

From Table 3.3, this *K*-value corresponds approximately to a design speed of 110 km/h ($K = 55$). For a 110-km/h design speed, the required stopping sight distance is 220 m. Since the curve length is greater than the required SSD (385 > 220), Eq. 3.29 applies:

$$L = \frac{A \times \text{SSD}^2}{800\left(H_c - 1.5\right)}$$

Rearranging this equation to solve for the clearance height, $H_c$, and substituting $A = 7\%$, SSD = 220 m, and $L = 385$ m gives

$$H_c = \frac{A \times \text{SSD}^2}{800L} + 1.5$$

$$= \frac{7 \times 220^2}{800\left(385\right)} + 1.5$$

$$= 2.6 \text{ m}$$

Although only 2.6 m is needed for SSD requirements, AASHTO [2011] recommends a minimum clearance height of 4.4 m to take maximum vehicle height into account. Thus, the bottom of the overpass should be placed at least 4.4 m above the roadway surface (at the *PVI*), but desirably at a height of 5.0 m according to AASHTO [2011].

## 3.4 HORIZONTAL ALIGNMENT

The critical aspect of horizontal alignment is the horizontal curve, with the focus on design of the directional transition of the roadway in a horizontal plane. Stated differently, a horizontal curve provides a transition between two straight (or tangent) sections of roadway. A key concern in this directional transition is the ability of the vehicle to negotiate a horizontal curve. (Provision of adequate drainage is also important, but is not discussed in this book; see [AASHTO 2011].) As was the case with the straight-line vehicle performance characteristics discussed at length in Chapter 2, the highway engineer must design a horizontal alignment to accommodate the cornering capabilities of a variety of vehicles, ranging from nimble sports cars to ponderous trucks. A theoretical assessment of vehicle cornering at the level of detail given to straight-line performance in Chapter 2 is beyond the scope of this book (see [Campbell 1978] and [Wong 2008]). Instead, vehicle cornering performance is viewed only at the practical design-oriented level, with equations simplified in a manner similar to that used for the stopping-distance equation discussed in Section 2.9.5.

### 3.4.1 Vehicle Cornering

Figure 3.12 illustrates the forces acting on a vehicle during cornering.

**Figure 3.12** Vehicle cornering forces.

$R_v$ = radius defined to the vehicle's traveled path in m,

$\alpha$ = angle of incline in degrees,

$e$ = number of vertical m of rise per 100 m of horizontal distance,

$W$ = weight of the vehicle in N,

$W_n$ = vehicle weight normal to the roadway surface in N,

$W_p$ = vehicle weight parallel to the roadway surface in N,

$F_f$ = side frictional force (centripetal, in N),

$F_c$ = centripetal force (lateral acceleration × mass, in N),

$F_{cp}$ = centripetal force acting parallel to the roadway surface in N, and

$F_{cn}$ = centripetal force acting normal to the roadway surface in N.

Some basic horizontal curve relationships can be derived by noting that

$$W_p + F_f = F_{cp} \tag{3.31}$$

From basic physics this equation can be written as [with $F_f = f_s(W_n + F_{cn})$]

$$W \sin \alpha + f_s \left( W \cos \alpha + \frac{WV^2}{gR_v} \sin \alpha \right) = \frac{WV^2}{gR_v} \cos \alpha \qquad (3.32)$$

where

$f_s$ = coefficient of side friction (unitless),
$V$ = vehicle speed in m/s,
$g$ = gravitational constant, 9.807 m/s², and

Other terms are as defined in Fig 3.12.

Dividing both sides of Eq. 3.32 by $W \cos \alpha$ gives

$$\tan \alpha + f_s = \frac{V^2}{gR_v}(1 - f_s \tan \alpha) \qquad (3.33)$$

The term $\tan \alpha$ indicates the superelevation of the curve (banking) and can be expressed in percent; it is denoted $e$ ($e = 100 \tan \alpha$). In words, the superelevation is the number of vertical meters of rise per 100 meters of horizontal distance (see Fig. 3.12). The term $f_s \tan \alpha$ in Eq. 3.33 is conservatively set equal to zero for practical applications due to the small values that $f_s$ and $\alpha$ typically assume (this is equivalent to ignoring the normal component of centripetal force). With $e = 100 \tan \alpha$, Eq. 3.33 can be arranged as

$$R_v = \frac{V^2 \left( 1 - \right)}{g \left( f_s + \dfrac{e}{100} \right)} \qquad (3.34)$$

EXAMPLE 3.14    SUPERELEVATION ON HORIZONTAL CURVES

A roadway is being designed for a speed of 110 km/h. At one horizontal curve, it is known that the superelevation is 8.0% and the coefficient of side friction is 0.11. Determine the minimum radius of curve (measured to the traveled path) that will provide for safe vehicle operation.

SOLUTION

The application of Eq. 3.34 gives [with 0.2778 (1000/3600) converting from km/h to m/s]

$$R_v = \frac{V^2}{g \left( f_s + \dfrac{e}{100} \right)} = \frac{(110 \times 0.2778)^2}{9.807(0.11 + 0.08)} = \underline{\underline{501.141 \text{ m}}}$$

This value is the minimum radius, because radii smaller than 501.141 m will generate centripetal forces higher than those capable of being safely supported by the superelevation and the side frictional force.

In the actual design of a horizontal curve, the engineer must select appropriate values of $e$ and $f_s$. The value selected for superelevation, $e$, is critical because high rates of superelevation can cause vehicle steering problems on the horizontal curve, and in cold climates, ice on the roadway can reduce $f_s$ such that vehicles traveling at less than the design speed on an excessively superelevated curve could slide inward off the curve due to gravitational forces. AASHTO provides general guidelines for the selection of $e$ and $f_s$ for horizontal curve design, as shown in Table 3.5. The values presented in this table are grouped by five values of maximum $e$. The selection of any one of these five maximum $e$ values is dependent on the type of road (for example, higher maximum $e$'s are permitted on freeways than on arterials and local roads) and local design practice. Limiting values of $f_s$ are simply a function of design speed. Table 3.5 also presents calculated radii (given $V$, $e$, and $f_s$) by applying Eq. 3.34.

### 3.4.2   Horizontal Curve Fundamentals

In connecting straight (tangent) sections of roadway with a horizontal curve, several options are available. The most obvious of these is the simple circular curve, which is just a curve with a single, constant radius. Other options include reverse curves, compound curves, and spiral curves. Reverse curves generally consist of two consecutive curves that turn in opposite directions. They are used to shift the alignment of a highway laterally. The curves used are usually circular and have equal radii. Reverse curves, however, are not recommended because drivers may find it difficult to stay within their lane as a result of sudden changes to the alignment. Compound curves consist of two or more curves, usually circular, in succession. Compound curves are used to fit horizontal curves to very specific alignment needs, such as interchange ramps, intersection curves, or difficult topography. In designing compound curves, care must be taken not to have successive curves with widely different radii, as this will make it difficult for drivers to maintain their lane position as they transition from one curve to the next. Spiral curves are curves with a continuously changing radius. Spiral curves are sometimes used to transition a tangent section of roadway to a circular curve. In such a case, the radius of the spiral curve is equal to infinity where it connects to the tangent section and ends with the radius value of the connecting circular curve at the other end. Because motorists usually create their own transition paths between tangent sections and circular curves by utilizing the full lane width available, spiral curves are not often used. However, there are exceptions. Spiral curves are sometimes used on high-speed roadways with sharp horizontal curves and also to gradually introduce the superelevation of an upcoming horizontal curve. To illustrate the basic principles involved in horizontal curve design, this book will focus only on the single simple circular curve. For detailed information regarding these additional horizontal curve types, refer to standard route-surveying texts, such as Hickerson [1964].  Figure 3.13 shows the basic elements of a simple horizontal curve.

**Table 3.5**  Minimum Radius Using Limiting Values of $e$ and $f_s$

| Design speed (km/h) | Maximum $e$ (%) | Limiting values of $f_s$ | Total ($e/100 + f_s$) | Calculated radius, $R_v$ (m) | Rounded radius, $R_v$ (m) | Design speed (km/h) | Maximum $e$ (%) | Limiting values of $f_s$ | Total ($e/100 + f_s$) | Calculated radius, $R_v$ (m) | Rounded radius, $R_v$ (m) |
|---|---|---|---|---|---|---|---|---|---|---|---|
| 15 | 4.0 | 0.40 | 0.44 | 4.0 | 4 | 15 | 10.0 | 0.40 | 0.50 | 3.5 | 4 |
| 20 | 4.0 | 0.35 | 0.39 | 8.1 | 8 | 20 | 10.0 | 0.35 | 0.45 | 7.0 | 7 |
| 30 | 4.0 | 0.28 | 0.32 | 22.1 | 22 | 30 | 10.0 | 0.28 | 0.38 | 18.6 | 19 |
| 40 | 4.0 | 0.23 | 0.27 | 46.7 | 47 | 40 | 10.0 | 0.23 | 0.33 | 38.2 | 38 |
| 50 | 4.0 | 0.19 | 0.23 | 85.6 | 86 | 50 | 10.0 | 0.19 | 0.29 | 67.9 | 68 |
| 60 | 4.0 | 0.17 | 0.21 | 135.0 | 135 | 60 | 10.0 | 0.17 | 0.27 | 105.0 | 105 |
| 70 | 4.0 | 0.15 | 0.19 | 203.1 | 203 | 70 | 10.0 | 0.15 | 0.25 | 154.3 | 154 |
| 80 | 4.0 | 0.14 | 0.18 | 280.0 | 280 | 80 | 10.0 | 0.14 | 0.24 | 210.0 | 210 |
| 90 | 4.0 | 0.13 | 0.17 | 375.2 | 375 | 90 | 10.0 | 0.13 | 0.23 | 277.3 | 277 |
| 100 | 4.0 | 0.12 | 0.16 | 492.1 | 492 | 100 | 10.0 | 0.12 | 0.22 | 357.9 | 358 |
|  |  |  |  |  |  | 110 | 10.0 | 0.11 | 0.21 | 453.7 | 454 |
|  |  |  |  |  |  | 120 | 10.0 | 0.09 | 0.19 | 596.8 | 597 |
| 15 | 6.0 | 0.40 | 0.46 | 3.9 | 4 | 130 | 10.0 | 0.08 | 0.18 | 739.3 | 739 |
| 20 | 6.0 | 0.35 | 0.41 | 7.7 | 8 |  |  |  |  |  |  |
| 30 | 6.0 | 0.28 | 0.34 | 20.8 | 21 |  |  |  |  |  |  |
| 40 | 6.0 | 0.23 | 0.29 | 43.4 | 43 | 15 | 12.0 | 0.40 | 0.52 | 3.4 | 3 |
| 50 | 6.0 | 0.19 | 0.25 | 78.7 | 79 | 20 | 12.0 | 0.35 | 0.47 | 6.7 | 7 |
| 60 | 6.0 | 0.17 | 0.23 | 123.2 | 123 | 30 | 12.0 | 0.28 | 0.40 | 17.7 | 18 |
| 70 | 6.0 | 0.15 | 0.21 | 183.7 | 184 | 40 | 12.0 | 0.23 | 0.35 | 36.0 | 36 |
| 80 | 6.0 | 0.14 | 0.20 | 252.0 | 252 | 50 | 12.0 | 0.19 | 0.31 | 63.5 | 64 |
| 90 | 6.0 | 0.13 | 0.19 | 335.7 | 336 | 60 | 12.0 | 0.17 | 0.29 | 97.7 | 98 |
| 100 | 6.0 | 0.12 | 0.18 | 437.4 | 437 | 70 | 12.0 | 0.15 | 0.27 | 142.9 | 143 |
| 110 | 6.0 | 0.11 | 0.17 | 560.4 | 560 | 80 | 12.0 | 0.14 | 0.26 | 193.8 | 194 |
| 120 | 6.0 | 0.09 | 0.15 | 755.9 | 756 | 90 | 12.0 | 0.13 | 0.25 | 255.1 | 255 |
| 130 | 6.0 | 0.08 | 0.14 | 950.5 | 951 | 100 | 12.0 | 0.12 | 0.24 | 328.1 | 328 |
|  |  |  |  |  |  | 110 | 12.0 | 0.11 | 0.23 | 414.2 | 414 |
|  |  |  |  |  |  | 120 | 12.0 | 0.09 | 0.21 | 539.9 | 540 |
| 15 | 8.0 | 0.40 | 0.48 | 3.7 | 4 | 130 | 12.0 | 0.08 | 0.20 | 665.4 | 665 |
| 20 | 8.0 | 0.35 | 0.43 | 7.3 | 7 |  |  |  |  |  |  |
| 30 | 8.0 | 0.28 | 0.36 | 19.7 | 20 |  |  |  |  |  |  |
| 40 | 8.0 | 0.23 | 0.31 | 40.6 | 41 |  |  |  |  |  |  |
| 50 | 8.0 | 0.19 | 0.27 | 72.9 | 73 |  |  |  |  |  |  |
| 60 | 8.0 | 0.17 | 0.25 | 113.4 | 113 |  |  |  |  |  |  |
| 70 | 8.0 | 0.15 | 0.23 | 167.8 | 168 |  |  |  |  |  |  |
| 80 | 8.0 | 0.14 | 0.22 | 229.1 | 229 |  |  |  |  |  |  |
| 90 | 8.0 | 0.13 | 0.21 | 303.7 | 304 |  |  |  |  |  |  |
| 100 | 8.0 | 0.12 | 0.20 | 393.7 | 394 |  |  |  |  |  |  |
| 110 | 8.0 | 0.11 | 0.19 | 501.5 | 501 |  |  |  |  |  |  |
| 120 | 8.0 | 0.09 | 0.17 | 667.0 | 667 |  |  |  |  |  |  |
| 130 | 8.0 | 0.08 | 0.16 | 831.7 | 832 |  |  |  |  |  |  |

*Note*: In recognition of safety considerations, use of $e_{max} = 4.0\%$ should be limited to urban conditions.

*Source*: American Association of State Highway and Transportation Officials, *A Policy on Geometric Design of Highways and Streets*, 6th Edition, Washington, DC, 2011. Used by permission.

**Figure 3.13** Elements of a simple circular horizontal curve.

$R$ = radius, usually measured to the centerline of the road, in m,

$\Delta$ = central angle of the curve in degrees,

$T$ = tangent length in m,

$E$ = external distance in m,

$M$ = middle ordinate in m,

$PC$ = point of curve (the beginning point of the horizontal curve),

$PI$ = point of tangent intersection,

$PT$ = point of tangent (the ending point of the horizontal curve), and

$L$ = length of curve in m.

Another important term is the degree of curve, which can be defined as the angle subtended by a 100-m arc along the horizontal curve. It is a measure of the sharpness of the curve and is frequently used instead of the radius in the construction of the curve. The degree of curve is directly related to the radius of the horizontal curve by

$$D = \frac{100\left(\dfrac{180}{\pi}\right)}{R} = \frac{18000}{\pi R} \qquad (3.35)$$

*where*

$D$ = degree of curve [angle subtended by a 100-m arc along the horizontal curve], and

Other terms are as defined in Fig. 3.13.

Note that the quantity $180/\pi$ converts from radians to degrees.

Geometric and trigonometric analyses of Fig. 3.13 reveal the following relationships:

$$T = R \, \tan \frac{\Delta}{2} \qquad (3.36)$$

$$E = R\left(\frac{1}{\cos(\Delta/2)} - 1\right) \qquad \text{(3.37)}$$

$$M = R\left(1 - \cos\frac{\Delta}{2}\right) \qquad \text{(3.38)}$$

$$L = \frac{\pi}{180} R\Delta \qquad \text{(3.39)}$$

where all terms are as defined in Fig. 3.13.

It is important to note that horizontal curve stationing, curve length, and curve radius ($R$) are usually measured to the centerline of the road. In contrast, the radius determined on the basis of vehicle forces ($R_v$ in Eq. 3.34) is measured from the innermost vehicle path, which is assumed to be the midpoint of the innermost vehicle lane. Thus, a slight correction for lane width is required in equating the $R_v$ of Eq. 3.34 with the $R$ in Eqs. 3.35 through 3.39.

**EXAMPLE 3.15    STATIONING ON HORIZONTAL CURVES**

A horizontal curve is designed with a 610-m radius. The curve has a tangent length of 120 m and the *PI* is at station 3 + 140. Determine the station of the *PT*.

*SOLUTION*

Equation 3.36 is applied to determine the central angle, $\Delta$:

$$T = R\tan\frac{\Delta}{2}$$

$$120 = 610\tan\frac{\Delta}{2}$$

$$\Delta = 22.26°$$

So, from Eq. 3.39, the length of the curve is

$$L = \frac{\pi}{180} R\Delta$$

$$L = \frac{3.1416}{180} 610(22.26) = 236.990 \text{ m}$$

Given that the tangent length is 120 m,

$$\text{station of } PC = 3 + 140 \text{ minus } 0 + 120 = 3 + 020$$

Since horizontal curve stationing is measured along the alignment of the road,

$$\text{station of } PT = \text{station of } PC + L$$

$$= 3 + 020 \text{ plus } 0 + 236.990 = \underline{\underline{3 + 256.990}}$$

### 3.4.3 Stopping Sight Distance and Horizontal Curve Design

As is the case for vertical curve design, adequate stopping sight distance must be provided in the design of horizontal curves. Sight distance restrictions on horizontal curves occur when obstructions are present, as shown in Fig. 3.14. Such obstructions are frequently encountered in highway design due to the cost of right-of-way acquisition or the cost of moving earthen materials, such as rock outcroppings. When such an obstruction exists, the stopping sight distance is measured along the horizontal curve from the center of the traveled lane (the assumed location of the driver's eyes). As shown in Fig. 3.14, for a specified stopping distance, some distance $M_s$ (the middle ordinate of a curve that has an arc length equal to the stopping sight distance) must be visually cleared so that the line of sight is such that sufficient stopping sight distance is available.

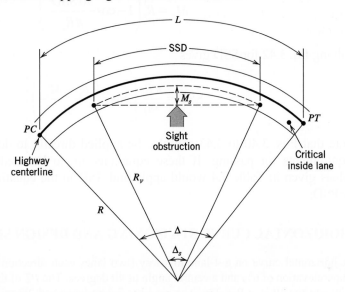

**Figure 3.14** Stopping sight distance considerations for horizontal curves.

$R$ = radius measured to the centerline of the road in m,

$R_v$ = radius to the vehicle's traveled path (usually measured to the center of the innermost lane of the road) in m,

$\Delta$ = central angle of the curve in degrees,

$\Delta_s$ = angle (in degrees) subtended by an arc equal in length to the required stopping sight distance (SSD),

$L$ = length of curve in m,

$M_s$ = middle ordinate necessary to provide adequate stopping sight distance (SSD) in m.

SSD = stopping sight distance in m,

$PC$ = point of curve (the beginning point of the horizontal curve), and

$PT$ = point of tangent (the ending point of the horizontal curve).

Equations for computing stopping sight distance (SSD) relationships for horizontal curves can be derived by first determining the angle, $\Delta_s$, for an arc length equal to the required stopping sight distance (see Fig. 3.14 and note that this is not the central angle, $\Delta$ of the horizontal curve whose arc length is equal to $L$). Assuming that the length of the horizontal curve exceeds the required SSD (as shown in Fig. 3.14), we have (as with Eq. 3.39)

$$SSD = \frac{\pi}{180} R_v \Delta_s \qquad (3.40)$$

Rearranging terms,

$$\Delta_s = \frac{180\, SSD}{\pi R_v} \qquad (3.41)$$

Substituting this into the general equation for the middle ordinate of a simple horizontal curve (Eq. 3.38) to get an expression for $M_s$ gives

$$M_s = R_v \left( 1 - \cos \frac{90\, SSD}{\pi R_v} \right) \qquad (3.42)$$

Solving Eq. 3.42 for SSD gives

$$SSD = \frac{\pi R_v}{90} \left[ \cos^{-1} \left( \frac{R_v - M_s}{R_v} \right) \right] \qquad (3.43)$$

Note that Eqs. 3.40 to 3.43 can also be applied directly to determine sight distance requirements for passing. If these equations are to be used for passing, distance values given in Table 3.4 would apply and SSD in the equations would be replaced by PSD.

## EXAMPLE 3.16  HORIZONTAL CURVE STATIONING AND DESIGN SPEED

A horizontal curve on a 4-lane highway (two lanes each direction with no median) has a superelevation of 6% and a central angle of 40 degrees. The *PT* of the curve is at station 9 + 830 and the *PI* is at 9 + 756. The road has 3.1-m lanes and 2.8-m shoulders on both sides with high retaining walls going up immediately next to the shoulders. What is the highest safe speed of this curve (highest in 10 km/h increments) and what is the station of the *PC*?

### SOLUTION

The tangent will be equal to the $PI - PC$ so $T = 9 + 756 - PC$. The length of the curve will be equal to the $PT - PC$ so $L = 9 + 830 - PC$. With the equations for the tangent and length of curve both put in terms of the *PC*, Eq. 3.36 and Eq. 3.39 can be rearranged respectively as,

$$R = \frac{T}{\tan \dfrac{\Delta}{2}} \quad \text{and} \quad R = \frac{L}{\dfrac{\pi}{180}\Delta} \quad \text{so that} \quad \frac{T}{\tan \dfrac{\Delta}{2}} = \frac{L}{\dfrac{\pi}{180}\Delta}$$

Substituting the previous tangent and length-of-curve equations ($T = 9756 - PC$ and $L = 9830 - PC$),

$$\frac{9756 - PC}{\tan \dfrac{40}{2}} = \frac{9830 - PC}{\dfrac{\pi}{180} 40}$$

Which gives $\underline{PC = 9 + 675.367}$. Using this value of $PC$, the tangent can be computed as $T = 9756 - 9675.367 = 80.633$ m. This value of $T$ can then be used to determine $R$ (see equations above),

$$R = \frac{T}{\tan \dfrac{\Delta}{2}} = \frac{80.633}{\tan \dfrac{40}{2}} = 221.537 \text{ m}$$

Because the curve radius is usually taken to the centerline of the roadway and there are two 3.1-m lanes before the centerline (working from the inside of the curve to the outside), $R_v = R - 3.1 - 3.1/2 = 221.537 - 4.65 = 216.887$ m. From Table 3.5 with a superelevation of 6%, at 70 km/h a radius of 184 m is needed; and at 80 km/h a radius of 252 m is needed. Therefore the highest deign speed for centripetal force is 70 km/h (since $216.887 > 184$, the design is acceptable for 70 km/h because more than the needed radius is available, but with $252 > 216.887$ the design is not acceptable for 80 km/h since insufficient radius is available).

To check for adequate sight distance, $M_s$ is going to be the shoulder width plus half of the inside lane width or $2.8 + 3.1/2 = 4.35$ m. Consider the stopping sight distance required at 70 km/h. At 70 km/h the required stopping sight distance (SSD) is 105 m (from Table 3.1). Applying Eq. 3.42 gives,

$$M_s = R_v \left( 1 - \cos \frac{90 \text{ SSD}}{\pi R_v} \right) = 216.887 \left( 1 - \cos \frac{90 \, (105)}{\pi (216.887)} \right) = 6.326 \text{ m}$$

Because 6.326 m is greater than the 4.35 m of available $M_s$, 70 km/h is too fast. Consider a speed of 60 km/h which gives SSD = 85 m (from Table 3.1). The application of Eq. 3.42 then gives,

$$M_s = R_v \left( 1 - \cos \frac{90 \text{ SSD}}{\pi R_v} \right) = 216.887 \left( 1 - \cos \frac{90 \, (85)}{\pi (216.887)} \right) = 4.15 \text{ m}$$

Because 4.15 m is less than 4.35 m, the highway is safe at 60 km/h. Considering both the maximum safe speeds for centripetal force (70 km/h) and sight distance (60 km/h), the lower of the two speeds will govern. Thus $\underline{60 \text{ km/h}}$ (the highest safe speed for sight distance) is the lower of the two speeds and is the highest safe speed for this curve.

## 3.5 COMBINED VERTICAL AND HORIZONTAL ALIGNMENT

Thus far the discussion on highway alignment has treated vertical and horizontal curves independently. The combination of vertical and horizontal curves, however, is quite common in geometric design, and often necessary. Obvious examples are highways through mountainous terrain and freeway interchange ramp roadways, which typically have to make significant changes in direction and elevation over a relatively short distance.

As previously mentioned, the design of an alignment that consists of a vertical and horizontal curve in combination usually consists of two two-dimensional alignment problems. The following examples illustrate this process.

### EXAMPLE 3.17   COMBINED HORIZONTAL/VERTICAL ALIGNMENT—DESIGN ADEQUACY

A two-lane highway (two 3.6-m lanes) has a posted speed limit of 80 km/h and, on one section, has both horizontal and vertical curves, as shown in Fig. 3.15. A recent daytime crash (driver traveling eastbound and striking a stationary roadway object) resulted in a fatality and a lawsuit alleging that the 80-km/h posted speed limit is an unsafe speed for the curves in question and was a major cause of the crash. Evaluate and comment on the roadway design.

**Figure 3.15**  Horizontal and vertical alignment for Example 3.17.

*SOLUTION*

Begin with an assessment of the horizontal alignment. Two concerns must be considered: the adequacy of the curve radius and the superelevation, and the adequacy of the sight distance on the eastbound (inside) lane. For the curve radius, note from Fig. 3.15 that

$$L = \text{station of } PT - \text{station of } PC$$

$$L = 4 + 600 \text{ minus } 4 + 160 = 440 \text{ m}$$

Rearranging Eq. 3.39, we get

$$R = \frac{180}{\pi\Delta}L = \frac{180}{\pi(80)}(440) = 315.126 \text{ m}$$

Using the posted speed limit of 80 km/h with $e = 8.0\%$, we find that Eq. 3.34 can be rearranged to give (with the vehicle traveling in the middle of the inside lane, $R_v = R$ − half the lane width, or $R_v = 315.126 - 1.8 = 313.326$ m)

$$f_s = \frac{V^2}{gR_v} - e = \frac{(80 \times 0.2778)^2}{9.807(313.326)} - 0.08 = 0.0807$$

From Table 3.5, the maximum $f_s$ for 80 km/h is 0.14. Since 0.0807 does not exceed 0.14, the radius and superelevation are sufficient for the 80-km/h design speed. For sight distance, the available $M_s$ is 6 m plus the 1.8-m distance to the center of the eastbound (inside) lane, or 7.8 m. Application of Eq. 3.43 gives

$$\text{SSD} = \frac{\pi R_v}{90}\left[\cos^{-1}\left(\frac{R_v - M_s}{R_v}\right)\right]$$

$$= \frac{\pi(313.326)}{90}\left[\cos^{-1}\left(\frac{313.326 - 7.8}{313.326}\right)\right]$$

$$= 140.119 \text{ m}$$

From Table 3.1, the required SSD at 80 km/h is 130 m, so the 140.119 m of SSD provided is sufficient. Turning to the sag vertical curve, the length of curve is

$$L = \text{station of } PVT - \text{station of } PVC$$

$$L = 4 + 290 \text{ minus } 4 + 140 = 150 \text{ m}$$

Using $A = 6\%$ (from Fig. 3.15) and applying Eq. 3.10, we obtain

$$K = \frac{L}{A} = \frac{150}{6} = 25$$

For the 80-km/h design speed, Table 3.3 indicates a necessary $K$-value of 30. Thus the $K$-value of 25 reveals that the curve is inadequate for the 80-km/h speed. However, because the crash occurred in daylight and sight distances on sag vertical curves are governed by nighttime conditions, this design did not contribute to the crash.

## EXAMPLE 3.18   DESIGN OF A COMBINED HORIZONTAL/VERTICAL ALIGNMENT

A new highway is to be constructed over an existing highway. The two highways will intersect at right angles and are to be grade-separated. Both highways are level grade (constant elevation). The new highway will run east-west and the existing highway runs north-south at elevation of 172 m. The proposed bridge structure for the new highway is such that the bridge girder thickness is 2 m (measured from the road surface to the bottom of the girder). A single-lane ramp is to be constructed to allow eastbound traffic to go southbound. A single horizontal curve, with a central angle of 90 degrees, is to be used. With a design speed of 70 km/h and a required superelevation of 4%, determine the following: the stationing of the *PC*, *PI*, and *PT*, assuming the curve begins at station 1 + 219; the stationing and elevation of all key points along the vertical alignment; the distance that must be cleared from the inside of the horizontal curve so that the line of sight is sufficient to provide sufficient stopping sight distance. Fig. 3.16 displays the horizontal and vertical alignments.

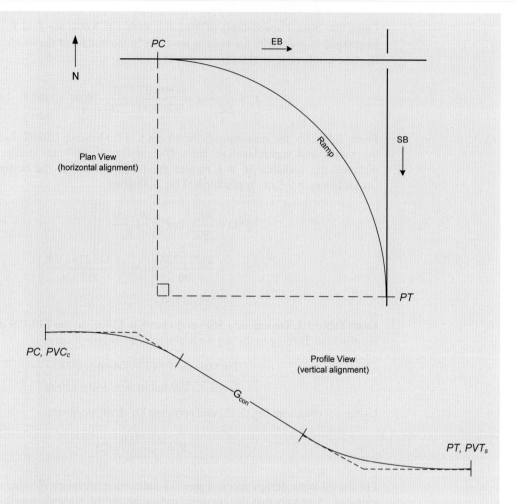

**Figure 3.16** Horizontal and vertical alignment for Example 3.18.

## SOLUTION

To begin, the required radius to the vehicle path $(R_v)$ is determined to be 203 m from Table 3.5 with a 70 km/h design speed and 4% superelevation. Because the ramp is a single lane, the horizontal curve radius will be equal to the radius to the vehicle path $(R = R_v)$. Applying Eq. 3.39 gives the length of the horizontal curve as (with $R = 203$ m and $\Delta = 90$ degrees):

$$L = \frac{\pi}{180} R \Delta$$

$$= \frac{3.1416}{180} 203(90) = 318.872 \text{ m}$$

Also, by inspection of Fig. 3.16 (or application of Eq. 3.36), the tangent length $T = R = 203$. The stationing for the horizontal curve is as follows:

$$\text{station of } PC = \underline{\underline{1 + 219}}$$

$$\text{station of } PI = \text{station of } PC + T$$

$$= 1 + 219 \text{ plus } 0 + 203 = \underline{\underline{1 + 422}}$$

$$\text{station of } PT = \text{station of } PC + L$$

$$= 1 + 219 \text{ plus } 0 + 318.872 = \underline{\underline{1 + 537.872}}$$

For the vertical alignment, both a sag and crest curve will be needed and adequate clearance must be provided over the existing highway. As shown in Section 3.3.6, AASHTO [2011] specifies a desirable clearance height of 5 m. The bridge girder thickness is given as 2 m so the total elevation difference between the two highways is 7 m (5 + 2).

For the vertical alignment, the elevation change will be the final offsets of the sag and crest curves plus the change in elevation resulting from the constant-grade section connecting the two curves. The constant-grade section is included because the available distance of 318.872 m (known from the length of the horizontal curve) is likely to be more than sufficient to affect a 7 m elevation difference at a 70 km/h design speed. Because both the sag and crest curves connect to a level grade at one end and have the constant grade in common at the other end, the $A$ value will be the same for both curves and will also be the grade for the constant-grade section. That is,

$$A_s = A_c = A$$

With this information, the equation that will solve the vertical alignment for this problem is

$$\frac{AL_s}{200} + \frac{AL_c}{200} + \frac{A(318.872 - L_s - L_c)}{100} = 7$$

where the third term accounts for the elevation difference attributable to the constant-grade section connecting the sag and crest curves (see also Example 3.9 for comparison). Using $L = KA$, we have

$$\frac{A^2 K_s}{200} + \frac{A^2 K_c}{200} + \frac{A(318.872 - K_s A - K_c A)}{100} = 7$$

From Table 3.2 for 70 km/h, $K_c = 17$, and from Table 3.3 for 70 km/h, $K_s = 23$. Putting these values in the above equation gives

$$0.2A^2 + 3.18872A - 0.4 A^2 = 7$$

$$-0.2A^2 + 3.18872A - 7 = 0$$

Solving this gives $A = 2.629$ and $A = 13.315$; $A = 2.629\%$ is chosen because we want to minimize the grade. For this value of $A$, the curve lengths are

$$L_s = K_s A = 23(2.629) = \underline{\underline{60.467 \text{ m}}}$$

$$L_c = K_c A = 17(2.629) = \underline{\underline{44.693 \text{ m}}}$$

and the length of the constant-grade section ($L_{con}$) will be 213.712 m (318.872 − 60.467 − 44.693). This means that about 5.618 m of the elevation difference will occur in the constant-grade section, with the remainder of the elevation difference attributable to the final curve offsets.

The stationing and elevation of the key points along the vertical alignment can now be calculated:

$$\text{station of } PVC_c = \text{station of } PC$$
$$= \underline{1 + 219}$$

$$\text{station of } PVI_c = 1 + 219 \text{ plus } L_c / 2 = 1 + 219 \text{ plus } (44.493 / 2) = \underline{1 + 241.247}$$

$$\text{station of } PVT_c = \text{station of } PVI + L_c$$
$$= 1 + 219 \text{ plus } (44.493) = \underline{1 + 263.493}$$

$$\text{station of } PVC_s = \text{station of } PVT_c + L_{con}$$
$$= 1 + 263.493 \text{ plus } 213.712 = \underline{1 + 477.205}$$

$$\text{station of } PVI_s = \text{station of } PVC_s + L_s / 2$$
$$= 1 + 477.205 \text{ plus } (60.467) / 2 = \underline{1 + 507.439}$$

$$\text{station of } PVT_s = \text{station of } PT$$
$$= \underline{1 + 537.872}$$

The elevation of the new east-west road will be 179 m which is determined by adding 7 m (the 5 m clearance plus the 2 m girder depth) above the north-south road elevation of 172 m. The PC and $PVC_c$ are both at station 1 + 219 and elevation 179 m (by inspection, the $PVI_c$ will also be at this elevation)

$$\text{elevation } PVT_c = \text{elev } PVI_c - \frac{AL_c}{200}$$
$$= 179 - \frac{2.629(44.493)}{200} = \underline{178.415 \text{ m}}$$

$$\text{elevation } PVC_s = \text{elev } PVT_c - \frac{AL_{con}}{100}$$
$$= 178.415 - \frac{2.629(213.712)}{100} = \underline{172.797 \text{ m}}$$

$$\text{elevation } PVT_s = \text{elev } PVC_s - \frac{AL_s}{200}$$
$$= 172.797 - \frac{2.629(60.467)}{200} = \underline{172 \text{ m}}$$

The PT and $PVT_s$ are both at station 1 + 537.872 and will thus both be at 172 m (by inspection, the $PVI_s$ will also be at this elevation).

Finally, the distance that must be cleared from the inside of the horizontal curve to provide sufficient stopping sight distance is determined by applying Equation 3.42:

$$M_s = R_v \left( 1 - \cos \frac{90 \, \text{SSD}}{\pi R_v} \right)$$

With $R_v = 203$ and SSD $= 105$ m (from Table 3.1 at 70 km/h),

$$M_s = 203 \left( 1 - \cos \frac{90(105)}{3.1416(203)} \right)$$

$$= \underline{\underline{6.751 \text{ m}}}$$

Thus, a distance of at least 6.751 m must be cleared from the center of the ramp's lane to the nearest sight obstruction on the inside of the curve.

## NOMENCLATURE FOR CHAPTER 3

| | |
|---|---|
| $A$ | absolute value of the algebraic difference in grades (in percent) |
| $a$ | coefficient in the parabolic curve equation or the deceleration in the stopping distance equation |
| $b$ | coefficient in the parabolic curve equation |
| $c$ | elevation of the $PVC$ |
| $D$ | degree of curvature |
| $e$ | rate of superelevation |
| $F_f$ | frictional side force |
| $F_c$ | centripetal force |
| $F_{cn}$ | centripetal force normal to the roadway surface |
| $F_{cp}$ | centripetal force parallel to the roadway surface |
| $f_s$ | coefficient of side friction |
| $G$ | grade |
| $G_1$ | initial roadway grade |
| $G_2$ | final roadway grade |
| $g$ | gravitational constant |
| $H$ | height of vehicle headlights |
| $H_c$ | clearance height of structure above sag curve |
| $H_1$ | height of driver's eye |
| $H_2$ | height of roadway object for stopping, height of oncoming car for passing |
| $K$ | horizontal distance required to effect a 1% change in slope |
| $L$ | length of curve |
| $L_m$ | minimum length of curve |
| $M$ | middle ordinate |
| $M_s$ | middle ordinate for stopping sight distance |
| $PC$ | initial point of horizontal curve |

| | |
|---|---|
| $PI$ | point of tangent intersection (horizontal curve) |
| PSD | passing sight distance |
| $PT$ | final point of horizontal curve |
| $PVC$ | initial point of vertical curve |
| $PVI$ | point of tangent intersection (vertical curve) |
| $PVT$ | final point of vertical curve |
| $R$ | radius of curve measured to the roadway centerline |
| $R_v$ | radius of curve to the vehicle's traveled path |
| $S$ | sight distance |
| SSD | stopping sight distance |
| $T$ | tangent length |
| $V$ | vehicle speed |
| $V_1$ | initial vehicle speed |
| $W$ | vehicle weight |
| $W_n$ | vehicle weight normal to the roadway surface |
| $W_p$ | vehicle weight parallel to the roadway surface |
| $x$ | distance from the beginning of the vertical curve to specified point |
| $x_{hl}$ | distance from the beginning of the vertical curve to high or low point |
| $Y$ | vertical curve offset |
| $Y_f$ | end-of-curve offset (vertical curve) |
| $Y_m$ | midcurve offset (vertical curve) |
| $\alpha$ | angle of superelevation |
| $\beta$ | upward angle of headlight beam |
| $\Delta$ | central angle |
| $\Delta_s$ | angle subtended by the stopping sight distance (SSD) arc |

# REFERENCES

AASHTO (American Association of State Highway and Transportation Officials). *A Policy on Geometric Design of Highways and Streets*, 6th ed. Washington, DC, 2011.

Campbell, C. *The Sports Car: Its Design and Performance*. Cambridge, MA: Robert Bently, 1978.

Hickerson, T. F. *Route Location and Design*, 5th ed. New York: McGraw-Hill, 1964.

Wong, J. Y. *Theory of Ground Vehicles*. New York: John Wiley & Sons, 2008.

# PROBLEMS

**3.1** A 490-m–long sag vertical curve (equal tangent) has a *PVC* at station 3 + 700 and elevation 460 m. The initial grade is −3.5% and the final grade is +6.5%. Determine the elevation and stationing of the low point, *PVI*, and *PVT*.

**3.2** A 150-m–long equal-tangent crest vertical curve connects tangents that intersect at station 10 + 360 and elevation 400 m. The initial grade is +4.0% and the final grade is −2.5%. Determine the elevation and stationing of the high point, *PVC*, and *PVT*.

**3.3** Consider Example 3.4. Solve this problem with the parabolic equation (Eq. 3.1) rather than by using offsets.

**3.4** Again consider Example 3.4. Does this curve provide sufficient stopping sight distance for a speed of 100 km/h?

**3.5** An equal-tangent sag vertical curve is designed with the *PVC* at station 3 + 320 and elevation 290 m, the *PVI* at station 3 + 375 and elevation 288.74 m, and the low point at station 3 + 365. Determine the design speed of the curve.

**3.6** An equal-tangent vertical curve was designed in 2012 (to 2011 AASHTO guidelines) for a design speed of 110 km/h to connect grades $G_1 = +1.0\%$ and $G_2 = -2.0\%$. The curve is to be redesigned for a 110-km/h design speed in the year 2025. Vehicle braking technology has advanced such that the recommended design deceleration rate is 25% greater than the 2004 value used to develop Table 3.1, but due to the higher percentage of older persons in the driving population, design reaction times have increased by 20%. Also, vehicles have become smaller so that the driver's eye height is assumed to be 900 mm above the pavement, and roadway objects are assumed to be 300 mm above the pavement surface. Compute the difference in design curve lengths for the 2012 and 2025 designs.

**3.7** A 365-m equal-tangent crest vertical curve is currently designed for 80 km/h. A civil engineering student contends that 100 km/h is safe in a van because of the higher driver's-eye height. If all other design inputs are standard, what must the driver's-eye height (in the van) be for the student's claim to be valid?

**3.8** A highway reconstruction project is being undertaken to reduce crash rates. The reconstruction involves a major realignment of the highway such that a 100-km/h design speed is attained. At one point on the highway, a 245-m equal-tangent crest vertical curve exists. Measurements show that at 0 + 107.3 stations from the *PVC*, the vertical curve offset is 1 m. Assess the adequacy of this existing curve in light of the reconstruction design speed of 100 km/h and, if the existing curve is inadequate, compute a satisfactory curve length.

**3.9** Two level sections of an east-west highway ($G = 0$) are to be connected. Currently, the two sections of highway are separated by a 1200-m (horizontal distance), 2% grade. The westernmost section of highway is the higher of the two and is at elevation 30 m. If the highway has a 100-km/h design speed, determine, for the crest and sag vertical curves required, the stationing and elevation of the *PVCs* and *PVTs* given that the *PVC* of the crest curve (on the westernmost level highway section) is at station 0 + 000 and elevation 30 m. In solving this problem, assume that the curve *PVIs* are at the intersection of $G = 0$ and the 2% grade, that is, $A = 2$.

**3.10** Consider Problem 3.9. Suppose it is necessary to keep the entire alignment within the 1200 m that currently separate the two level sections. It is determined that the crest and sag curves should be connected (the *PVT* of the crest and *PVC* of the sag) with a constant-grade section that has the lowest grade possible. Again using a 100-km/h design speed, determine, for the crest and sag vertical curves, the stationing and elevation of the *PVCs* and *PVTs* given that the westernmost level section ends at station 0 + 000 and elevation 30 m. (Note that $A$ must now be determined and will not be equal to 2.)

**3.11** An equal-tangent crest vertical curve is designed for 100 km/h. The initial grade is +4.0% and the final grade is negative. What is the elevation difference between the *PVC* and the high point of the curve?

**3.12** An equal-tangent crest vertical curve has an 80-km/h design speed. The initial grade is +3%. The high point is at station 1 + 017.2 and the *PVT* is at station 1 + 133.3. What is the elevation difference between the high point and the *PVT*?

**3.13** A vertical curve is designed for 90 km/h and has an initial grade of +2.5% and a final grade of −1.0%. The *PVT* is at station 3 + 480. It is known that a point on the curve at station 3 + 440 is at elevation 75 m. What is the stationing and elevation of the *PVC*? What is the stationing and elevation of the high point on the curve?

**3.14** An equal-tangent crest curve connects a +1.0% and a −0.5% grade. The *PVC* is at station 1 + 670 and the *PVI* is at station 1 + 750. Is this curve long enough to provide passing sight distance for a 90-km/h design speed?

**3.15** Due to crashes at a railroad crossing, an overpass (with a roadway surface 7.5 m above the existing road) is to be constructed on an existing level highway. The existing highway has a design speed of 80 km/h. The overpass structure is to be level, centered above the railroad, and 60 m long. What length of the existing level highway must be reconstructed to provide an appropriate vertical alignment?

**3.16** A section of a freeway ramp has a +4.0% grade and ends at station 3 + 870 and elevation 42 m. It must be connected to another section of the ramp (which has a 0.0% grade) that is at station 4 + 940 and elevation 30 m. It is determined that the crest and sag curves required to connect the ramp should be connected (the *PVT* of the crest and *PVC* of the sag) with a constant-grade section that has the lowest grade possible. Design a vertical alignment to connect between these two stations using an 80-km/h design speed. Provide the lengths of the curves and constant-grade section.

**3.17** A tangent section of highway has a −1.0% grade and ends at station 0 + 145 and elevation 25 m. It must be connected to another section of highway that has a −1.0% grade and that begins at station 1 + 345 and elevation 40 m. The connecting alignment should consist of a sag curve, constant-grade section, and crest curve, and be designed for a speed of 80 km/h. What is the lowest grade possible for the constant-grade section that will complete this alignment?

**3.18** A roadway has a design speed of 80 km/h, and at station 3 + 045 a +3.0% grade roadway section ends and at station 3 + 960 a +2.0% grade roadway section

begins. The +3.0% grade section of highway (at station 3 + 045) is at a higher elevation than the +2.0% grade section of highway (at station 3 + 960). If a −5% constant-grade section is used to connect the crest and sag vertical curves that are needed to link the +3.0 and +2.0% grade sections, what is the elevation difference between stations 3 + 045 and 3 + 960? (The entire alignment, crest and sag curves, and constant-grade section must fit between stations 3 + 045 and 3 + 960.)

**3.19** A sag curve and crest curve connect a −3.0% tangent section of highway (to the west) with a +2.0% tangent section of highway (to the east). The +2.0% tangent section is at a higher elevation than the −3.0% tangent section. The two tangent sections are separated by 390 m of horizontal distance. If the design speed of the curves is 80 km/h, what is the common grade between the sag and crest curves ($G_2$ of sag and $G_1$ of crest, from west to east), and what is the elevation difference between $PVC_s$ and $PVT_c$?

**3.20** An overpass is being built over the *PVI* of an existing equal-tangent sag curve. The sag curve has a 110-km/h design speed, and $G_1 = -6\%$, $G_2 = +3\%$. Determine the minimum necessary clearance height of the overpass and the resultant elevation of the bottom of the overpass over the *PVI*. (Ignore the cross-sectional width of the overpass.)

**3.21** An equal-tangent sag curve has its *PVI* at station 0 + 305 and elevation at 42 m. Directly above the *PVI*, the bottom of an overpass structure is at elevation 50 m. The *PVC* is at station 0 + 120. If the initial grade is −4%, what is the highest possible value of the final grade given that a 110-km/h design speed is to be provided in daytime conditions? What is the highest possible final grade in nighttime conditions? (*Note*: Be careful of units of *A*, and ignore the cross-sectional width of the overpass.)

**3.22** An existing highway-railway at-grade crossing is being redesigned as grade separated to improve traffic operations. The railway must remain at the same elevation. The highway is being reconstructed to travel under the railway. The underpass will be a sag curve that connects to 2% tangent sections on both ends, and the *PVI* will be centered under the railway (a symmetrical alignment). The sag curve design speed is 70 km/h. How many meters below the railway should the curve *PVI* be located?

**3.23** You are asked to design a horizontal curve for a two-lane road. The road has 3.6-m lanes. Due to expensive excavation, it is determined that a maximum of 10 m can be cleared from the road's centerline toward the inside lane to provide for stopping sight distance. Also, local guidelines dictate a maximum superelevation of 0.08 m/m. What is the highest possible design speed for this curve?

**3.24** A horizontal curve on a single-lane highway has its PC at station 3 + 780 and its PI at station 4 + 000. The curve has a superelevation of 0.06 m/m and is designed for 110 km/h. What is the station of the PT?

**3.25** A horizontal curve is being designed through mountainous terrain for a four-lane road with lanes that are 3 m wide. The central angle (Δ) is known to be 40 degrees, the tangent distance is 155 m, and the stationing of the tangent intersection (PI) is 82 + 300. Under specified conditions and vehicle speed, the roadway surface is determined to have a coefficient of side friction of 0.08, and the curve's superelevation is 0.09 m/m. What is the stationing of the PC and PT and what is the safe vehicle speed?

**3.26** A new interstate highway is being built with a design speed of 110 km/h. For one of the horizontal curves, the radius (measured to the innermost vehicle path) is tentatively planned as 275 m. What rate of superelevation is required for this curve?

**3.27** A developer is having a single-lane raceway constructed with a 160-km/h design speed. A curve on the raceway has a radius of 305 m, a central angle of 30 degrees, and PI stationing at 34 + 300. If the design coefficient of side friction is 0.20, determine the superelevation required at the design speed (do not ignore the normal component of the centripetal force). Also, compute the degree of curve, length of curve, and stationing of the PC and PT.

**3.28** A horizontal curve is being designed for a new two-lane highway (3.6-m lanes). The PI is at station 7 + 640, the design speed is 105 km/h, and a maximum superelevation of 0.08 m/m is to be used. If the central angle of the curve is 35 degrees, design a curve for the highway by computing the radius and stationing of the PC and PT.

**3.29** You are asked to design a horizontal curve with a 40-degree central angle (Δ = 40) for a two-lane road with 3-m lanes. The design speed is 110 km/h and superelevation is limited to 0.06 m/m. Give the radius, degree of curvature, and length of curve that you would recommend.

**3.30** A horizontal curve on a single-lane freeway ramp is 120 m long, and the design speed of the ramp is 70 km/h. If the superelevation is 10% and the station of the PC is 0 + 530, what is the station of the PI and how much distance must be cleared from the center of the lane to provide adequate stopping sight distance?

**3.31** A freeway exit ramp has a single lane and consists entirely of a horizontal curve with a central angle of 90 degrees and a length of 190 m. If the distance cleared from the centerline for sight distance is 5.9 m, what design speed was used?

**3.32** A horizontal curve on a two-lane highway (3.6-m lanes) has PC at station 3 + 765 and PT at station 3 + 940. The central angle is 34 degrees, the superelevation is 0.08, and 6 m are cleared (for sight distance) from the inside edge of the innermost lane. Determine a maximum safe speed (assuming current design standards) to the nearest 10 km/h.

**3.33** For the horizontal curve in Problem 3.29, what distance must be cleared from the inside edge of the inside lane to provide adequate stopping sight distance?

**3.34** A horizontal curve was designed for a four-lane highway for adequate SSD. Lane widths are 3.6 m, and the superelevation is 0.06 and was set assuming maximum $f_s$. If the necessary sight distance required 16 m of lateral clearance from the roadway centerline, what design speed was used for the curve?

**3.35** A section of highway has vertical and horizontal curves with the same design speed. A vertical curve on this highway connects a +1% and a +3% grade and is 124 m long. If a horizontal curve on this highway is on a two-lane section with 3.6-m lanes, has a central angle of 37 degrees, and has a superelevation of 6%, what is the length of the horizontal curve?

**3.36** A section of a two-lane highway (3.6-m lanes) is designed for 120 km/h. At one point a vertical curve connects a −2.5% and a +1.5% grade. The PVT of this curve is at station 0 + 765. It is known that a horizontal curve starts (has PC) 90 m before the vertical curve's PVC. If the superelevation of the horizontal curve is 0.08 and the central angle is 38 degrees, what is the station of the PT?

**3.37** Two straight sections of freeway cross at a right angle. At the point of crossing, the east-west highway is at elevation 45 m and has a constant +5.0% grade (upgrade in the east direction), and the north-south highway is at elevation 38 m and has a constant −3.0% grade (downgrade in the north direction). Design a 90-degree ramp that connects the northbound direction of travel to the eastbound direction of travel. Design the ramp for the highest design speed (to the nearest 10 km/h) with the constraint that the minimum allowable value of D is 8.0. (Assume that the PC of the horizontal curve is at station 0 + 460, and the vertical curve PVIs are at the PC and PT.) Give the stationing and elevations of the PC, PT, PVCs, and PVTs.

# Chapter 4

# Pavement Design

## 4.1 INTRODUCTION

Pavements are among the costliest items associated with highway construction and maintenance, and are largely responsible for making the U.S. highway system the most expensive public works project undertaken by any society. Because the pavement and associated shoulder structures are the most expensive items to construct and maintain, it is important for highway engineers to have a basic understanding of pavement design principles.

In the United States, there are over 4.8 million kilometers of highways. About 45% of these roads are lower-volume roads that are not paved (these roads generally have a gravel surface or are composed of a stabilized material consisting of an aggregate bound together with a cementing agent such as portland cement, lime fly ash, or asphaltic cement). Highways that carry higher volumes of traffic with heavy axle loads require surfaces with asphalt concrete or portland cement concrete to provide for all-weather operations and prevent permanent deformation of the highway surface. These types of pavements can cost upward of several million dollars per mile to construct. Some states, such as California, Illinois, New York, Pennsylvania, and Texas, have pavement construction, maintenance, and rehabilitation budgets that easily exceed a billion dollars per year. Given the magnitude of this pavement-asset investment, it is easy to understand why the construction, maintenance and rehabilitation of pavement infrastructure must be done in a cost-effective manner.

Fundamentally, a paved surface performs two basic functions. First, it helps guide drivers by giving them a visual perspective of the horizontal and vertical alignment of the traveled path—thus giving drivers information relating to the driving task and the steering control of the vehicle. The second function of pavement is to support vehicle loads, and this second function is the focus of this chapter.

## 4.2 PAVEMENT TYPES

In general, there are two types of pavement structures: flexible pavements and rigid pavements. There are, however, many variations of these pavement types. Composite pavements (which are made of both rigid and flexible pavement layers), continuously reinforced pavements, and post-tensioned pavements are other types, which usually require specialized designs and are not covered in this chapter.

As with any structure, the underlying soil must ultimately carry the load that is placed on it. A pavement's function is to distribute the traffic load stresses to the soil (subgrade) at a magnitude that will not shear or distort the soil. Typical soil-bearing capacities can be less than 345 kPa (50 lb/in$^2$) and in some cases as low as 14 to 21 kPa (2 to 3 lb/in$^2$). When soil is saturated with water, the bearing capacity can be very low, and in these cases it is very important for pavement to distribute tire loads to the soil in such a way as to prevent failure of the pavement structure.

A typical automobile weighs approximately 15.5 kN (3500 lb), with tire pressures around 241 kPa (35 lb/in$^2$). These loads are small compared with a typical tractor–semi-trailer truck, which can weigh up to 355.8 kN (80,000 lb)—the legal limit, in many states, on five axles with tire pressures of 690 kPa (100 lb/in$^2$) or higher. Truck loads such as these represent the standard type of loading used in pavement design. In this chapter, attention is directed toward an accepted procedure that can be used to design pavement structures for high–traffic-volume highway facilities subjected to heavy truck traffic. The design of lower-volume facilities, which may have stabilized-soil and gravel-surfaced pavements, is discussed elsewhere [Yoder and Witczak, 1975].

### 4.2.1 Flexible Pavements

A flexible pavement is constructed with asphaltic cement and aggregates and usually consists of several layers, as shown in Fig. 4.1. The lower layer is called the subgrade (the soil itself). The upper 152 to 203 mm (6 to 8 inches) of the subgrade is usually scarified and blended to provide a uniform material before it is compacted to maximum density. The next layer is the subbase, which usually consists of crushed aggregate (rock). This material has better engineering properties (higher modulus values) than the subgrade material in terms of its bearing capacity. The next layer is the base layer and is also often made of crushed aggregates (of a higher strength than those used in the subbase), which are either unstabilized or stabilized with a cementing material. The cementing material can be portland cement, lime fly ash, or asphaltic cement.

The top layer of a flexible pavement is referred to as the wearing surface. It is usually made of asphaltic concrete, which is a mixture of asphalt cement and aggregates. The purpose of the wearing layer is to protect the base layer from wheel abrasion and to waterproof the entire pavement structure. It also provides a skid-resistant surface that is important for safe vehicle stops. Typical thicknesses of the individual layers are shown in Fig. 4.1. These thicknesses vary with the type of axle loading, available materials, and expected pavement design life, which is the number of years the pavement is expected to provide adequate service before it must undergo major rehabilitation.

**Figure 4.1** Typical flexible-pavement cross section.

### 4.2.2   Rigid Pavements

A rigid pavement is constructed with portland cement concrete (PCC) and aggregates, as shown in Fig. 4.2. As with flexible pavements, the subgrade (the lower layer) is often scarified, blended, and compacted to maximum density. In rigid pavements, the base layer (see Fig. 4.2) is optional, depending on the engineering properties of the subgrade. If the subgrade soil is poor and erodable, then it is advisable to use a base layer. However, if the soil has good engineering properties and drains well, a base layer need not be used. The top layer (wearing surface) is the portland cement concrete slab. Slab length varies from a spacing of 3 to 4 m (10 to 13 ft) to a spacing of 12.2 m (40 ft) or more.

Transverse contraction joints are built into the pavement to control cracking due to shrinkage of the concrete during the curing process. Load transfer devices, such as dowel bars, are placed in the joints to minimize deflections and reduce stresses near the edges of the slabs. Slab thicknesses for PCC highway pavements usually vary from 200 to 300 mm (roughly 8 to 12 inches), as shown in Fig. 4.2.

## 4.3   PAVEMENT SYSTEM DESIGN: PRINCIPLES FOR FLEXIBLE PAVEMENTS

The primary function of the pavement structure is to reduce and distribute the surface stresses (contact tire pressure) to an acceptable level at the subgrade (to a level that prevents permanent deformation). A flexible pavement reduces the stresses by distributing the traffic wheel loads over greater and greater areas, through the individual layers, until the stress at the subgrade is at an acceptably low level. The traffic loads are transmitted to the subgrade by aggregate-to-aggregate particle contact. Confining pressures (lateral forces due to material weight) in the subbase and base layers increase the bearing strength of these materials. A cone of distributed loads reduces and spreads the stresses to the subgrade, as shown in Fig. 4.3.

**Figure 4.2** Typical rigid-pavement cross section.

**Figure 4.3** Distribution of load on a flexible pavement.

## 4.4   TRADITIONAL AASHTO FLEXIBLE-PAVEMENT DESIGN PROCEDURE

There are several accepted flexible-pavement design procedures available, including the Asphalt Institute method, the National Stone Association procedure, the Shell procedure, and the Mechanistic-Empirical Pavement Design Guide (which is discussed later in this chapter). Most of the procedures have been field verified and used by highway agencies for several years. The selection of one procedure over another is usually based on a highway agency's experience and satisfaction with design results.

A traditional and widely accepted flexible-pavement design procedure is presented in the *AASHTO Guide for Design of Pavement Structures*, which is published by the American Association of State Highway and Transportation Officials (AASHTO). The procedure was first published in 1972, with the latest revisions in 1993. The 1993 AASHTO design procedure is the same as the 1986 AASHTO procedure, except the new procedure has a revised section for overlay designs. Test data, used for the development of the design procedure, were collected at the AASHO Road Test in Illinois from 1958 to 1960 (AASHO, which stands for American Association of State Highway Officials, was the prior name of AASHTO).

A pavement can be subjected to a number of detrimental effects, including fatigue failures (cracking), which are the result of repeated loading caused by traffic passing over the pavement. The pavement is also placed in an uncontrolled environment that produces temperature extremes and moisture variations. The combination of the environment, traffic loads, material variations, and construction variations requires a comparatively complex set of design procedures to incorporate all of the variables. The AASHTO pavement design procedure meets most of the demands placed on a flexible-pavement design procedure. It considers environment, load, and materials in a methodology that is relatively easy to use. The AASHTO pavement design procedure has been widely accepted throughout the United States and around the world. Details of this procedure are presented in the following sections.

### 4.4.1   Serviceability Concept

Prior to the AASHO Road Test, there was no real consensus on the definition of pavement failure. In the eyes of an engineer, pavement failure occurred whenever cracking, rutting, or other surface distresses became visible. In contrast, the motoring public usually associated pavement failure with poor ride quality. Pavement engineers conducting the AASHO Road Test were faced with the task of combining the two failure definitions so that a single design procedure could be used to satisfy both critics. The Pavement Serviceability-Performance Concept was developed by Carey and Irick [1962] to handle the question of pavement failure. Carey and Irick considered pavement performance histories and noted that pavements usually begin their service life in excellent condition and deteriorate as traffic loading is applied in conjunction with prevailing environmental conditions. The performance curve is the historical record of the performance of the pavement. Pavement performance, at any point in time, is known as the present serviceability index, or PSI. Examples of pavement performance (or PSI trends) are shown in Fig. 4.4.

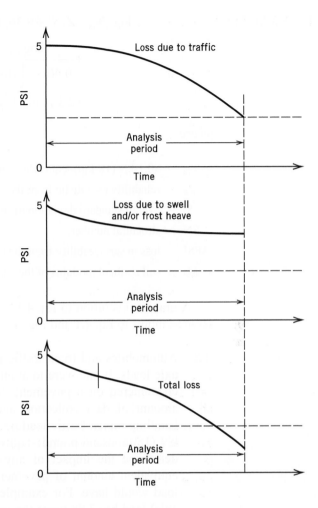

**Figure 4.4** Pavement performance trends.

Redrawn from *AASHTO Guide for Design of Pavement Structures*, Washington, DC, The American Association of State Highway and Transportation Officials, 1993. Used by permission.

At any time, the present serviceability index of a pavement can be measured. This is usually done by a panel of raters who drive over the pavement section and rate the pavement performance on a scale of 1 to 5, with 5 being the smoothest ride. The accumulation of traffic loads causes the pavement to deteriorate, and, as expected, the serviceability rating drops. At some point, a terminal serviceability index (TSI) is reached and the pavement is in need of rehabilitation or replacement.

It has been found that new pavements usually have an initial PSI rating of approximately 4.2 to 4.5. The point at which pavements are considered to have failed (the TSI) varies by type of highway. Highway facilities such as interstate highways or principal arterials usually have TSIs of 2.5 or 3.0, whereas local roads can have TSIs of 2.0.

## 4.4.2   Flexible-Pavement Design Equation

The basic equation for flexible-pavement design given in the 1993 AASHTO design guide permits engineers to determine a structural number necessary to carry a designated traffic loading. The AASHTO equation is

$$\log_{10}W_{18} = Z_R S_o + 9.36 \left[ \log_{10}(SN+1) \right] - 0.20$$

$$+ \frac{\log_{10}\left[ \Delta PSI / 2.7 \right]}{0.40 + \left[ 1094 \big/ (SN+1)^{5.19} \right]} \tag{4.1}$$

$$+ 2.32 \log_{10}M_R - 8.07$$

*where*

$W_{18}$ = 80.1-kN (18-kip)–equivalent single-axle load (1 kip = 1,000 lb or 4448 N),

$Z_R$ = reliability ($z$-statistic from the standard normal curve),

$S_o$ = overall standard deviation of traffic,

SN = structural number,

$\Delta PSI$ = loss in serviceability from the time the pavement is new until it reaches its TSI, and

$M_R$ = soil resilient modulus of the subgrade in lb/in$^2$.

A graphical solution to Eq. 4.1 is shown in Fig. 4.5. Details on the variables that serve as inputs to Eq. 4.1 and Fig. 4.5 are as follows:

$W_{18}$    Automobiles and truck traffic provide a wide range of vehicle axle types and axle loads. If one were to attempt to account for the variety of traffic loadings encountered on a pavement, this input variable would require a significant amount of data collection and design evaluation. Instead, the problem of handling mixed traffic loading is solved with the adoption of a standard 80.1-kN (18-thousand-pound)–equivalent single-axle load: (ESAL). The idea is to determine the impact of any axle load on the pavement in terms of the equivalent amount of pavement impact that an 80.1-kN (18-kip) single-axle load would have. For example, if a 195.7-kN (44-kip) tandem-axle (double-axle) load has 2.88 times the impact on pavement structure as an 80.1-kN (18-kip) single-axle load, 2.88 would be the $W_{18}$ value assigned to this tandem-axle load. The AASHO Road Test also found that the 80.1-kN (18-kip)–equivalent axle load is a function of the terminal serviceability index of the pavement structure. The axle-load equivalency factors for flexible pavement design, with a TSI of 2.5, are presented in Tables 4.1 (for single axles), 4.2 (for tandem axles), and 4.3 (for triple axles).

$Z_R$    Represents the probability that serviceability will be maintained at adequate levels from a user's point of view throughout the design life of the facility. This factor estimates the likelihood that the pavement will perform at or above the TSI level during the design period, and takes into account the inherent uncertainty in design. Equation 4.1 uses the $z$-statistic, which is obtained from the cumulative probabilities of the standard normal distribution (a normal distribution with mean equal to 0 and variance equal to 1). The $z$-statistics corresponding to various probability levels are given in Table 4.4. In the flexible-pavement-design nomograph (Fig. 4.5), the probabilities (in percent) are used directly (instead of $Z_R$ as in the case of Eq. 4.1), and these percent probabilities are denoted $R$, the reliability (see Table 4.4).

**Figure 4.5** Design chart for flexible pavements based on the use of mean values for each input.
Redrawn from *AASHTO Guide for Design of Pavement Structures*, Washington, DC, The American Association of State Highway and Transportation Officials, 1993. Used by permission.

*Example:*

$W_{18} = 5 \times 10^6$
$R = 95\%$
$S_o = 0.35$
$M_R = 5000 \ \text{lb/in}^2$
$\Delta \text{PSI} = 1.9$

*Solution:* SN = 5.0

Highways such as interstates and major arterials, which are costly to reconstruct (have their pavements rehabilitated) because of resulting traffic delay and disruption, require a high reliability level, whereas local roads, which will have lower impacts on users in the event of pavement rehabilitation, do not. Typical reliability values for interstate highways are 90% or higher, whereas local roads can have a reliability as low as 50%.

$S_o$    The overall standard deviation, $S_o$, takes into account the designers' inability to accurately estimate the variation in future 80.1-kN (18-kip)–equivalent axle loads, and the statistical error in the equations resulting from variability in materials and construction practices. Typical values of $S_o$ are on the order of 0.30 to 0.50.

SN    The structural number, SN, represents the overall structural requirement needed to sustain the design's traffic loadings. The structural number is discussed further in Section 4.4.3.

ΔPSI    The amount of serviceability loss over the life of the pavement, ΔPSI, is determined during the pavement design process. The engineer must decide on the final PSI level for a particular pavement. Loss of serviceability is caused by pavement roughness, cracking, patching, and rutting. As pavement distress increases, serviceability decreases. If the design is for a pavement with heavy traffic loads, such as an interstate highway, then the serviceability loss may only be 1.2 (an initial PSI of 4.2 and a TSI of 3.0), whereas a low-volume road can be allowed to deteriorate further, with a possible total serviceability loss of 2.7 or more.

$M_R$    The soil resilient modulus, $M_R$, is used to reflect the engineering properties of the subgrade (the soil). Each time a vehicle passes over pavement, stresses are developed in the subgrade. After the load passes, the subgrade soil relaxes and the stress is relieved. The resilient modulus test is used to determine the properties of the soil under this repeated load. The resilient modulus can be determined by AASHTO test method T274. Measurement of the resilient modulus is not performed by all transportation agencies; therefore, a relationship between $M_R$ and the California bearing ratio (CBR) has been determined. The CBR has been widely used to determine the supporting characteristics of soils since the mid-1930s, and a significant amount of historical information is available. The CBR is the ratio of the load-bearing capacity of the soil to the load-bearing capacity of a high-quality aggregate, multiplied by 100. The relationship, used to provide a very basic approximation of $M_R$ (in lb/in$^2$) from a known CBR, is

$$M_R = 1500 \times CBR \qquad \textbf{(4.2)}$$

The coefficient of 1500 in Eq. 4.2 is used for CBR values less than 10. Caution must be exercised in applying this equation to higher CBRs because the coefficient (the value 1500 shown in Eq. 4.2) has a range of 750 to 3000.

**Table 4.1**  Axle-Load Equivalency Factors for Flexible Pavements, Single Axles, and TSI = 2.5

| Axle load (kips) | Pavement structural number (SN) | | | | | |
|---|---|---|---|---|---|---|
| | 1 | 2 | 3 | 4 | 5 | 6 |
| 2 | 0.0004 | 0.0004 | 0.0003 | 0.0002 | 0.0002 | 0.0002 |
| 4 | 0.003 | 0.004 | 0.004 | 0.003 | 0.002 | 0.002 |
| 6 | 0.011 | 0.017 | 0.017 | 0.013 | 0.010 | 0.009 |
| 8 | 0.032 | 0.047 | 0.051 | 0.041 | 0.034 | 0.031 |
| 10 | 0.078 | 0.102 | 0.118 | 0.102 | 0.088 | 0.080 |
| 12 | 0.168 | 0.198 | 0.229 | 0.213 | 0.189 | 0.176 |
| 14 | 0.328 | 0.358 | 0.399 | 0.388 | 0.360 | 0.342 |
| 16 | 0.591 | 0.613 | 0.646 | 0.645 | 0.623 | 0.606 |
| 18 | 1.00 | 1.00 | 1.00 | 1.00 | 1.00 | 1.00 |
| 20 | 1.61 | 1.57 | 1.49 | 1.47 | 1.51 | 1.55 |
| 22 | 2.48 | 2.38 | 2.17 | 2.09 | 2.18 | 2.30 |
| 24 | 3.69 | 3.49 | 3.09 | 2.89 | 3.03 | 3.27 |
| 26 | 5.33 | 4.99 | 4.31 | 3.91 | 4.09 | 4.48 |
| 28 | 7.49 | 6.98 | 5.90 | 5.21 | 5.39 | 5.98 |
| 30 | 10.3 | 9.5 | 7.9 | 6.8 | 7.0 | 7.8 |
| 32 | 13.9 | 12.8 | 10.5 | 8.8 | 8.9 | 10.0 |
| 34 | 18.4 | 16.9 | 13.7 | 11.3 | 11.2 | 12.5 |
| 36 | 24.0 | 22.0 | 17.7 | 14.4 | 13.9 | 15.5 |
| 38 | 30.9 | 28.3 | 22.6 | 18.1 | 17.2 | 19.0 |
| 40 | 39.3 | 35.9 | 28.5 | 22.5 | 21.1 | 23.0 |
| 42 | 49.3 | 45.0 | 35.6 | 27.8 | 25.6 | 27.7 |
| 44 | 61.3 | 55.9 | 44.0 | 34.0 | 31.0 | 33.1 |
| 46 | 75.5 | 68.8 | 54.0 | 41.4 | 37.2 | 39.3 |
| 48 | 92.2 | 83.9 | 65.7 | 50.1 | 44.5 | 46.5 |
| 50 | 112.0 | 102.0 | 79.0 | 60.0 | 53.0 | 55.0 |

Source: *AASHTO Guide for Design of Pavement Structures*, The American Association of State Highway and Transportation Officials, Washington, DC, 1993. Used by permission.

**Table 4.2** Axle-Load Equivalency Factors for Flexible Pavements, Tandem Axles, and TSI = 2.5

| Axle load (kips) | Pavement structural number (SN) | | | | | |
|---|---|---|---|---|---|---|
| | 1 | 2 | 3 | 4 | 5 | 6 |
| 2 | 0.0001 | 0.0001 | 0.0001 | 0.0000 | 0.0000 | 0.0000 |
| 4 | 0.0005 | 0.0005 | 0.0004 | 0.0003 | 0.0003 | 0.0002 |
| 6 | 0.002 | 0.002 | 0.002 | 0.001 | 0.001 | 0.001 |
| 8 | 0.004 | 0.006 | 0.005 | 0.004 | 0.003 | 0.003 |
| 10 | 0.008 | 0.013 | 0.011 | 0.009 | 0.007 | 0.006 |
| 12 | 0.015 | 0.024 | 0.023 | 0.018 | 0.014 | 0.013 |
| 14 | 0.026 | 0.041 | 0.042 | 0.033 | 0.027 | 0.024 |
| 16 | 0.044 | 0.065 | 0.070 | 0.057 | 0.047 | 0.043 |
| 18 | 0.070 | 0.097 | 0.109 | 0.092 | 0.077 | 0.070 |
| 20 | 0.107 | 0.141 | 0.162 | 0.141 | 0.121 | 0.110 |
| 22 | 0.160 | 0.198 | 0.229 | 0.207 | 0.180 | 0.166 |
| 24 | 0.231 | 0.273 | 0.315 | 0.292 | 0.260 | 0.242 |
| 26 | 0.327 | 0.370 | 0.420 | 0.401 | 0.364 | 0.342 |
| 28 | 0.451 | 0.493 | 0.548 | 0.534 | 0.495 | 0.470 |
| 30 | 0.611 | 0.648 | 0.703 | 0.695 | 0.658 | 0.633 |
| 32 | 0.813 | 0.843 | 0.889 | 0.887 | 0.857 | 0.834 |
| 34 | 1.06 | 1.08 | 1.11 | 1.11 | 1.09 | 1.08 |
| 36 | 1.38 | 1.38 | 1.38 | 1.38 | 1.38 | 1.38 |
| 38 | 1.75 | 1.73 | 1.69 | 1.68 | 1.70 | 1.73 |
| 40 | 2.21 | 2.16 | 2.06 | 2.03 | 2.08 | 2.14 |
| 42 | 2.76 | 2.67 | 2.49 | 2.43 | 2.51 | 2.61 |
| 44 | 3.41 | 3.27 | 2.99 | 2.88 | 3.00 | 3.16 |
| 46 | 4.18 | 3.98 | 3.58 | 3.40 | 3.55 | 3.79 |
| 48 | 5.08 | 4.80 | 4.25 | 3.98 | 4.17 | 4.49 |
| 50 | 6.12 | 5.76 | 5.03 | 4.64 | 4.86 | 5.28 |
| 52 | 7.33 | 6.87 | 5.93 | 5.38 | 5.63 | 6.17 |
| 54 | 8.72 | 8.14 | 6.95 | 6.22 | 6.47 | 7.15 |
| 56 | 10.3 | 9.6 | 8.1 | 7.2 | 7.4 | 8.2 |
| 58 | 12.1 | 11.3 | 9.4 | 8.2 | 8.4 | 9.4 |
| 60 | 14.2 | 13.1 | 10.9 | 9.4 | 9.6 | 10.7 |
| 62 | 16.5 | 15.3 | 12.6 | 10.7 | 10.8 | 12.1 |
| 64 | 19.1 | 17.6 | 14.5 | 12.2 | 12.2 | 13.7 |
| 66 | 22.1 | 20.3 | 16.6 | 13.8 | 13.7 | 15.4 |
| 68 | 25.3 | 23.3 | 18.9 | 15.6 | 15.4 | 17.2 |
| 70 | 29.0 | 26.6 | 21.5 | 17.6 | 17.2 | 19.2 |
| 72 | 33.0 | 30.3 | 24.4 | 19.8 | 19.2 | 21.3 |
| 74 | 37.5 | 34.4 | 27.6 | 22.2 | 21.3 | 23.6 |
| 76 | 42.5 | 38.9 | 31.1 | 24.8 | 23.7 | 26.1 |
| 78 | 48.0 | 43.9 | 35.0 | 27.8 | 26.2 | 28.8 |
| 80 | 54.0 | 49.4 | 39.2 | 30.9 | 29.0 | 31.7 |
| 82 | 60.6 | 55.4 | 43.9 | 34.4 | 32.0 | 34.8 |
| 84 | 67.8 | 61.9 | 49.0 | 38.2 | 35.3 | 38.1 |
| 86 | 75.7 | 69.1 | 54.5 | 42.3 | 38.8 | 41.7 |
| 88 | 84.3 | 76.9 | 60.6 | 46.8 | 42.6 | 45.6 |
| 90 | 93.7 | 85.4 | 67.1 | 51.7 | 46.8 | 49.7 |

Source: *AASHTO Guide for Design of Pavement Structures*, The American Association of State Highway and Transportation Officials, Washington, DC, 1993. Used by permission.

**Table 4.3**  Axle-Load Equivalency Factors for Flexible Pavements, Triple Axles, and TSI = 2.5

| Axle load (kips) | Pavement structural number (SN) | | | | | |
|---|---|---|---|---|---|---|
| | 1 | 2 | 3 | 4 | 5 | 6 |
| 2 | 0.0000 | 0.0000 | 0.0000 | 0.0000 | 0.0000 | 0.0000 |
| 4 | 0.0002 | 0.0002 | 0.0002 | 0.0001 | 0.0001 | 0.0001 |
| 6 | 0.0006 | 0.0007 | 0.0005 | 0.0004 | 0.0003 | 0.0003 |
| 8 | 0.001 | 0.002 | 0.001 | 0.001 | 0.001 | 0.001 |
| 10 | 0.003 | 0.004 | 0.003 | 0.002 | 0.002 | 0.002 |
| 12 | 0.005 | 0.007 | 0.006 | 0.004 | 0.003 | 0.003 |
| 14 | 0.008 | 0.012 | 0.010 | 0.008 | 0.006 | 0.006 |
| 16 | 0.012 | 0.019 | 0.018 | 0.013 | 0.011 | 0.010 |
| 18 | 0.018 | 0.029 | 0.028 | 0.021 | 0.017 | 0.016 |
| 20 | 0.027 | 0.042 | 0.042 | 0.032 | 0.027 | 0.024 |
| 22 | 0.038 | 0.058 | 0.060 | 0.048 | 0.040 | 0.036 |
| 24 | 0.053 | 0.078 | 0.084 | 0.068 | 0.057 | 0.051 |
| 26 | 0.072 | 0.103 | 0.114 | 0.095 | 0.080 | 0.072 |
| 28 | 0.098 | 0.133 | 0.151 | 0.128 | 0.109 | 0.099 |
| 30 | 0.129 | 0.169 | 0.195 | 0.170 | 0.145 | 0.133 |
| 32 | 0.169 | 0.213 | 0.247 | 0.220 | 0.191 | 0.175 |
| 34 | 0.219 | 0.266 | 0.308 | 0.281 | 0.246 | 0.228 |
| 36 | 0.279 | 0.329 | 0.379 | 0.352 | 0.313 | 0.292 |
| 38 | 0.352 | 0.403 | 0.461 | 0.436 | 0.393 | 0.368 |
| 40 | 0.439 | 0.491 | 0.554 | 0.533 | 0.487 | 0.459 |
| 42 | 0.543 | 0.594 | 0.661 | 0.644 | 0.597 | 0.567 |
| 44 | 0.666 | 0.714 | 0.781 | 0.769 | 0.723 | 0.692 |
| 46 | 0.811 | 0.854 | 0.918 | 0.911 | 0.868 | 0.838 |
| 48 | 0.979 | 1.015 | 1.072 | 1.069 | 1.033 | 1.005 |
| 50 | 1.17 | 1.20 | 1.24 | 1.25 | 1.22 | 1.20 |
| 52 | 1.40 | 1.41 | 1.44 | 1.44 | 1.43 | 1.41 |
| 54 | 1.66 | 1.66 | 1.66 | 1.66 | 1.66 | 1.66 |
| 56 | 1.95 | 1.93 | 1.90 | 1.90 | 1.91 | 1.93 |
| 58 | 2.29 | 2.25 | 2.17 | 2.16 | 2.20 | 2.24 |
| 60 | 2.67 | 2.60 | 2.48 | 2.44 | 2.51 | 2.58 |
| 62 | 3.09 | 3.00 | 2.82 | 2.76 | 2.85 | 2.95 |
| 64 | 3.57 | 3.44 | 3.19 | 3.10 | 3.22 | 3.36 |
| 66 | 4.11 | 3.94 | 3.61 | 3.47 | 3.62 | 3.81 |
| 68 | 4.71 | 4.49 | 4.06 | 3.88 | 4.05 | 4.30 |
| 70 | 5.38 | 5.11 | 4.57 | 4.32 | 4.52 | 4.84 |
| 72 | 6.12 | 5.79 | 5.13 | 4.80 | 5.03 | 5.41 |
| 74 | 6.93 | 6.54 | 5.74 | 5.32 | 5.57 | 6.04 |
| 76 | 7.84 | 7.37 | 6.41 | 5.88 | 6.15 | 6.71 |
| 78 | 8.83 | 8.28 | 7.14 | 6.49 | 6.78 | 7.43 |
| 80 | 9.92 | 9.28 | 7.95 | 7.15 | 7.45 | 8.21 |
| 82 | 11.1 | 10.4 | 8.8 | 7.9 | 8.2 | 9.0 |
| 84 | 12.4 | 11.6 | 9.8 | 8.6 | 8.9 | 9.9 |
| 86 | 13.8 | 12.9 | 10.8 | 9.5 | 9.8 | 10.9 |
| 88 | 15.4 | 14.3 | 11.9 | 10.4 | 10.6 | 11.9 |
| 90 | 17.1 | 15.8 | 13.2 | 11.3 | 11.6 | 12.9 |

Source: *AASHTO Guide for Design of Pavement Structures*, The American Association of State Highway and Transportation Officials, Washington, DC, 1993. Used by permission.

**Table 4.4** Cumulative Percent Probabilities of Reliability, $R$, of the Standard Normal Distribution, and Corresponding $Z_R$

| $R$ | 0 | 1 | 2 | 3 | 4 | 5 | 6 | 7 | 8 | 9 | 9.5 | 9.9 |
|---|---|---|---|---|---|---|---|---|---|---|---|---|
| 90 | −1.282 | −1.341 | −1.405 | −1.476 | −1.555 | −1.645 | −1.751 | −1.881 | −2.054 | −2.326 | −2.576 | −3.080 |
| 80 | −0.842 | −0.878 | −0.915 | −0.954 | −0.994 | −1.036 | −1.080 | −1.126 | −1.175 | −1.227 | −1.253 | −1.272 |
| 70 | −0.524 | −0.553 | −0.583 | −0.613 | −0.643 | −0.675 | −0.706 | −0.739 | −0.772 | −0.806 | −0.824 | −0.838 |
| 60 | −0.253 | −0.279 | −0.305 | −0.332 | −0.358 | −0.385 | −0.412 | −0.440 | −0.468 | −0.496 | −0.510 | −0.522 |
| 50 | 0 | −0.025 | −0.050 | −0.075 | −0.100 | −0.125 | −0.151 | −0.176 | −0.202 | −0.228 | −0.241 | −0.251 |

*Example*: To be 95% confident that the pavement will remain at or above its TSI ($R = 95$ for use in Fig. 4.7), a $Z_R$ value of −1.645 would be used in Eq. 4.1 (and in Eq. 4.4).

### 4.4.3 Structural Number

The objective of Eq. 4.1 and the nomograph in Fig. 4.5 is to determine a required structural number for given axle loadings, reliability, overall standard deviation, change in PSI, and soil resilient modulus. As previously mentioned, there are many pavement material combinations and thicknesses that will provide satisfactory pavement service life. The following equation can be used to relate individual material types and thicknesses to the structural number:

$$SN = a_1 D_1 + a_2 D_2 M_2 + a_3 D_3 M_3 \tag{4.3}$$

*where*

$a_1, a_2, a_3$ = structural-layer coefficients of the wearing surface, base, and subbase layers, respectively,

$D_1, D_2, D_3$ = thickness of the wearing surface, base, and subbase layers in inches, respectively, and

$M_2, M_3$ = drainage coefficients for the base and subbase, respectively.

Values for the structural-layer coefficients for various types of material are presented in Table 4.5. Drainage coefficients are used to modify the thickness of the lower pavement layers (base and subbase) to take into account a material's drainage characteristics. A value of 1.0 for a drainage coefficient represents a material with good drainage characteristics (a sandy material). A soil such as clay does not drain very well and, consequently, will have a lower drainage coefficient (less than 1.0) than a sandy material. The reader is referred to [AASHTO 1993] for further information on drainage coefficients.

Because there are many combinations of structural-layer coefficients and thicknesses that solve Eq. 4.3, some guidelines are used to narrow the number of solutions. Experience has shown that wearing layers are typically 51 to 102 mm (2 to 4 inches) thick, whereas subbases and bases range from 102 to 254 mm (4 to 10 inches) thick. Knowing which of the materials is the most costly per inch of depth will assist in the determination of an initial layer thickness.

**Table 4.5** Structural-Layer Coefficients

| Pavement component | Coefficient |
|---|---|
| Wearing surface | |
| Sand-mix asphaltic concrete | 0.35 |
| Hot-mix asphaltic (HMA) concrete | 0.44 |
| Base | |
| Crushed stone | 0.14 |
| Dense-graded crushed stone | 0.18 |
| Soil cement | 0.20 |
| Emulsion/aggregate-bituminous | 0.30 |
| Portland cement/aggregate | 0.40 |
| Lime-pozzolan/aggregate | 0.40 |
| Hot-mix asphaltic (HMA) concrete | 0.40 |
| Subbase | |
| Crushed stone | 0.11 |

**EXAMPLE 4.1   FLEXIBLE PAVEMENT DESIGN—STRUCTURAL NUMBER DETERMINATION**

A pavement is to be designed to last 10 years. The initial PSI is 4.2 and the TSI (the final PSI) is determined to be 2.5. The subgrade has a soil resilient modulus of 103,430 kPa (15,000 lb/in$^2$). Reliability is 95% with an overall standard deviation of 0.4. For design, the daily car, pickup truck, and light van traffic is 30,000, and the daily truck traffic consists of 1000 passes of single-unit trucks with two single axles and 350 passes of tractor semi-trailer trucks with single, tandem, and triple axles. The axle weights are

$$
\begin{aligned}
\text{cars, pickups, light vans} &= \text{two 8.9-kN (2000-lb) single axles} \\
\text{single-unit truck} &= \text{35.6-kN (8000-lb) steering, single axle} \\
&= \text{97.9-kN (22,000-lb) drive, single axle} \\
\text{tractor semi-trailer truck} &= \text{44.5-kN (10,000-lb) steering, single axle} \\
&= \text{71.2-kN (16,000-lb) drive, tandem axle} \\
&= \text{195.7-kN (44,000-lb) trailer, triple axle}
\end{aligned}
$$

$M_2$ and $M_3$ are equal to 1.0 for the materials in the pavement structure. Four inches of hot-mix asphalt (HMA) is to be used as the wearing surface and 10 inches of crushed stone as the subbase. Determine the thickness required for the base if soil cement is the material to be used.

*SOLUTION*

Because the axle-load equivalency factors presented in Tables 4.1, 4.2, and 4.3 are a function of the structural number (SN), we have to assume an SN to start the problem (later we will arrive at a structural number and check to make sure that it is consistent with our assumed value). A typical assumption is to let SN = 4. Given this, the 18-kip–equivalent single-axle load for cars, pickups, and light vans is

2-kip single-axle equivalent = 0.0002 (Table 4.1)

This gives an 18-kip ESAL total of 0.0004 for each vehicle. For single-unit trucks,

8-kip single-axle equivalent = 0.041 (Table 4.1)

22-kip single-axle equivalent = 2.090 (Table 4.1)

This gives an 18-kip ESAL total of 2.131 for single-unit trucks. For tractor semi-trailer trucks,

10-kip single-axle equivalent = 0.102 (Table 4.1)

16-kip tandem-axle equivalent = 0.057 (Table 4.2)

44-kip triple-axle equivalent = 0.769 (Table 4.3)

This gives an 18-kip ESAL total of 0.928 for tractor semi-trailer trucks. Note the comparatively small effect of cars and other light vehicles in terms of the 18-kip ESAL. This small effect underscores the nonlinear relationship between axle loads and pavement damage. For example, from Table 4.2 with SN = 4, a 36-kip single-axle load has 14.4 times the impact on pavement as an 18-kip single-axle load (twice the weight has 14.4 times the impact).

Given the computed 18-kip ESAL, the daily traffic on this highway produces an 18-kip ESAL total of 2467.8 (0.0004 ×30,000 + 2.131 × 1000 + 0.928 × 350). Traffic (total axle accumulations) over the 10-year design period will be

$$2467.8 \times 365 \times 10 = 9{,}007{,}470 \text{ 18-kip ESAL}$$

With an initial PSI of 4.2 and a TSI of 2.5, $\Delta$PSI = 1.7. Solving Eq. 4.3 for SN (using an equation solver on a calculator or computer) with $Z_R$ = −1.645 (which corresponds to R = 95%, as shown in Table 4.4) gives SN = 3.94 (Fig. 4.5 can also be used to arrive at an approximate solution for SN). Note that this is very close to the value that was assumed (SN = 4.0) to get the load equivalency factors from Tables 4.1, 4.2, and 4.3. If Eq. 4.1 gave SN = 5, we would go back and recompute total axle accumulations using the SN of 5 to read the axle-load equivalency factors in Tables 4.1, 4.2, and 4.3. Usually one iteration of this type is all that is needed. Later, Examples 4.3 and 4.5 will demonstrate this type of iteration.

Given that SN = 3.94, Eq. 4.3 can be applied with $a_1$ = 0.44 (surface course, hot-mix asphalt, Table 4.5), $a_2$ = 0.20 (base course, soil cement, Table 4.5), and $a_3$ = 0.11 (subbase, crushed stone, Table 4.5), $M_2$ = 1.0 (given), $M_3$ = 1.0 (given), $D_1$ = 4.0 inches (given), and $D_3$ = 10.0 inches (given). We have

$$SN = a_1 D_1 + a_2 D_2 M_2 + a_3 D_3 M_3$$

$$3.94 = 0.44(4) + 0.20 D_2 (1.0) + 0.11(10.0)(1.0)$$

Solving for $D_2$ gives $D_2$ = 5.4 inches (137 mm). Using $D_2$ = <u>5.5 inches (140 mm)</u> would be a conservative estimate and allow for variations in construction. Rounding up to the nearest 0.5 inch is a safe practice.

**EXAMPLE 4.2   FLEXIBLE PAVEMENT DESIGN—RELIABILITY ASSESSMENT**

A flexible pavement is constructed with 102 mm (4 inches) of hot-mix asphalt (HMA) wearing surface, 203 mm (8 inches) of emulsion/aggregate-bituminous base, and 203 mm (8 inches) of crushed stone subbase. The subgrade has a soil resilient modulus of 68,950 kPa (10,000 lb/in²), and $M_2$ and $M_3$ are equal to 1.0 for the materials in the pavement structure. The overall standard deviation is 0.5, the initial PSI is 4.5, and the TSI is 2.5. The daily traffic has 1080 89.0-kN (20-kip) single axles, 400 106.8-kN (24-kip) single axles, and 680 177.9-kN (40-kip) tandem axles. How many years would you estimate this pavement would last (how long before its PSI drops below a TSI of 2.5) if you wanted to be 90% confident that your estimate was not too high, and if you wanted to be 99% confident that your estimate was not too high?

**SOLUTION**

The pavement's structural number is determined from Eq. 4.3 , using Table 4.5 to find $a_1 = 0.44$, $a_2 = 0.30$ and $a_3 = 0.11$, and with $D_1 = 4$, $D_2 = 8$, $D_3 = 8$, $M_2 = M_3 = 1.0$ (all given) as

$$SN = a_1 D_1 + a_2 D_2 M_2 + a_3 D_3 M_3$$

$$SN = 0.44(4) + 0.30(8)(1.0) + 0.11(8.0)(1.0) = 5.04$$

For the daily axle loads, the equivalency factors (reading axle equivalents from Tables 4.1 and 4.2 while using SN = 5, which is very close to the 5.04 computed above) are

20-kip single-axle equivalent = 1.51 (Table 4.1)

24-kip single-axle equivalent = 3.03 (Table 4.1)

40-kip tandem-axle equivalent = 2.08 (Table 4.2)

Thus the total daily 18-kip ESAL is

Daily $W_{18}$ = 1.51(1080) + 3.03(400) + 2.08(680) = 4257.2 18-kip ESAL

Applying Eq. 4.1, with $S_o = 0.5$, SN = 5.04, $\Delta$PSI = 2.0 (4.5 − 2.5), and $M_R = 10,000$ lb/in², we find that at R = 90% ($Z_R = -1.282$ for purposes of Eq. 4.1, as shown in Table 4.4), $W_{18}$ is 26,128,077. Therefore, the number of years is

$$\text{years} = \frac{26,128,077}{365 \times 4257.2}$$

$$= 16.82 \text{ years}$$

Similarly, with R = 99% ($Z_R = -2.326$ for purposes of Eq. 4.1, as shown in Table 4.4), $W_{18}$ is 7,854,299, so the number of years is

$$\text{years} = \frac{7,854,299}{365 \times 4257.2}$$

$$= 5.05 \text{ years}$$

These results show that one can be 99% confident that the pavement will last (have a PSI above 2.5) at least 5.05 years, and one can be 90% confident that it will have a PSI above 2.5 for 16.82 years. This example demonstrates the large impact that the chosen reliability value can have on pavement design.

## 4.5   PAVEMENT SYSTEM DESIGN: PRINCIPLES FOR RIGID PAVEMENTS

Rigid pavements distribute wheel loads by the beam action of the portland cement concrete (PCC) slab, which is made of a material that has a high modulus of elasticity, on the order of 27.6 to 34.5 GPa (4 to 5 million lb/in$^2$). This beam action (see Fig. 4.6) distributes the wheel loads over a large area of the pavement, thus reducing the high stresses experienced at the surface of the pavement to a level that is acceptable to the subgrade soil.

**Figure 4.6** Beam action of a rigid pavement.

## 4.6   TRADITIONAL AASHTO RIGID-PAVEMENT DESIGN PROCEDURE

The design procedure for rigid pavements presented in the AASHTO design guide is also based on the field results of the AASHO Road Test. The AASHTO design procedure is applicable to jointed plain concrete pavements, jointed reinforced concrete pavements, and continuously reinforced concrete pavements. Jointed plain concrete pavements (JPCP) do not have slab-reinforcing material and can have doweled joints (steel bars to transfer loads between slabs as shown in Fig. 4.2) or undoweled joints. The traverse joints between slabs are spaced at about 3 to 4 m (10 to 13 ft). Jointed reinforced concrete pavements (JRCP) have steel reinforced slabs with joints that are 12.2 m (40 ft) or more apart. Finally, continuously reinforced concrete pavements (CRCP) do not have traverse expansion/contraction joints, necessitating the use of extensive steel-bar reinforcement in the slab. The idea with both jointed-reinforced and continuously-reinforced pavements is to permit slab cracking but to provide sufficient slab reinforcement to hold the cracks tightly together to ensure load transfer. It is important to note that faulting, which is a distress characterized by different slab elevations, was not a failure consideration in the AASHO Road Test, and thus the design of nondoweled joints must be checked with a procedure other than that presented here (more information on faulting is provided in Section 4.7.5).

The design procedure for rigid pavements is based on a selected reduction in serviceability and is similar to the procedure for flexible pavements. However, instead of measuring pavement strength by using a structural number, the thickness of the PCC slab is the measure of strength. The regression equation that is used (in U.S. Customary units) to determine the thickness of a rigid-pavement PCC slab is

$$\log_{10} W_{18} = Z_R S_o + 7.35 \left[ \log_{10} (D+1) \right] - 0.06$$

$$+ \frac{\log_{10} \left[ \Delta PSI / 3.0 \right]}{1 + \left[ 1.624 \times 10^7 / (D+1)^{8.46} \right]} \tag{4.4}$$

$$+ (4.22 - 0.32 \text{TSI}) \log_{10} \left( \frac{S'_c C_d \left[ D^{0.75} - 1.132 \right]}{215.63 J \left\{ D^{0.75} - \left[ 18.42 / (E_c / k)^{0.25} \right] \right\}} \right)$$

*where*

$W_{18}$ = 80.1-kN (18-kip)–equivalent single-axle loads,

$Z_r$ = Reliability ($z$-statistic from the standard normal curve),

$S_o$ = Overall standard deviation of traffic,

$D$ = PCC slab thickness in inches

TSI = Pavement's terminal serviceability index,

$\Delta PSI$ = Loss in serviceability from the time when the pavement is new until it reaches its TSI,

$S'_c$ = Concrete modulus of rupture in lb/in$^2$,

$C_d$ = Drainage coefficient,

$J$ = Load transfer coefficient,

$E_c$ = Concrete modulus of elasticity in lb/ in$^2$, and

$k$ = Modulus of subgrade reaction in lb/in$^3$.

A graphic solution to Eq. 4.4 is shown in Figs. 4.7 and 4.8. The terms used in Eq. 4.4 and Figs. 4.8 and 4.9 are defined as follows:

$W_{18}$   The 80.1-kN (18-kip)–equivalent single-axle load is the same concept as discussed for the flexible-pavement design procedure. However, instead of being a function of the structural number, this value is a function of slab thickness. The axle-load equivalency factors used in rigid-pavement design are presented in Tables 4.6 (for single axles), 4.7 (for tandem axles), and 4.8 (for triple axles).

$Z_R$   As in flexible-pavement design, the reliability, $Z_R$, is defined as the probability that serviceability will be maintained at adequate levels from a user's point of view throughout the design life of the facility (the PSI will stay above the TSI). In the rigid-pavement design nomograph (Figs. 4.7 and 4.8), the probabilities (in percent) are used directly (instead of $Z_R$ as in Eq. 4.4), and these percent probabilities are denoted $R$ (see Table 4.4, which still applies).

$S_o$   As in flexible-pavement design, the overall standard deviation, $S_o$, takes into account designers' inability to accurately estimate future 80.1-kN (18-kip)–equivalent axle loads and the statistical error in the equations resulting from variability in materials and construction practices.

**Table 4.6**  Axle-Load Equivalency Factors for Rigid Pavements, Single Axles, and TSI = 2.5

| Axle load (kips) | Slab thickness, $D$ (inches) | | | | | | | | |
|---|---|---|---|---|---|---|---|---|---|
| | 6 | 7 | 8 | 9 | 10 | 11 | 12 | 13 | 14 |
| 2 | 0.0002 | 0.0002 | 0.0002 | 0.0002 | 0.0002 | 0.0002 | 0.0002 | 0.0002 | 0.0002 |
| 4 | 0.003 | 0.002 | 0.002 | 0.002 | 0.002 | 0.002 | 0.002 | 0.002 | 0.002 |
| 6 | 0.012 | 0.011 | 0.010 | 0.010 | 0.010 | 0.010 | 0.010 | 0.010 | 0.010 |
| 8 | 0.039 | 0.035 | 0.033 | 0.032 | 0.032 | 0.032 | 0.032 | 0.032 | 0.032 |
| 10 | 0.097 | 0.089 | 0.084 | 0.082 | 0.081 | 0.080 | 0.080 | 0.080 | 0.080 |
| 12 | 0.203 | 0.189 | 0.181 | 0.176 | 0.175 | 0.174 | 0.174 | 0.174 | 0.173 |
| 14 | 0.376 | 0.360 | 0.347 | 0.341 | 0.338 | 0.337 | 0.336 | 0.336 | 0.336 |
| 16 | 0.634 | 0.623 | 0.610 | 0.604 | 0.601 | 0.599 | 0.599 | 0.599 | 0.598 |
| 18 | 1.00 | 1.00 | 1.00 | 1.00 | 1.00 | 1.00 | 1.00 | 1.00 | 1.00 |
| 20 | 1.51 | 1.52 | 1.55 | 1.57 | 1.58 | 1.58 | 1.59 | 1.59 | 1.59 |
| 22 | 2.21 | 2.20 | 2.28 | 2.34 | 2.38 | 2.40 | 2.41 | 2.41 | 2.41 |
| 24 | 3.16 | 3.10 | 3.22 | 3.36 | 3.45 | 3.50 | 3.53 | 3.54 | 3.55 |
| 26 | 4.41 | 4.26 | 4.42 | 4.67 | 4.85 | 4.95 | 5.01 | 5.04 | 5.05 |
| 28 | 6.05 | 5.76 | 5.92 | 6.29 | 6.61 | 6.81 | 6.92 | 6.98 | 7.01 |
| 30 | 8.16 | 7.67 | 7.79 | 8.28 | 8.79 | 9.14 | 9.35 | 9.46 | 9.52 |
| 32 | 10.8 | 10.1 | 10.1 | 10.7 | 11.4 | 12.0 | 12.3 | 12.6 | 12.7 |
| 34 | 14.1 | 13.0 | 12.9 | 13.6 | 14.6 | 15.4 | 16.0 | 16.4 | 16.5 |
| 36 | 18.2 | 16.7 | 16.4 | 17.1 | 18.3 | 19.5 | 20.4 | 21.0 | 21.3 |
| 38 | 23.1 | 21.1 | 20.6 | 21.3 | 22.7 | 24.3 | 25.6 | 26.4 | 27.0 |
| 40 | 29.1 | 26.5 | 25.7 | 26.3 | 27.9 | 29.9 | 31.6 | 32.9 | 33.7 |
| 42 | 36.2 | 32.9 | 31.7 | 32.2 | 34.0 | 36.3 | 38.7 | 40.4 | 41.6 |
| 44 | 44.6 | 40.4 | 38.8 | 39.2 | 41.0 | 43.8 | 46.7 | 49.1 | 50.8 |
| 46 | 54.5 | 49.3 | 47.1 | 47.3 | 49.2 | 52.3 | 55.9 | 59.0 | 61.4 |
| 48 | 66.1 | 59.7 | 56.9 | 56.8 | 58.7 | 62.1 | 66.3 | 70.3 | 73.4 |
| 50 | 79.4 | 71.7 | 68.2 | 67.8 | 69.6 | 73.3 | 78.1 | 83.0 | 87.1 |

*Source*: *AASHTO Guide for Design of Pavement Structures*, The American Association of State Highway and Transportation Officials, Washington, DC, 1993. Used by permission.

**Table 4.7** Axle-Load Equivalency Factors for Rigid Pavements, Tandem Axles, and TSI = 2.5

| Axle load (kips) | Slab thickness, $D$ (inches) | | | | | | | | |
|---|---|---|---|---|---|---|---|---|---|
| | 6 | 7 | 8 | 9 | 10 | 11 | 12 | 13 | 14 |
| 2 | 0.0001 | 0.0001 | 0.0001 | 0.0001 | 0.0001 | 0.0001 | 0.0001 | 0.0001 | 0.0001 |
| 4 | 0.0006 | 0.0006 | 0.0005 | 0.0005 | 0.0005 | 0.0005 | 0.0005 | 0.0005 | 0.0005 |
| 6 | 0.002 | 0.002 | 0.002 | 0.002 | 0.002 | 0.002 | 0.002 | 0.002 | 0.002 |
| 8 | 0.007 | 0.006 | 0.006 | 0.005 | 0.005 | 0.005 | 0.005 | 0.005 | 0.005 |
| 10 | 0.015 | 0.014 | 0.013 | 0.013 | 0.012 | 0.012 | 0.012 | 0.012 | 0.012 |
| 12 | 0.031 | 0.028 | 0.026 | 0.026 | 0.025 | 0.025 | 0.025 | 0.025 | 0.025 |
| 14 | 0.057 | 0.052 | 0.049 | 0.048 | 0.047 | 0.047 | 0.047 | 0.047 | 0.047 |
| 16 | 0.097 | 0.089 | 0.084 | 0.082 | 0.081 | 0.081 | 0.080 | 0.080 | 0.080 |
| 18 | 0.155 | 0.143 | 0.136 | 0.133 | 0.132 | 0.131 | 0.131 | 0.131 | 0.131 |
| 20 | 0.234 | 0.220 | 0.211 | 0.206 | 0.204 | 0.203 | 0.203 | 0.203 | 0.203 |
| 22 | 0.340 | 0.325 | 0.313 | 0.308 | 0.305 | 0.304 | 0.303 | 0.303 | 0.303 |
| 24 | 0.475 | 0.462 | 0.450 | 0.444 | 0.441 | 0.440 | 0.439 | 0.439 | 0.439 |
| 26 | 0.644 | 0.637 | 0.627 | 0.622 | 0.620 | 0.619 | 0.618 | 0.618 | 0.618 |
| 28 | 0.855 | 0.854 | 0.852 | 0.850 | 0.850 | 0.850 | 0.849 | 0.849 | 0.849 |
| 30 | 1.11 | 1.12 | 1.13 | 1.14 | 1.14 | 1.14 | 1.14 | 1.14 | 1.14 |
| 32 | 1.43 | 1.44 | 1.47 | 1.49 | 1.50 | 1.51 | 1.51 | 1.51 | 1.51 |
| 34 | 1.82 | 1.82 | 1.87 | 1.92 | 1.95 | 1.96 | 1.97 | 1.97 | 1.97 |
| 36 | 2.29 | 2.27 | 2.35 | 2.43 | 2.48 | 2.51 | 2.52 | 2.52 | 2.53 |
| 38 | 2.85 | 2.80 | 2.91 | 3.03 | 3.12 | 3.16 | 3.18 | 3.20 | 3.20 |
| 40 | 3.52 | 3.42 | 3.55 | 3.74 | 3.87 | 3.94 | 3.98 | 4.00 | 4.01 |
| 42 | 4.32 | 4.16 | 4.30 | 4.55 | 4.74 | 4.86 | 4.91 | 4.95 | 4.96 |
| 44 | 5.26 | 5.01 | 5.16 | 5.48 | 5.75 | 5.92 | 6.01 | 6.06 | 6.09 |
| 46 | 6.36 | 6.01 | 6.14 | 6.53 | 6.90 | 7.14 | 7.28 | 7.36 | 7.40 |
| 48 | 7.64 | 7.16 | 7.27 | 7.73 | 8.21 | 8.55 | 8.75 | 8.86 | 8.92 |
| 50 | 9.11 | 8.50 | 8.55 | 9.07 | 9.68 | 10.14 | 10.42 | 10.58 | 10.66 |
| 52 | 10.8 | 10.0 | 10.0 | 10.6 | 11.3 | 11.9 | 12.3 | 12.5 | 12.7 |
| 54 | 12.8 | 11.8 | 11.7 | 12.3 | 13.2 | 13.9 | 14.5 | 14.8 | 14.9 |
| 56 | 15.0 | 13.8 | 13.6 | 14.2 | 15.2 | 16.2 | 16.8 | 17.3 | 17.5 |
| 58 | 17.5 | 16.0 | 15.7 | 16.3 | 17.5 | 18.6 | 19.5 | 20.1 | 20.4 |
| 60 | 20.3 | 18.5 | 18.1 | 18.7 | 20.0 | 21.4 | 22.5 | 23.2 | 23.6 |
| 63 | 23.5 | 21.4 | 20.8 | 21.4 | 22.8 | 24.4 | 25.7 | 26.7 | 27.3 |
| 64 | 27.0 | 24.6 | 23.8 | 24.4 | 25.8 | 27.7 | 29.3 | 30.5 | 31.3 |
| 66 | 31.0 | 28.1 | 27.1 | 27.6 | 29.2 | 31.3 | 33.2 | 34.7 | 35.7 |
| 68 | 35.4 | 32.1 | 30.9 | 31.3 | 32.9 | 35.2 | 37.5 | 39.3 | 40.5 |
| 70 | 40.3 | 36.5 | 35.0 | 35.3 | 37.0 | 39.5 | 42.1 | 44.3 | 45.9 |
| 72 | 45.7 | 41.4 | 39.6 | 39.8 | 41.5 | 44.2 | 47.2 | 49.8 | 51.7 |
| 74 | 51.7 | 46.7 | 44.6 | 44.7 | 46.4 | 49.3 | 52.7 | 55.7 | 58.0 |
| 76 | 58.3 | 52.6 | 50.2 | 50.1 | 51.8 | 54.9 | 58.6 | 62.1 | 64.8 |
| 78 | 65.5 | 59.1 | 56.3 | 56.1 | 57.7 | 60.9 | 65.0 | 69.0 | 72.3 |
| 80 | 73.4 | 66.2 | 62.9 | 62.5 | 64.2 | 67.5 | 71.9 | 76.4 | 80.2 |
| 82 | 82.0 | 73.9 | 70.2 | 69.6 | 71.2 | 74.7 | 79.4 | 84.4 | 88.8 |
| 84 | 91.4 | 82.4 | 78.1 | 77.3 | 78.9 | 82.4 | 87.4 | 93.0 | 98.1 |
| 86 | 102.0 | 92.0 | 87.0 | 86.0 | 87.0 | 91.0 | 96.0 | 102.0 | 108.0 |
| 88 | 113.0 | 102.0 | 96.0 | 95.0 | 96.0 | 100.0 | 105.0 | 112.0 | 119.0 |
| 90 | 125.0 | 112.0 | 106.0 | 105.0 | 106.0 | 110.0 | 115.0 | 123.0 | 130.0 |

Source: *AASHTO Guide for Design of Pavement Structures*, The American Association of State Highway and Transportation Officials, Washington, DC, 1993. Used by permission.

**Table 4.8** Axle-Load Equivalency Factors for Rigid Pavements, Triple Axles, and TSI = 2.5

| Axle load (kips) | Slab thickness, D (inches) | | | | | | | | |
|---|---|---|---|---|---|---|---|---|---|
| | 6 | 7 | 8 | 9 | 10 | 11 | 12 | 13 | 14 |
| 2 | 0.0001 | 0.0001 | 0.0001 | 0.0001 | 0.0001 | 0.0001 | 0.0001 | 0.0001 | 0.0001 |
| 4 | 0.0003 | 0.0003 | 0.0003 | 0.0003 | 0.0003 | 0.0003 | 0.0003 | 0.0003 | 0.0003 |
| 6 | 0.001 | 0.001 | 0.001 | 0.001 | 0.001 | 0.001 | 0.001 | 0.001 | 0.001 |
| 8 | 0.003 | 0.002 | 0.002 | 0.002 | 0.002 | 0.002 | 0.002 | 0.002 | 0.002 |
| 10 | 0.006 | 0.005 | 0.005 | 0.005 | 0.005 | 0.005 | 0.005 | 0.005 | 0.005 |
| 12 | 0.011 | 0.010 | 0.010 | 0.009 | 0.009 | 0.009 | 0.009 | 0.009 | 0.009 |
| 14 | 0.020 | 0.018 | 0.017 | 0.017 | 0.016 | 0.016 | 0.016 | 0.016 | 0.016 |
| 16 | 0.033 | 0.030 | 0.029 | 0.028 | 0.027 | 0.027 | 0.027 | 0.027 | 0.027 |
| 18 | 0.053 | 0.048 | 0.045 | 0.044 | 0.044 | 0.043 | 0.043 | 0.043 | 0.043 |
| 20 | 0.080 | 0.073 | 0.069 | 0.067 | 0.066 | 0.066 | 0.066 | 0.066 | 0.066 |
| 22 | 0.116 | 0.107 | 0.101 | 0.099 | 0.098 | 0.097 | 0.097 | 0.097 | 0.097 |
| 24 | 0.163 | 0.151 | 0.144 | 0.141 | 0.139 | 0.139 | 0.138 | 0.138 | 0.138 |
| 26 | 0.222 | 0.209 | 0.200 | 0.195 | 0.194 | 0.193 | 0.192 | 0.192 | 0.192 |
| 28 | 0.295 | 0.281 | 0.271 | 0.265 | 0.263 | 0.262 | 0.262 | 0.262 | 0.262 |
| 30 | 0.384 | 0.371 | 0.359 | 0.354 | 0.351 | 0.350 | 0.349 | 0.349 | 0.349 |
| 32 | 0.490 | 0.480 | 0.468 | 0.463 | 0.460 | 0.459 | 0.458 | 0.458 | 0.458 |
| 34 | 0.616 | 0.609 | 0.601 | 0.596 | 0.594 | 0.593 | 0.592 | 0.592 | 0.592 |
| 36 | 0.765 | 0.762 | 0.759 | 0.757 | 0.756 | 0.755 | 0.755 | 0.755 | 0.755 |
| 38 | 0.939 | 0.941 | 0.946 | 0.948 | 0.950 | 0.951 | 0.951 | 0.951 | 0.951 |
| 40 | 1.14 | 1.15 | 1.16 | 1.17 | 1.18 | 1.18 | 1.18 | 1.18 | 1.18 |
| 42 | 1.38 | 1.38 | 1.41 | 1.44 | 1.45 | 1.46 | 1.46 | 1.46 | 1.46 |
| 44 | 1.65 | 1.65 | 1.70 | 1.74 | 1.77 | 1.78 | 1.78 | 1.78 | 1.78 |
| 46 | 1.97 | 1.96 | 2.03 | 2.09 | 2.13 | 2.15 | 2.16 | 2.16 | 2.16 |
| 48 | 2.34 | 2.31 | 2.40 | 2.49 | 2.55 | 2.58 | 2.59 | 2.60 | 2.60 |
| 50 | 2.76 | 2.71 | 2.81 | 2.94 | 3.02 | 3.07 | 3.09 | 3.10 | 3.11 |
| 52 | 3.24 | 3.15 | 3.27 | 3.44 | 3.56 | 3.62 | 3.66 | 3.68 | 3.68 |
| 54 | 3.79 | 3.66 | 3.79 | 4.00 | 4.16 | 4.26 | 4.30 | 4.33 | 4.34 |
| 56 | 4.41 | 4.23 | 4.37 | 4.63 | 4.84 | 4.97 | 5.03 | 5.07 | 5.09 |
| 58 | 5.12 | 4.87 | 5.00 | 5.32 | 5.59 | 5.76 | 5.85 | 5.90 | 5.93 |
| 60 | 5.91 | 5.59 | 5.71 | 6.08 | 6.42 | 6.64 | 6.77 | 6.84 | 6.87 |
| 63 | 6.80 | 6.39 | 6.50 | 6.91 | 7.33 | 7.62 | 7.79 | 7.88 | 7.93 |
| 64 | 7.79 | 7.29 | 7.37 | 7.82 | 8.33 | 8.70 | 8.92 | 9.04 | 9.11 |
| 66 | 8.90 | 8.28 | 8.33 | 8.83 | 9.42 | 9.88 | 10.17 | 10.33 | 10.42 |
| 68 | 10.1 | 9.4 | 9.4 | 9.9 | 10.6 | 11.2 | 11.5 | 11.7 | 11.9 |
| 70 | 11.5 | 10.6 | 10.6 | 11.1 | 11.9 | 12.6 | 13.0 | 13.3 | 13.5 |
| 72 | 13.0 | 12.0 | 11.8 | 12.4 | 13.3 | 14.1 | 14.7 | 15.0 | 15.2 |
| 74 | 14.6 | 13.5 | 13.2 | 13.8 | 14.8 | 15.8 | 16.5 | 16.9 | 17.1 |
| 76 | 16.5 | 15.1 | 14.8 | 15.4 | 16.5 | 17.6 | 18.4 | 18.9 | 19.2 |
| 78 | 18.5 | 16.9 | 16.5 | 17.1 | 18.2 | 19.5 | 20.5 | 21.1 | 21.5 |
| 80 | 20.6 | 18.8 | 18.3 | 18.9 | 20.2 | 21.6 | 22.7 | 23.5 | 24.0 |
| 82 | 23.0 | 21.0 | 20.3 | 20.9 | 22.2 | 23.8 | 25.2 | 26.1 | 26.7 |
| 84 | 25.6 | 23.3 | 22.5 | 23.1 | 24.5 | 26.2 | 27.8 | 28.9 | 29.6 |
| 86 | 28.4 | 25.8 | 24.9 | 25.4 | 26.9 | 28.8 | 30.5 | 31.9 | 32.8 |
| 88 | 31.5 | 28.6 | 27.5 | 27.9 | 29.4 | 31.5 | 33.5 | 35.1 | 36.1 |
| 90 | 34.8 | 31.5 | 30.3 | 30.7 | 32.2 | 34.4 | 36.7 | 38.5 | 39.8 |

Source: *AASHTO Guide for Design of Pavement Structures*, The American Association of State Highway and Transportation Officials, Washington, DC, 1993. Used by permission.

TSI The pavement's terminal serviceability index, TSI, is the point at which the pavement can no longer perform in a serviceable manner, as discussed previously for the flexible-pavement design procedure.

ΔPSI The amount of serviceability loss, ΔPSI, over the life of the pavement is the difference between the initial PSI and the TSI, as discussed for the flexible-pavement design procedure.

$S'_c$ The concrete modulus of rupture, $S'_c$, is a measure of the tensile strength of the concrete and is determined by loading a beam specimen, at the third points, to failure. The test method is ASTM C78, Flexural Strength of Concrete. Because concrete gains strength with age, the average 28-day strength is used for design purposes. Typical values are 3450 to 8270 kPa (500 to 1200 lb/in$^2$).

$C_d$ The drainage coefficient, $C_d$, is slightly different from the value used in flexible-pavement design. In rigid-pavement design, it accounts for the drainage characteristics of the subgrade. A value of 1.0 for the drainage coefficient represents a material with good drainage characteristics (such as a sandy material). Soils with less-than-ideal drainage characteristics will have drainage coefficients less than 1.0.

$J$ The load transfer coefficient, $J$, is a factor that is used to account for the ability of pavement to transfer a load from one PCC slab to another across the slab joints. Many rigid pavements have dowel bars across the joints to transfer loads between slabs. Pavements with dowel bars at the joints are typically designed with a $J$ value of 3.2.

$E_c$ The concrete modulus of elasticity, $E_c$, is derived from the stress-strain curve as taken in the elastic region. The modulus of elasticity is also known as Young's modulus. Typical values of $E_c$ for portland cement concrete are between 20.7 and 48.3 GPa (3 and 7 million lb/in$^2$).

$k$ The modulus of subgrade reaction, $k$, depends upon several different factors, including the moisture content and density of the soil. It should be noted that most highway agencies do not perform testing to measure the $k$ value of the soil. At best, the agency will have a CBR value for the subgrade. Typical values for $k$ range from 0.0271 to 0.2165 N/mm$^3$ (100 to 800 lb/in$^3$). Table 4.9 indicates the relationship between CBR and $k$ values.

**Table 4.9** Relationship Between California Bearing Ratio (CBR) and Modulus of Subgrade Reaction, $k$

| CBR | $k$, lb/in$^3$ |
|---|---|
| 2 | 100 |
| 10 | 200 |
| 20 | 250 |
| 25 | 290 |
| 40 | 420 |
| 50 | 500 |
| 75 | 680 |
| 100 | 800 |

**Figure 4.7** Segment 1 of the design chart for rigid pavement based on the use of mean values for each input variable.

Redrawn from *AASHTO Guide for Design of Pavement Structures*, The American Association of State Highway and Transportation Officials, Washington, DC, 1993. Used by permission.

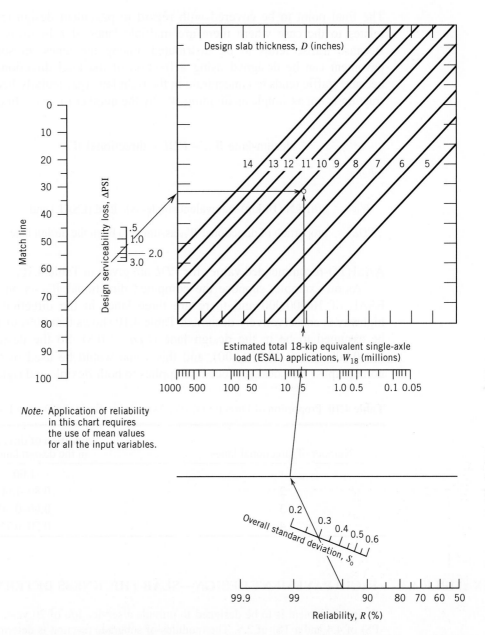

**Figure 4.8** Segment 2 of the design chart for rigid pavements based on the use of mean values for each input variable.

Redrawn from *AASHTO Guide for Design of Pavement Structures*, The American Association of State Highway and Transportation Officials, Washington, DC, 1993. Used by permission.

The final point to be covered with regard to pavement design (rigid and flexible) relates to the case where there are multiple lanes of a highway in one direction. Because traffic tends to be distributed among the lanes, in some instances the pavement can be designed using a fraction of the total directional $W_{18}$. However, because traffic tends to concentrate in the right lane (particularly heavy vehicles), this fraction is not as simple as dividing $W_{18}$ by the number of lanes. In equation form,

$$\text{design-lane } W_{18} = PDL \times \text{directional } W_{18} \tag{4.5}$$

*where*

$W_{18}$ = 80.1-kN (18-kip)–equivalent single-axle load (ESAL) and

$PDL$ = proportion of directional $W_{18}$ assumed to be in the design lane.

AASHTO-recommended values for *PDL* are given in Table 4.10.

As an example, suppose the computed directional $W_{18}$ is an 80.1-kN (18-kip) ESAL of 10,000,000 and there are three lanes in the direction of travel. If the highway is conservatively designed, Table 4.10 shows that 80% of the axle loads can be assumed to be in the design lane (*PDL* = 0.8). So the design $W_{18}$ would be 8,000,000 (0.8 × 10,000,000), and this value would be used in the equations and nomographs. This design procedure applies to both flexible and rigid pavements.

**Table 4.10** Proportion of Directional $W_{18}$ Assumed to Be in the Design Lane

| Number of directional lanes | Proportion of directional $W_{18}$ in the design lane (*PDL*) |
|:---:|:---:|
| 1 | 1.00 |
| 2 | 0.80–1.00 |
| 3 | 0.60–0.80 |
| 4 | 0.50–0.75 |

**EXAMPLE 4.3    RIGID PAVEMENT DESIGN—SLAB THICKNESS DETERMINATION**

A rigid pavement is to be designed to provide a service life of 20 years and has an initial PSI of 4.4 and a TSI of 2.5. The modulus of subgrade reaction is determined to be 0.0813 N/mm³ (300 lb/in³). For design, the daily car, pickup truck, and light van traffic is 20,000; and the daily truck traffic consists of 200 passes of single-unit trucks with single and tandem axles, and 410 passes of tractor semi-trailer trucks with single, tandem, and triple axles. The axle weights are

| | |
|---|---|
| cars, pickups, light vans | = two 8.9-kN (2000-lb) single axles |
| single-unit trucks | = 44.5-kN (10,000-lb) steering, single axle |
| | = 97.9-kN (22,000-lb) drive, tandem axle |

tractor semi-trailer trucks   = 53.4-kN (12,000-lb) steering, single axle

= 80.1-kN (18,000-lb) drive, tandem axle

= 222.4-kN (50,000-lb) trailer, triple axle

Reliability is 95%, the overall standard deviation is 0.45, the concrete's modulus of elasticity is 31.03 GPa (4.5 million lb/in$^2$), the concrete's modulus of rupture is 6210 kPa (900 lb/in$^2$), the load transfer coefficient is 3.2, and the drainage coefficient is 1.0. Determine the required slab thickness.

**SOLUTION**

Because the axle-load equivalency factors presented in Tables 4.6, 4.7, and 4.8 are a function of the slab thickness ($D$), we have to assume a $D$ value to start the problem (later we will arrive at a slab thickness and check to make sure that it is consistent with our assumed value). A typical assumption is to let $D = 10$ inches. Given this, the 18-kip–equivalent single-axle load (18-kip ESAL) for cars, pickups, and light vans is

2-kip single-axle equivalent = 0.0002 (Table 4.6)

This gives an 18-kip ESAL total of 0.0004 for each vehicle. For single-unit trucks,

10-kip single-axle equivalent   = 0.081 (Table 4.6)

22-kip tandem-axle equivalent   = 0.305 (Table 4.7)

This gives an 18-kip ESAL total of 0.386 for single-unit trucks. For tractor semi-trailer trucks,

12-kip single-axle equivalent   = 0.175 (Table 4.6)

18-kip tandem-axle equivalent   = 0.132 (Table 4.7)

50-kip triple-axle equivalent   = 3.020 (Table 4.8)

This gives an 18-kip ESAL total of 3.327 for tractor semi-trailer trucks.

Given the computed 18-kip ESAL, the daily traffic on this highway produces an 18-kip ESAL total of 1449.27 ($0.0004 \times 20,000 + 0.386 \times 200 + 3.327 \times 410$). Traffic (total axle accumulations) over the 20-year design period will be

$$1449.27 \times 365 \times 20 = 10,579,671 \text{ 18-kip ESAL}$$

With an initial PSI of 4.4 and a TSI of 2.5, $\Delta$PSI = 1.9. Solving Eq. 4.4 for $D$ (using an equation solver on a calculator or computer) with $Z_R = -1.645$ (which corresponds to $R = 95\%$, as shown in Table 4.4) gives $D = 9.21$ inches (233.9 mm). (Figs. 4.7 and 4.8 can also be used to arrive at an approximate solution for $D$.) Note that this value differs from the slab thickness assumed to derive the axle-load equivalency factors. Recomputing the axle-load equivalency factors with $D = 9$ inches (228.6 mm) (for Tables 4.6, 4.7, and 4.8) for cars, pickups, and light vans gives

2-kip single-axle equivalent = 0.0002 (Table 4.6)

This gives an 18-kip ESAL total of 0.0004 (same as before) for each vehicle. For single-unit trucks,

$$10\text{-kip single-axle equivalent} = 0.082 \text{ (Table 4.6)}$$

$$22\text{-kip tandem-axle equivalent} = 0.308 \text{ (Table 4.7)}$$

This gives an 18-kip ESAL total of 0.390 (up from 0.386) for single-unit trucks. For tractor semi-trailer trucks,

$$12\text{-kip single-axle equivalent} = 0.176 \text{ (Table 4.6)}$$

$$18\text{-kip tandem-axle equivalent} = 0.133 \text{ (Table 4.7)}$$

$$50\text{-kip triple-axle equivalent} = 2.940 \text{ (Table 4.8)}$$

This gives an 18-kip ESAL total of 3.249 (down from 3.327) for tractor semi-trailer trucks. Given the computed 18-kip ESAL, the daily traffic on this highway produces an 18-kip ESAL of 1418.09 ($0.0004 \times 20{,}000 + 0.390 \times 200 + 3.249 \times 410$). Traffic (total axle accumulations) over the 20-year design period will be

$$1418.09 \times 365 \times 20 = 10{,}352{,}057 \text{ 18-kip ESAL}$$

Again, solving Eq. 4.4 for $D$ gives $D = 9.17$ inches (232.9 mm). (Figs. 4.7 and 4.8 can also be used to arrive at an approximate solution for $D$.) This is very close to the assumed $D = 9.0$ inches (229 mm) and is only a minor change from the 9.21 inches (234 mm) previously obtained. To be conservative, we would round up to the nearest 0.5 inch and make the slab 9.5 inches (240 mm).

## EXAMPLE 4.4  RIGID PAVEMENT DESIGN WITH TRAFFIC DISTRIBUTION BY LANE

In 1996, a rigid pavement on a northbound section of interstate highway was designed with a 304.8-mm (12-inch) PCC slab, an $E_c$ of 41.37 GPa ($6 \times 10^6$ lb/in$^2$), a concrete modulus of rupture of 5,520 kPa (800 lb/in$^2$), a load transfer coefficient of 3.0, an initial PSI of 4.5, and a terminal serviceability index of 2.5. The overall standard deviation was 0.45, the modulus of subgrade reaction was 0.05149 N/mm$^3$ (190 lb/in$^3$), and a reliability of 95% was used along with a drainage coefficient of 1.0. The pavement was designed for a 20-year life, and traffic was assumed to be composed entirely of tractor semi-trailer trucks with one 71.2-kN (16-kip) single axle, one 88.9-kN (20-kip) single axle, and one 155.7-kN (35-kip) tandem axle (the effect of all other vehicles was ignored). The interstate has four northbound lanes and was conservatively designed. How many tractor semi-trailer trucks, per day, were assumed to be traveling in the northbound direction?

### SOLUTION

Given that the slab thickness $D$ is 12 inches, for the tractor semi-trailer trucks we have

$$16\text{-kip single-axle equivalent} = 0.599 \text{ (Table 4.6)}$$

$$20\text{-kip single-axle equivalent} = 1.590 \text{ (Table 4.6)}$$

$$35\text{-kip tandem-axle equivalent} = 2.245 \text{ (Table 4.7)}$$

Note that the value of 2.245 for the 35,000-lb tandem-axle linear interpolation uses 34-kip and 36-kip values [(1.97 + 2.52)/2]. Summing these axle equivalents gives 4.434 18-kip ESAL per truck.

With an initial PSI of 4.5 and a TSI of 2.5, $\Delta$PSI = 2.0. Solving Eq. 4.4 for $W_{18}$ with $Z_R = -1.645$ (which corresponds to $R = 95\%$ as shown in Table 4.4) gives $W_{18} = 39{,}740{,}309$ 18-kip ESAL (Figs. 4.7 and 4.8 can also be used to arrive at an approximate solution for $W_{18}$). Thus the total daily truck traffic in the design lane is

$$\text{traffic} = \frac{39{,}740{,}309\ 18\text{-kip ESAL}}{365\ \text{days / year} \times 20\ \text{years} \times 4.434\ 18\text{-kip ESAL / truck}}$$

$$= 1227.76\ \text{trucks / day}$$

To determine the total directional volume (total number of northbound trucks), we note from Table 4.10 that the *PDL* for a conservative design on a four-lane highway is 0.75, and the application of Eq. 4.5 gives

$$\text{directional } W_{18} = \frac{\text{design - lane } W_{18}}{PDL}$$

$$= \frac{1227.76}{0.75}$$

$$= 1637.01\ \text{trucks / day}$$

## EXAMPLE 4.5  FLEXIBLE AND RIGID PAVEMENT DESIGN-LIFE EQUIVALENCE

A new flexible pavement was designed for four lanes of traffic (conservatively designed for load distribution among the lanes). The design is for a total directional daily traffic of 967 44.48-kN (10-kip) single axles and 1935 133.4-kN (30-kip) tandem axles. The pavement has an 203 mm (8-in) hot-mix asphaltic (HMA) surface, 254 mm (10-in) dense-graded crushed stone base and a 254 mm (10-in) crushed-stone subbase (the drainage coefficients are 0.9 for the base and 0.78 for the subbase). The soil CBR is 2, the reliability used was 95%, the overall standard deviation was 0.4, initial PSI was 4.5, and the TSI was 2.5. Determine the required slab thickness for a rigid pavement designed to last the same number of years as the flexible pavement, but with only three lanes (instead of four) in the design direction (again, conservatively designing for the load distribution among the lanes). The design is to use the same truck traffic, reliability, soil, initial PSI, TSI, and overall standard deviations as the flexible pavement. In addition, the rigid pavement is to have a modulus of rupture of 5516 kPa (800 lb/in$^2$), a concrete modulus of elasticity of 37.92 GPa (5.5 × 10$^6$ lb/in$^2$), a load transfer coefficient of 3.0, and a drainage coefficient of 0.9.

### SOLUTION

We will first need to determine the design life of the existing flexible pavement and then use this value to determine the required slab thickness of the rigid pavement.

The flexible pavement's structural number is determined from Eq. 4.3, using Table 4.5 to find $a_1 = 0.44$, $a_2 = 0.18$ and $a_3 = 0.11$, and with $D_1 = 8$, $D_2 = 10$, $D_3 = 10$, $M_2 = 0.9$ and $M_3 = 0.78$ (all given) as

$$SN = a_1 D_1 + a_2 D_2 M_2 + a_3 D_3 M_3$$

$$SN = 0.44(8) + 0.18(10)(0.9) + 0.11(10)(0.9) = 5.998 \approx 6$$

With a CBR = 2 (given), Eq. 4.2 is applied to give $M_R$ = 3000 (1500 × 2). For other elements required to solve Eq. 4.1 (or alternatively Fig. 4.5), $\Delta$PSI = 2 (initial PSI of 4.5 minus the TSI of 2.5), $S_o$ = 0.4 (given), and $Z_R$ = −1.645 (which corresponds to R = 95%, as shown in Table 4.4). With these values, applying Eq. 4.1 (or using Fig. 4.5) gives $W_{18}$ = 5,703,439.

For the daily axle loads, the equivalency factors (reading axle equivalents from Tables 4.1 and 4.2 while using SN = 6) are

10-kip single-axle equivalent = 0.08 (Table 4.1)

30-kip tandem-axle equivalent = 0.633 (Table 4.2)

Thus the total daily 18-kip ESAL is

Daily $W_{18}$ = 0.08(967) + 0.633(1935) = 1302.215 18-kip ESAL/day

From Table 4.10, the *PDL* for a conservative design on a four-lane highway is 0.75, so applying Eq. 4.4 gives

design - lane $W_{18}$ = PDL × directional $W_{18}$ = 976.66 $W_{18}$ / day

or 356,481 $W_{18}$/yr (976.66 × 365). So the design life of the flexible pavement is:

$$\text{design life (in years)} = \frac{\text{design } W_{18}}{\text{daily } W_{18}}$$

$$= \frac{5,703,439}{356,481}$$

$$= 16 \text{ years}$$

For the rigid pavement, we have that all of the design parameters are the same and are given $S'_c$ = 800 lb/in$^2$, $E_c$ = 5.5 × 10$^6$ lb/in$^2$, J = 3.0, $C_d$ = 0.9. In addition, the modulus of subgrade reaction, k, is 100 lb/in$^3$ from Table 4.9 (with CBR = 2). Note that the accumulated $W_{18}$ for the rigid pavement will not be the same as the daily axle loads for the flexible pavement because the load equivalency factors will be different. To determine the accumulated $W_{18}$ for the rigid pavement, we assume a slab thickness of 10 inches (D = 10 inches) and read axle equivalents from Tables 4.6 and 4.7 as

10-kip single-axle equivalent = 0.08 (Table 4.6)

30-kip tandem-axle equivalent = 1.14 (Table 4.7)

Thus the total daily 18-kip ESAL is

Daily $W_{18}$ = 0.08(967) + 1.14(1935) = 2283.26 18-kip ESAL/day

From Table 4.10, the *PDL* for a conservative design on a three-lane highway is 0.8, so applying Eq. 4.4 gives a required slab thickness of <u>10.28 inches (261 mm)</u> (Figs. 4.7 and 4.8 can also be used to arrive at an approximate solution for slab thickness). For design, this can be rounded up to 10.5 inches (270 mm). It can be shown by the reader that going back to the load equivalency factors and assuming 10.5 inches (270 mm) instead of the previously assumed 10 inches (254 mm) will result in the same, correct slab-thickness solution of 10.5 inches (270 mm).

**EXAMPLE 4.6**   **FLEXIBLE AND RIGID PAVEMENT DESIGN-LIFE COMPARISON**

A roadway has 400 80.1-kN (18-kip) single axles, 200 106.8-kN (24-kip) tandem axles and 100 177.9-kN (40-kip) triple axles per day.  The subgrade CBR is 2 and the roadway pavement is designed for an overall standard deviation of 0.4, a reliability of 99 percent and the initial PSI is 4.5 and the TSI is 2.5. One newly constructed section of this roadway is a rigid pavement designed with a 229-mm (9-inch) slab with a modulus of rupture of 4827 kPa (700 lb/in$^2$) a modulus of elasticity of 27.58 GPa ($4.0 \times 10^6$ lb/in$^2$), and a joint transfer coefficient of 3.0. Another newly constructed section of the same roadway is a flexible pavement with a 127-mm (5-in) hot-mix asphalt (HMA) surface, 254-mm (10-in) dense-graded crushed stone base and a 229-mm (9-in) crushed-stone subbase. If the roadway has four lanes in each direction and is conservatively designed, which of the pavement sections will last longer and by how many years (all drainage coefficients are 1.0)?

**SOLUTION**

We first determine the total design $W_{18}$ for the flexible pavement. To do this, the flexible pavement's structural number is computed from Eq. 4.3, using Table 4.5 to find $a_1 = 0.44$, $a_2 = 0.18$ and $a_3 = 0.11$, and with $D_1 = 5$, $D_2 = 10$, $D_3 = 9$, $M_2 = 1.0$ and $M_3 = 1.0$ (all given) as

$$SN = a_1 D_1 + a_2 D_2 M_2 + a_3 D_3 M_3$$

$$SN = 0.44(5) + 0.18(10)(1.0) + 0.11(9)(1.0) = 4.99 \approx 5$$

With a CBR = 2 (given), Eq. 4.2 is applied to give $M_R = 3000$ ($1500 \times 2$). For other elements required to solve Eq. 4.1 (or alternatively Fig. 4.5), $S_o = 0.4$ (given), $Z_R = -2.326$ (which corresponds to $R = 99\%$, as shown in Table 4.4), and $\Delta PSI = 2$ (initial PSI of 4.5 minus the TSI of 2.5). With these values, applying Eq. 4.1 (or using Fig. 4.5) gives $W_{18} = 775{,}133$. To determine the total truck traffic, the equivalency factors (reading axle equivalents from Tables 4.1, 4.2 and 4.3 while using SN = 5) are

18-kip single-axle equivalent  =  1.0 (Table 4.1)
24-kip tandem-axle equivalent  =  0.260 (Table 4.1)
40-kip triple-axle equivalent  =  0.487 (Table 4.2)

which gives a total $W_{18}$ of 500.7 18-kip ESAL/day ($1.0 \times 400 + 0.260 \times 200 + 0.487 \times 100$). From Table 4.10, the *PDL* for a conservative design on a four-lane highway is 0.75, so applying Eq. 4.4 gives the design-lane $W_{18} = 375.53$ 18-kip ESAL/day ($0.75 \times 500.7$). So the design life for the flexible pavement is

$$\text{design life (in years)} = \frac{\text{total design - lane } W_{18}}{\text{daily design - lane } W_{18} \times 365 \text{ days/yr}}$$

$$= \frac{775{,}133}{375.53 \times 365}$$

$$= 5.655 \text{ years}$$

For the rigid pavement, we are given a slab thickness of 9 inches ($D = 9$ inches) $S'_c = 900$ lb/in$^2$, $E_c = 4.0 \times 10^6$ lb/in$^2$, $J = 3.0$, $C_d = 1.0$ and the modulus of subgrade reaction, $k$, is 100 lb/in$^3$ from Table 4.9 (with CBR = 2), and all other design parameters are the same as those for the flexible pavement. With these values, applying Eq. 4.4 (or using Figs. 4.7 and

4.8) gives $W_{18}$ = 2,364,522. To determine the total truck traffic, the equivalency factors (reading axle equivalents from Tables 4.7, 4.8, and 4.9 while using $D = 9$) are

$$18\text{-kip single-axle equivalent} = 1.0 \text{ (Table 4.6)}$$
$$24\text{-kip tandem-axle equivalent} = 0.444 \text{ (Table 4.7)}$$
$$40\text{-kip triple-axle equivalent} = 1.17 \text{ (Table 4.8)}$$

which gives a total $W_{18}$ of 605.8 18-kip ESAL/day ($1.0 \times 400 + 0.444 \times 200 + 1.17 \times 100$). As in the flexible-pavement case, from Table 4.10, the *PDL* for a conservative design on a four-lane highway is 0.75, so applying Eq. 4.4 gives the design-lane $W_{18}$ = 454.35 18-kip ESAL/day ($0.75 \times 605.8$). So the design life for the rigid pavement is

$$\text{design life (in years)} = \frac{\text{total design-lane } W_{18}}{\text{daily design-lane } W_{18} \times 365 \text{ days/yr}}$$

$$= \frac{2,364,522}{454.35 \times 365}$$

$$= 14.258 \text{ years}$$

So the rigid pavement will last <u>8.603 years</u> longer ($14.258 - 5.655$).

## 4.7 MEASURING PAVEMENT QUALITY AND PERFORMANCE

The design procedure for pavements originally focused on the pavement serviceability index (PSI) as a measure of pavement quality. However, the pavement serviceability index is based on the opinions of a panel of experts (as discussed in Section 4.4.1), which can introduce some variability into their determination. As a result, efforts have been undertaken to develop quantitative measures of pavement condition that provide additional insights into pavement quality and performance and that correlate with the traditional pavement serviceability index. Some factors that are regularly measured by highway pavement agencies now include the International Roughness Index, friction measurements, and rut depth.

### 4.7.1 International Roughness Index

The International Roughness Index (IRI) has become the most popular measure for evaluating the condition of pavements. The IRI evolved out of a study commissioned by the World Bank [Sayers et al., 1986] to establish uniformity of the physical measurement of pavement roughness. The IRI is determined by measuring vertical movements in a standardized vehicle's suspension per unit length of roadway. Units of IRI are reported in meters per kilometer (m/km). The higher the value of the IRI, the rougher the road. To get some sense for how the IRI relates to pavement condition assessments and PSI, Tables 4.11 and 4.12 provide IRI and PSI values corresponding to what is considered poor, mediocre, fair, good, and very good for Interstate and non-Interstate highways [Federal Highway Administration, 2006]. Note that, due to the higher design standards and performance expectations, interstate highways are held to a higher standard for fair, mediocre, and poor pavement assessments.

**Table 4.11** Relationship Between the International Roughness Index (IRI) and Perceptions of Pavement Quality for Interstate Highways

| Pavement condition | Present Serviceability Index | Measured International Roughness Index m/km |
|---|---|---|
| Very good | ≥ 4.0 | < 0.95 |
| Good | 3.5 – 3.9 | 0.96–1.52 |
| Fair | 3.1 – 3.4 | 1.53–1.88 |
| Mediocre | 2.6 – 3.0 | 1.89–2.68 |
| Poor | ≤ 2.5 | > 2.68 |

**Table 4.12** Relationship Between the International Roughness Index (IRI) and Perceptions of Pavement Quality for Non-Interstate Highways

| Pavement condition | Present Serviceability Index | Measured International Roughness Index m/km |
|---|---|---|
| Very good | ≥ 4.0 | < 0.95 |
| Good | 3.5 – 3.9 | 0.96–1.52 |
| Fair | 3.1 – 3.4 | 1.53–2.68 |
| Mediocre | 2.6 – 3.0 | 1.69–3.47 |
| Poor | ≤ 2.5 | > 3.47 |

### 4.7.2   Friction Measurements

Another important measurement of pavement performance is the surface friction. This is critical because low friction values can increase stopping distances and the probability of accidents. Given the variability of pavement surfaces, weather conditions, and tire characteristics, determining pavement friction over the range of possible values is not an easy task. To estimate friction, a standardized test is conducted under wet conditions using either a treaded or smooth tire. Although other speeds are sometimes used, the standard test is generally conducted at 40 mi/h using a friction-testing trailer in which the wheel is locked on the wetted road surface, and the torque developed from this wheel locking is used to measure a friction number. The friction number resulting from this test gives an approximation of the coefficient of road adhesion under wet conditions (as shown in Table 2.4) and is multiplied by 100 to produce a value between 0 and 100. The friction number with a treaded tire ($FN_t$) attempts to measure pavement microtexture, which is a function of the aggregate quality and composition. The friction number with a smooth tire ($FN_s$) provides a measure of pavement macrotexture, which is critical in providing a water drainage escape path between the pavement and tire.

A number of factors influence the friction number, such as changes in traffic volumes or traffic composition, surface age (friction has been found to increase quickly after construction, then as time passes, to level off and eventually decline), seasonal changes (in northern states, the friction number tends to be highest in the spring and lowest in the fall), and speed (the measured value tends to decrease as the test speed increases).

Also, the friction number measured with the treaded tire tends to be greater than that measured with the smooth tire (usually by a value of about 20), but the difference decreases as the surface texture becomes rougher [Li et al., 2003].

In terms of safety, the amount of friction needed to minimize safety-related problems depends on prevailing traffic and geometric conditions. Guidelines used by some states suggest that values of $FN_t < 30$ or $FN_s < 15$ indicate that poor friction may be contributing to wet-weather accidents. Other state agencies have simply put in place guidelines for minimum friction requirements. For example, in Indiana, the minimum friction value is based on the smooth tire test at 64 km/h (40 mi/h), and a pavement with $FN_s < 20$ is considered in need of surfacing work to improve friction (generally resurfacing).

### 4.7.3    Rut Depth

Rut depth, which is a measure of pavement surface deformation in the wheel paths, can affect roadway safety because the ruts accumulate water and increase the possibility of vehicle hydroplaning (which results in the tire skimming over a film of water, greatly reducing braking and steering effectiveness). Because of its potential impact on vehicle control, rut depths are regularly measured on many highways to determine if pavement rutting has reached critical values that would require resurfacing or other pavement treatments. Virtually all states measure rut depth using automated equipment that seeks to determine the difference in surface elevation of the pavement in the wheel path relative to the pavement that is not in the wheel path. The critical values of rut depth can vary from one highway agency to the next. Usually, rut depths are considered unacceptably high when their values reach between 13 and 25 mm (0.5 and 1.0 inches), indicating that corrective action is warranted.

### 4.7.4    Cracking

For flexible pavements, four types of cracking are usually monitored: longitudinal-fatigue cracking, transverse cracking, alligator cracking, and reflection cracking. Longitudinal-fatigue cracking is a surface-down cracking that occurs due to material fatigue in the wheel path. Such cracking can accelerate over time and require significant repairs to protect against water penetration into the flexible pavement structure. Transverse cracking is generally the result of low temperatures that cause fractures across the traffic lanes (resulting in an increase in pavement roughness). Alligator-fatigue cracking is a consequence of material fatigue in the wheel path, generally starting from the bottom of the asphalt layer. Such material fatigue creates a patch of connected cracks that resembles the skin of an alligator (as with other types of cracks, these can accelerate quickly over time and generate the need for maintenance to protect the integrity of the pavement structure). Finally, reflection cracking occurs when hot-mix asphalt (HMA) overlays are placed over exiting pavement structures that had alligator-fatigue cracking, or other indications of pavement distress, and these old distresses manifest themselves in new distresses in the overlay. This results in surface cracking that increases surface roughness and the need for maintenance to protect water intrusion into the pavement structure.

For rigid pavements, transverse cracking is a common measure of pavement distress. Such cracking can be the result of slab fatigue and can be initiated either at the surface or base of the slab. The spacing and width of transverse cracks, and the potential impact of severe cracking on the structural integrity of the pavement, are critical measures of rigid-pavement distress.

### 4.7.5   Faulting

For traditional JPCP (Jointed Plain Concrete Pavements) rigid pavements, joint faulting (characterized by different slab elevations) is a critical measure of pavement distress. Faulting is an indicator of erosion or fatigue of the layers beneath the slab and reflects a failure of the load-transfer ability of the pavement between adjacent slabs. Faulting is associated with increased roughness and will be reflected in International Roughness Index measurements.

### 4.7.6   Punchouts

For Continuously Reinforced Concrete Pavements (CRCP) rigid pavements (those built without expansion/contraction joints), fatigue damage at the top of the slab is often measured by punchouts, which occur when the close spacing of transverse cracks cause in high tensile stresses that result in portions of the slab being broken into pieces. Punchouts are associated with increased roughness and are reflected in International Roughness Index measurements.

## 4.8   MECHANISTIC-EMPIRICAL PAVEMENT DESIGN

The Mechanistic-Empirical Pavement Design Guide (AASHTO 2008) is one of the more recent tools for the design and rehabilitation of pavement structures. The Mechanistic-Empirical Pavement Design Guide was developed to improve on the traditional pavement design procedures presented earlier in this chapter (*AASHTO Guide for Design of Pavement Structures*, 1993) by providing the ability to predict multiple pavement-performance measures (such as rut depth, various types of cracking, joint faulting, International Roughness Index, etc.) and providing a direct link among pavement elements (materials, structural design, construction, traffic, climate and pavement management practices).

Unlike the traditional pavement-design procedures presented earlier in this chapter, the Mechanistic-Empirical Pavement Design Guide is quite complex and must be done using a software package (the software package is referred to simply as MEPDG, standing for Mechanistic-Empirical Pavement Design Guide). The design of pavements with MEPDG is an iterative process that can be summarized as follows:

1.   The design engineer first selects a pavement structure (layer thicknesses, etc.), often using the traditional AASHTO approach (*AASHTO Guide for Design of Pavement Structures*, 1993).

2.   Various inputs needed for MEPDG pavement assessment are then gathered and classified in the following broad topic groupings (please note that this is a much more time-intensive effort than the traditional AASHTO pavement design approach demonstrated earlier in this chapter):

   a.   General project information. For this, factors such as base/subbase construction month, pavement construction month, and month that the pavement is open to traffic are needed because these factors can affect pavement-performance criteria.

   b.   Design criteria and reliability level. For design criteria, the level of tolerable distress such as cracking, faulting, International Roughness

Index are needed (these criteria roughly replace the terminal serviceability index in the traditional AASHTO pavement design). The reliability level needed is similar to that currently used in the traditional AASHTO process.

c. Site conditions and factors. Here, information is needed on truck traffic (including axle-load distributions, speed limit to account for the effect of truck speed on pavement distress, and monthly and hourly distributions of truck-travel), climate (including hourly temperature, precipitation, wind speed, relative humidity and cloud cover), and detailed soil information (strength, variability, etc.).

d. Material properties. Detailed information on new-pavement material properties is needed. This information is along the lines of the structural coefficient values and concrete-strength measurements used in the traditional AASHTO pavement design (although at a significantly higher level of detail).

**3.** With the above, the MEPDG software can then be run and software outputs will include calculated changes in pavement layer properties, various distresses (such as rut depth, cracking, and faulting), and the International Roughness Index over the design life of the pavement. The designer can then determine if the criteria for a successful pavement design have been met (critical distresses do not cross values that can be considered a failure of the pavement over its design life). If these criteria are not met, the pavement design is altered and the process is continued until an acceptable pavement design is achieved.

Currently, the use of the mechanistic-empirical pavement design process and the MEPDG software is increasing; however, many highway and transportation agencies still use the traditional AASHTO pavement-design approach (*AASHTO Guide for Design of Pavement Structures*, 1993).

## NOMENCLATURE FOR CHAPTER 4

| | | | |
|---|---|---|---|
| $a_1, a_2, a_3$ | structural-layer coefficients for wearing surface, base, and subbase | $M_R$ | soil resilient modulus |
| $C_d$ | drainage coefficient for rigid-pavement design | $PDL$ | proportion of directional $W_{18}$ assumed in the design lane |
| CBR | California bearing ratio | PSI | present serviceability index |
| $D$ | slab thickness, AASHTO design equation | $R$ | reliability |
| $D_1, D_2, D_3$ | structural-layer thicknesses for wearing surface, base, and subbase | $S_c'$ | concrete modulus of rupture |
| $E_c$ | concrete modulus of elasticity | $S_o$ | overall standard deviation for AASHTO design equations structural number |
| $FN_s$ | friction number with a smooth tire | SN | structural number |
| $FN_t$ | friction number with a treaded tire | TSI | terminal serviceability index |
| IRI | International Roughness Index | $W_{18}$ | 80.1-kN (18-kip)–equivalent single-axle load |
| $k$ | modulus of subgrade reaction | $Z_R$ | reliability for AASHTO design equations |
| $M_2, M_3$ | drainage coefficients for base and subbase | $\Delta$PSI | change in the present serviceability index |

# REFERENCES

AASHTO (American Association of State Highway and Transportation Officials). *AASHTO Guide for Design of Pavement Structures*. Washington, DC: AASHTO, 1993.

AASHTO (American Association of State Highway and Transportation Officials). *Mechanistic-Empirical Pavement Design Guide —A Manual of Practice*. Washington, DC: AASHTO, 2008.

Carey, W., and P. Irick. *The Pavement Serviceability-Performance Concept*. Highway Research Board Special Report 61E, AASHO Road Test, 1962.

Federal Highway Administration. "2006 Status of the Nation's Highways, Bridges, and Transit: Conditions and Performance; Report to Congress",

United States Department of Transportation, Washington, DC, 2007.

Li, S., S. Noureldin and K. Zhu. *Upgrading the INDOT Pavement Friction Testing Program*. Final report FHWA/IN/JTRP-2003/23, West Lafayette, Indiana, 2003.

Sayers, M., T. Gillespie, and W. Paterson. *Guidelines for the Conducting and Calibrating Road Roughness Measurements*. World Bank Technical Paper No. 46. The World Bank, Washington, DC, 1986.

Yoder, E. J., and M. W. Witczak. *Principles of Pavement Design*, 2nd ed. New York: Wiley, 1975.

# PROBLEMS

**4.1** Truck A has two single axles. One axle weighs 53.4 kN (12,000 lb) and the other weighs 102.3 kN (23,000 lb). Truck B has an 35.6-kN (8000-lb) single axle and a 191.3-kN (43,000-lb) tandem axle. On a flexible pavement with a 76-mm (3-inch) hot-mix asphalt wearing surface, a 152-mm (6-inch) soil-cement base, and an 203-mm (8-inch) crushed stone subbase, which truck will cause more pavement damage? (Assume drainage coefficients are 1.0.)

**4.2** A flexible pavement has a 102-mm (4-inch) hot-mix asphalt wearing surface, a 178-mm (7-inch) dense-graded crushed stone base, and a 254-mm (10-inch) crushed stone subbase. The pavement is on a soil with a resilient modulus of 34,475 kPa (5000 lb/in$^2$). The pavement was designed with 90% reliability, an overall standard deviation of 0.4, and a $\Delta$PSI of 2.0 (a TSI of 2.5). The drainage coefficients are 0.9 and 0.8 for the base and subbase, respectively. How many 111.2-kN (25-kip) single-axle loads can be carried before the pavement reaches its TSI (with given reliability)?

**4.3** A highway has the following pavement design daily traffic: 300 single axles at 44.5 kN (10,000 lb) each, 120 single axles at 80.1 kN (18,000 lb) each, 100 single axles at 102.3 kN (23,000 lb) each, 100 tandem axles at 142.3 kN (32,000 lb) each, 30 single axles at 142.3 kN (32,000 lb) each, and 100 triple axles at 177.9 kN (40,000 lb) each. A flexible pavement is designed to have 102 mm (4 inches) of sand-mix asphalt wearing surface, 152 mm (6 inches) of soil-cement base, and 178 mm (7 inches) of crushed stone subbase. The pavement has a 10-year design life, a reliability of 85%, an overall standard deviation of 0.30, drainage coefficients of 1.0,

an initial PSI of 4.7, and a TSI of 2.5. What is the minimum acceptable soil resilient modulus?

**4.4** Consider the conditions in Problem 4.3. Suppose the state has relaxed its truck weight limits and the impact has been to reduce the number of 80.1-kN (18,000-lb) single-axle loads from 120 to 20 and increase the number of 142.3-kN (32,000-lb) single-axle loads from 30 to 90 (all other traffic is unaffected). Under these revised daily counts, what is the minimum acceptable soil resilient modulus?

**4.5** A flexible pavement was designed for the following daily traffic with a 12-year design life: 1300 single axles at 35.6 kN (8,000 lb) each, 900 tandem axles at 66.7 kN (15,000 lb) each, 20 single axles at 177.9 kN (40,000 lb) each, and 200 tandem axles at 177.9 kN (40,000 lb) each. The highway was designed with 102 mm (4 inches) of hot-mix asphalt wearing surface, 102 mm (4 inches) of hot-mix asphaltic base, and 204 mm (8 inches) of crushed stone subbase. The reliability was 70%, overall standard deviation was 0.5, $\Delta$PSI was 2.0 (with a TSI of 2.5), and all drainage coefficients were 1.0. What was the soil resilient modulus of the subgrade used in design?

**4.6** A flexible pavement has a structural number of 3.8 (all drainage coefficients are equal to 1.0). The initial PSI is 4.7 and the terminal serviceability is 2.5. The soil has a CBR of 9. The overall standard deviation is 0.40 and the reliability is 95%. The pavement is currently designed for 1800 equivalent 80.1-kN (18-kip) single-axle loads per day. If the number of 80.1-kN (18-kip) single-axle loads were to increase by 30%, by how many years would the pavement's design life be reduced?

**4.7** An engineer plans to replace the rigid pavement in Example 4.3 with a flexible pavement. The chosen design has 152 mm (6 inches) of sand-mix asphalt wearing surface, 229 (9 inches) of soil-cement base, and 254 mm (10 inches) of crushed stone subbase. All drainage coefficients are 1.0 and the soil resilient modulus is 34,475 kPa (5000 lb/in$^2$). If the highway's traffic is the same (same axle loadings per vehicle as in Example 4.3), for how many years could you be 95% sure that this pavement will last? (Assume that any parameters not given in this problem are the same as those given in Example 4.3.)

**4.8** You have been asked to design a flexible pavement, and the following daily traffic is expected for design: 5000 single axles at 44.5 kN (10,000 lb) each, 400 single axles at 106.8 kN (24,000 lb) each, 1000 tandem axles at 133.4 kN (30,000 lb) each, and 100 tandem axles at 222.4 kN (50,000 lb) each. There are three lanes in the design direction (conservative design is to be used). Reliability is 90%, overall standard deviation is 0.40, ΔPSI is 1.8, and the design life is 15 years. The soil has a resilient modulus of 94,806 kPa (13,750 lb/in$^2$). If the TSI is 2.5, what is the required structural number?

**4.9** A flexible pavement is designed with 127 mm (5 inches) of hot-mix asphalt wearing surface, 152 mm (6 inches) of hot-mix asphaltic base, and 254 mm (10 inches) of crushed stone subbase. All drainage coefficients are 1.0. Daily traffic is 200 passes of a 89.0-kN (20-kip) single axle, 200 passes of a 177.9 kN (40-kip) tandem axle, and 80 passes of a 97.9-kN (22-kip) single axle. If the initial minus the terminal PSI is 2.0 (the TSI is 2.5), the soil resilient modulus is 20,685 kPa (3000 lb/in$^2$), and the overall standard deviation is 0.6, what is the probability (reliability) that this pavement will last 20 years before reaching its terminal serviceability?

**4.10** A flexible pavement is designed with 102 mm (4 inches) of sand-mix asphalt wearing surface, 152 mm (6 inches) of dense-graded crushed stone base, and 203 mm (8 inches) of crushed stone subbase. All drainage coefficients are 1.0. The pavement is designed for 80.1-kN (18-kip) single-axle loads (1290 per day). The initial PSI is 4.5 and the TSI is 2.5. The soil has a resilient modulus of 82,740 kPa (12,000 lb/in$^2$). If the overall standard deviation is 0.40, what is the probability that this pavement will have a PSI greater than 2.5 after 20 years?

**4.11** A flexible pavement has a 102-mm (4-inch) sand-mix asphalt wearing surface, 254-mm (10-inch) soil cement base, and a 254-mm (10-inch) crushed stone subbase. It is designed to withstand 400 89.0-kN (20-kip) single-axle loads and 900 155.7-kN (35-kip) tandem-axle loads per day. The subgrade CBR is 8, the overall standard deviation is 0.45, the initial PSI is 4.2, and the final PSI is 2.5. What is the probability that this pavement will have a PSI above 2.5 after 25 years? (Drainage coefficients are 1.0.)

**4.12** A three-lane northbound section of interstate (with the design lane conservatively designed) has rigid pavement (PCC) and was designed with a 254-mm (10-inch) slab, 90% reliability, 4827 kPa (700 lb/in$^2$) concrete modulus of rupture, 31.03 GPa (4.5 million lb/in$^2$) modulus of elasticity, 3.0 load transfer coefficient, and an overall standard deviation of 0.35. The initial PSI is 4.6 and the TSI is 2.5. The CBR is 2 with a drainage coefficient of 1.0. The road was designed exclusively for trucks that have one 106.8-kN (24-kip) tandem axle and one 53.4-kN (12-kip) single axle. It is known from weigh-in-motion scales that there have been 13 million 80.1-kN (18-kip)–equivalent single-axle loads in the entire northbound direction of this freeway so far. If a section of flexible pavement is used to replace a section of the PCC that was removed for utility work, what structural number should be used so that the PCC and flexible pavements have the same life expectancy (the new life of the flexible pavement and the remaining life of the PCC)?

**4.13** Consider the two trucks in Problem 4.1. Which truck will cause more pavement damage on a rigid pavement with a 254-mm (10-inch) slab?

**4.14** A rigid pavement is designed with an 279-mm (11-inch) slab thickness, 90% reliability, $E_c$ = 27.58 GPa (4 million lb/in$^2$), modulus of rupture of 4137 kPa (600 lb/in$^2$) modulus of subgrade reaction of 0.0407 N/mm$^3$ (150 lb/in$^3$), a 2.8 load transfer coefficient, initial PSI of 4.8, final PSI of 2.5, overall standard deviation of 0.35, and a drainage coefficient of 0.8. The pavement has a 20-year design life. The pavement has three lanes and is conservatively designed for trucks that have one 89.0-kN (20-kip) single axle, one 115.6-kN (26-kip) tandem axle, and one 151.2-kN (34-kip) triple axle. What is the daily estimated truck traffic on the three lanes?

**4.15** A rigid pavement is on a highway with two lanes in one direction, and the pavement is conservatively designed. The pavement has an 279-mm (11-inch) slab with a modulus of elasticity of 34.48 GPa (5,000,000 lb/in$^2$) and a concrete modulus of rupture of 4827 kPa (700 lb/in$^2$), and it is on a soil with a CBR of 25. The design drainage coefficient is 1.0, the overall standard deviation is 0.3, and the load transfer coefficient is 3.0.

The pavement was designed to last 20 years (initial PSI of 4.7 and a final PSI of 2.5) with 95% reliability carrying trucks with one 18-kip single axle and one 28-kip tandem axle. However, after the pavement was designed, one more lane was added in the design direction (conservative design still used), and the weight limits on the trucks were increased to a 89.0-kN (20-kip) single and a 151.2-kN (34-kip) tandem axle (the slab thickness was unchanged from the original two-lane design with lighter trucks). If climate change has caused the drainage coefficient to drop to 0.8, how long will the pavement last with the new loading and the additional lane (same volume of truck traffic)?

**4.16** You have been asked to design the pavement for an access highway to a major truck terminal. The design daily truck traffic consists of the following: 80 single axles at 100.1 kN (22,500 lb) each, 570 tandem axles at 111.2 kN (25,000 lb) each, 50 tandem axles at 173.5 kN (39,000 lb) each, and 80 triple axles at 213.5 kN (48,000 lb) each. The highway is to be designed with rigid pavement having a modulus of rupture of 4137 kPa (600 lb/in$^2$) and a modulus of elasticity of 34.48 GPa (5,000,000 lb/in$^2$). The reliability is to be 95%, the overall standard deviation is 0.4, the drainage coefficient is 0.9, $\Delta$PSI is 1.7 (with a TSI of 2.5), and the load transfer coefficient is 3.2. The modulus of subgrade reaction is 0.0542 N/mm$^3$ (200 lb/in$^3$). If a 20-year design life is to be used, determine the required slab thickness.

**4.17** A rigid pavement is being designed with the same parameters as used in Problem 4.5. The modulus of subgrade reaction is 0.0813 N/mm$^3$ (300 lb/in$^3$) and the slab thickness is determined to be 216 mm (8.5 inches). The load transfer coefficient is 3.0, the drainage coefficient is 1.0, and the modulus of elasticity is 27.58 GPa (4 million lb/in$^2$). What is the design modulus of rupture? (Assume that any parameters not given in this problem are the same as those given in Problem 4.5.)

**4.18** A rigid pavement is designed with a 254-mm (10-inch) slab, an $E_c$ of 41.37 GPa (6 million lb/in$^2$), a concrete modulus of rupture of 2979 kPa (432 lb/in$^2$), a load transfer coefficient of 3.0, an initial PSI of 4.7, and a terminal serviceability index of 2.5. The overall standard deviation is 0.35, the modulus of subgrade reaction is 0.0515 N/mm$^3$ (190 lb/in$^3$), and a reliability of 90% is used along with a drainage coefficient of 0.8. The pavement is designed assuming traffic is composed entirely of trucks (100 per day). Each truck has one 89.0-kN (20-kip) single axle and one 186.8-kN (42-kip) tandem axle (the effect of all other vehicles is ignored). A section of this road is to be replaced (due to different subgrade characteristics) with a flexible pavement

having a structural number of 4 and is expected to last the same number of years as the rigid pavement. What is the assumed soil resilient modulus? (Assume all other factors are the same as for the rigid pavement.)

**4.19** A four-lane northbound section of interstate has rigid pavement and was designed with an 203-mm (8-inch) slab, 90% reliability, a 4827 kPa (700 lb/in$^2$) concrete modulus of rupture, a 34.48 GPa (5 million lb/in$^2$) modulus of elasticity, a 3.0 load transfer coefficient, and an overall standard deviation of 0.3. The initial PSI is 4.6 and the TSI is 2.5. The pavement was conservatively designed (assuming the upper limit of the $W_{18}$ design lane load) to last 20 years, and the CBR is 25 with a drainage coefficient of 1.0. A design mistake was made that ignored 1000 total northbound (daily) passes of trucks with 151.7 kN (22-kip) single and 133.4 kN (30-kip) tandem axles. What slab thickness should have been used?

**4.20** Consider the loading conditions in Problem 4.3. A rigid pavement is used with a modulus of subgrade reaction of 0.0542 N/mm$^3$ (200 lb/in$^3$), a slab thickness of 203 mm (8 inches), a load transfer coefficient of 3.2, a modulus of elasticity of 34.48 GPa (5 million lb/in$^2$) a modulus of rupture of 4137 kPa (600 lb/in$^2$), and a drainage coefficient of 1.0. How many years would the pavement be expected to last using the same reliability as in Problem 4.3? (Assume all other factors are as in Problem 4.3.)

**4.21** Consider Problem 4.20. How long would the rigid pavement be expected to last if you wanted to be 95% sure that the pavement would stay above the 2.5 TSI?

**4.22** Consider the traffic conditions in Example 4.3. Suppose a 10-inch slab was used and all other parameters are as described in Example 4.3. What would the design life be if the drainage coefficient was 0.8, and what would it be if it was 0.6?

**4.23** Consider the conditions in Example 4.4. Suppose all of the parameters are the same, but further soil tests found that the modulus of subgrade reaction was only 0.0407 N/mm$^3$ (150 lb/in$^3$). In light of this new soil finding, how would the design life of the pavement change?

**4.24** Consider the conditions in Example 4.4. Suppose all of the parameters are the same, but a quality control problem resulted in a modulus of rupture of 4137 kPa (600 lb/in$^2$) instead of 5516 kPa (800 lb/in$^2$). How would the design life of the pavement change?

**4.25** A flexible pavement is constructed with 127 mm (5 inches) of sand-mix asphaltic wearing surface, 229 mm (9 inches) of dense-graded crushed stone base, and

254 mm (10 inches) of crushed stone subbase. The base has a drainage coefficient of 0.90 while the subbase drainage coefficient is 1.0. Determine the structural number of the pavement.

**4.26** A flexible pavement is designed to last 10 years to withstand truck traffic that consists only of trucks with two 80.1-kN (18-kip) single axles. The pavement is designed for a soil CBR of 10, an initial PSI of 5.0, a TSI of 2.5, an overall standard deviation of 0.40 and a reliability of 90%, and the structural number was determined to be 6. On one section of this roadway, beneath an underpass, an engineer uses a 203-mm (8-inch) rigid pavement in an attempt to have it last longer before resurfacing. How many years will this rigid-pavement section last? (Given the same traffic conditions, modulus of rupture = 5516 kPa (800 lb/in$^2$), modulus of elasticity = 34.48 GPa (5,000,000 lb/in$^2$), load transfer coefficient of 3.0 and drainage coefficient of 1.0.).

**4.27** A flexible pavement at an access road to a sports stadium parking lot is designed with a 102-mm (4-inch) sand-mix asphaltic concrete surface, 127-mm (5-inch) aggregate bituminous emulsion base, and a 254-mm (10-inch) crushed stone subbase. There are 95 scheduled baseball and football games at the stadium per year. The access road to the parking lot is three lanes in each direction (conservatively designed). The pavement was designed for recreational vehicles with one 89.0-kN (20-kip) single axle and one 89.0-kN (20-kip) tandem axle. There are 9,000 recreational vehicles estimated at each event. Given that drainage coefficients are 1.0, the overall standard deviation of

traffic is 0.45, reliability is 90%, and the soil's resilient modulus is 103,425 kPa (15,000 lb/in$^2$), how many years will the access road last if the initial PSI is 4.0 and the terminal serviceability index is 2.5?

**4.28** A rigid pavement on a new interstate (3 lanes each direction) has been conservatively designed with a 305-mm (12-inch slab), an $E_c$ of 37.92 GPa (5.5 × 10$^6$ lb/in$^2$), a concrete modulus of rupture of 4827 kPa (700 lb/in$^2$), a load transfer coefficient of 3.0, an initial present serviceability index of 4.5, and a terminal serviceability index of 2.5. The overall standard deviation is 0.35, the subgrade CBR is 25, and the drainage coefficient is 0.9. The pavement was designed for 600 133.4-kN (30-kip) tandem axles per day and 1400 89.0-kN (20-kip) single axle loads per day. If the desired reliability was 90%, how long was this pavement designed to last?

**4.29** A flexible pavement was designed to have a 152-mm (6-inch) sand-mix asphaltic surface, 203-mm (8-inch) soil-cement base and a 533-mm (21-inch) crushed-stone subbase (all drainage coefficients are 1.0). The pavement was designed for 800 53.4-kN (12-kip) single axles and 1600 151.2-kN (34-kip) tandem axles per day in the design direction. The reliability used was 90%, the overall standard deviation was 0.35, initial PSI was 4.7, the TSI was 2.5 and the soil resilient modulus was 17803 kPa (2582 lb/in$^2$). If the road has three lanes in the design direction (and was conservatively designed), for how many years was the pavement designed to last?

# Chapter 5

# Fundamentals of Traffic Flow and Queuing Theory

## 5.1 INTRODUCTION

It is important to realize that the primary function of a highway is to provide mobility. This mobility must be provided with safety in mind while achieving an acceptable level of performance (such as acceptable vehicle speeds). Many of the safety-related aspects of highway design were discussed in Chapter 3, and focus is now shifted to measures of performance.

The analysis of vehicle traffic provides the basis for measuring the operating performance of highways. In undertaking such an analysis, the various dimensions of traffic, such as number of vehicles per unit time (flow), vehicle types, vehicle speeds, and the variation in traffic flow over time, must be addressed because they all influence highway design (the selection of the number of lanes, pavement types, and geometric design) and highway operations (selection of traffic control devices, including signs, markings, and traffic signals), both of which impact the performance of the highway. In light of this, it is important for the analysis of traffic to begin with theoretically consistent quantitative techniques that can be used to model traffic flow, speed, and temporal fluctuations. The intent of this chapter is to focus on models of traffic flow and queuing, thus providing the groundwork for quantifying measures of performance (and levels of service, which will be discussed in Chapters 6 and 7).

## 5.2 TRAFFIC STREAM PARAMETERS

Traffic streams can be characterized by a number of different operational performance measures. Before commencing a discussion of the specific measures, it is important to provide definitions for the contexts in which these measures apply. A traffic stream that operates free from the influence of such traffic control devices as signals and stop signs is classified as uninterrupted flow. This type of traffic flow is influenced primarily by roadway characteristics and the interactions of the vehicles in the traffic stream. Freeways, multilane highways, and two-lane highways often operate under uninterrupted flow conditions. Traffic streams that operate under the influence of signals and stop signs are classified as interrupted flow. Although all the concepts in this chapter are generally applicable to both types of flow, there are some additional complexities involved with the analysis of traffic flow at signalized and unsignalized intersections. Chapter 7 will address the additional complexities relating to the analysis of traffic flow at signalized intersections. For details on the analysis of traffic flow at unsignalized intersections, refer to other sources [Transportation Research Board 1975, 2010]. It should be noted that environmental conditions (day vs. night, sunny vs. rainy, etc.) can also affect the flow of traffic, but these issues are beyond the scope of this book.

## 5.2.1    Traffic Flow, Speed, and Density

Traffic flow, speed, and density are variables that form the underpinnings of traffic analysis. To begin the study of these variables, the basic definitions of traffic flow, speed, and density must be presented. Traffic flow is defined as

$$q = \frac{n}{t} \tag{5.1}$$

*where*

$\quad q$ = traffic flow in vehicles per unit time,

$\quad n$ = number of vehicles passing some designated roadway point during time $t$, and

$\quad t$ = duration of time interval.

Flow is often measured over the course of an hour, in which case the resulting value is typically referred to as volume. Thus, when the term "volume" is used, it is generally understood that the corresponding value is in units of vehicles per hour (veh/h). The definition of flow is more generalized to account for the measurement of vehicles over any period of time. In practice, the analysis flow rate is usually based on the peak 15-minute flow within the hour of interest. This aspect will be described in more detail in Chapter 6.

Aside from the total number of vehicles passing a point in some time interval, the amount of time between the passing of successive vehicles (or time between the arrival of successive vehicles) is also of interest. The time between the passage of the front bumpers of successive vehicles, at some designated highway point, is known as the time headway. The time headway is related to $t$, as defined in Eq. 5.1, by

$$t = \sum_{i=1}^{n} h_i \tag{5.2}$$

*where*

$\quad t$ = duration of time interval,

$\quad h_i$ = time headway of the $i$th vehicle (the elapsed time between the arrivals of vehicles $i$ and $i-1$), and

$\quad n$ = number of measured vehicle time headways at some designated roadway point.

Substituting Eq. 5.2 into Eq. 5.1 gives

$$q = \frac{n}{\sum\limits_{i=1}^{n} h_i} \tag{5.3}$$

or

$$q = \frac{1}{\bar{h}} \tag{5.4}$$

where $\bar{h}$ = average time headway ($\sum h_i / n$) in unit time per vehicle. The importance of time headways in traffic analysis will be given additional attention in later sections of this chapter.

The average traffic speed is defined in two ways. The first is the arithmetic mean of the vehicle speeds observed at some designated point along the roadway. This is referred to as the time-mean speed and is expressed as

$$\bar{u}_t = \frac{\sum\limits_{i=1}^{n} u_i}{n} \tag{5.5}$$

*where*

$\bar{u}_t$ = time-mean speed in unit distance per unit time,

$u_i$ = spot speed (the speed of the vehicle at the designated point on the highway, as might be obtained using a radar gun) of the $i$th vehicle, and

$n$ = number of measured vehicle spot speeds.

The second definition of speed is more useful in the context of traffic analysis and is determined on the basis of the time necessary for a vehicle to travel some known length of roadway. This measure of average traffic speed is referred to as the space-mean speed and is expressed as (assuming that the travel time for all vehicles is measured over the same length of roadway)

$$\bar{u}_s = \frac{l}{\bar{t}} \tag{5.6}$$

*where*

$\bar{u}_s$ = space-mean speed in unit distance per unit time,

$l$ = length of roadway used for travel time measurement of vehicles, and

$\bar{t}$ = average vehicle travel time, defined as

$$\bar{t} = \frac{1}{n} \sum_{i=1}^{n} t_i \tag{5.7}$$

*where*

$t_i$ = time necessary for vehicle $i$ to travel a roadway section of length $l$, and

$n$ = number of measured vehicle travel times.

Substituting Eq. 5.7 into Eq. 5.6 yields

$$\bar{u}_s = \frac{l}{\dfrac{1}{n}\displaystyle\sum_{i=1}^{n} t_i} \tag{5.8}$$

or

$$\bar{u}_s = \frac{1}{\dfrac{1}{n}\displaystyle\sum_{i=1}^{n}\left[\dfrac{1}{(l/t_i)}\right]} \tag{5.9}$$

which is the harmonic mean of speed (space-mean speed). Space-mean speed is the speed variable used in traffic models.

**EXAMPLE 5.1    TIME- AND SPACE-MEAN SPEEDS**

The speeds of five vehicles were measured (with radar) at the midpoint of a 0.8-kilometer section of roadway. The speeds for vehicles 1, 2, 3, 4, and 5 were 71, 68, 82, 79, and 74 km/h, respectively. Assuming all vehicles were traveling at constant speed over this roadway section, calculate the time-mean and space-mean speeds.

*SOLUTION*

For the time-mean speed, Eq. 5.5 is applied, giving

$$\bar{u}_t = \frac{\displaystyle\sum_{i=1}^{n} u_i}{n}$$

$$= \frac{71 + 68 + 82 + 79 + 74}{5}$$

$$= \underline{74.8\ \text{km/h}}$$

For the space-mean speed, Eq. 5.9 will be applied. This equation is based on travel time; however, because it is known that the vehicles were traveling at constant speed, we can rearrange this equation to utilize the measured speed, knowing that distance, $l$, divided by travel time, $t$, is equal to speed ($l/t_i = u$):

$$\bar{u}_s = \frac{1}{\dfrac{1}{n}\displaystyle\sum_{i=1}^{n}\left[\dfrac{1}{(l/t_i)}\right]} = \frac{1}{\dfrac{1}{n}\displaystyle\sum_{i=1}^{n}\left[\dfrac{1}{u_i}\right]}$$

$$= \cfrac{1}{\cfrac{1}{5}\left[\cfrac{1}{71} + \cfrac{1}{68} + \cfrac{1}{82} + \cfrac{1}{79} + \cfrac{1}{74}\right]}$$

$$= \frac{1}{0.013431}$$

$$= \underline{74.45 \ \text{km/h}}$$

Note that the space-mean speed will always be lower than the time-mean speed, unless all vehicles are traveling at exactly the same speed, in which case the two measures will be equal.

Finally, traffic density is defined as

$$k = \frac{n}{l} \tag{5.10}$$

*where*

 $k$ = traffic density in vehicles per unit distance,
 $n$ = number of vehicles occupying some length of roadway at some specified time, and
 $l$ = length of roadway.

The density can also be related to the individual spacing between successive vehicles (measured from front bumper to front bumper). The roadway length, $l$, in Eq. 5.10 can be defined as

$$l = \sum_{i=1}^{n} s_i \tag{5.11}$$

*where*

 $s_i$ = spacing of the $i$th vehicle (the distance between vehicles $i$ and $i-1$, measured from front bumper to front bumper), and

 $n$ = number of measured vehicle spacings.

Substituting Eq. 5.11 into Eq. 5.10 gives

$$k = \frac{n}{\displaystyle\sum_{i=1}^{n} s_i} \tag{5.12}$$

or

$$k = \frac{1}{s} \tag{5.13}$$

where $\bar{s}$ = average spacing ($\sum s_i / n$) in unit distance per vehicle.

Time headway and spacing are referred to as microscopic measures because they describe characteristics specific to individual pairs of vehicles within the traffic stream. Measures that describe the traffic stream as a whole, such as flow, average speed, and density, are referred to as macroscopic measures. As indicated by the preceding equations, the microscopic measures can be aggregated and related to the macroscopic measures.

Based on the definitions presented, a simple identity provides the basic relationship among traffic flow, speed (space-mean), and density (denoting space-mean speed, $\bar{u}_s$ as simply $u$ for notational convenience):

$$q = uk \qquad \textbf{(5.14)}$$

*where*

$q$ = flow, typically in units of veh/h,

$u$ = speed (space-mean speed), typically in units of km/h, and

$k$ = density, typically in units of veh/km.

## EXAMPLE 5.2    SPEED, FLOW, AND DENSITY

Vehicle time headways and spacings were measured at a point along a highway, from a single lane, over the course of an hour. The average values were calculated as 2.5 s/veh for headway and 60 m/veh for spacing. Calculate the average speed of the traffic.

## SOLUTION

To calculate the average speed of the traffic, the fundamental relationship in Eq. 5.14 is used. To begin, the flow and density need to be calculated from the headway and spacing data. Flow is determined from Eq. 5.4 as

$$q = \frac{1}{2.5 \text{ s/veh}}$$
$$= 0.40 \text{ veh/s}$$

or, because the data were collected for an hour,

$$q = 0.40 \text{ veh/s} \times 3600 \text{ s/h}$$
$$= 1440 \text{ veh/h}$$

Density is determined from Eq. 5.13 as

$$k = \frac{1}{60 \text{ m/veh}}$$
$$= 0.0167 \text{ veh/m}$$

or, applying this spacing over the course of one mile,

$$k = 0.0167 \text{ veh/m} \times 1000 \text{ m/km}$$
$$= 16.7 \text{ veh/km}$$

Now applying Eq. 5.14, after rearranging to solve for speed, gives

$$u = \frac{q}{k}$$

$$= \frac{1440 \text{ veh/h}}{16.7 \text{ veh/km}}$$

$$= \underline{\underline{86.4 \text{ km/h}}}$$

Note that the average speed of traffic can be determined directly from the average headway and spacing values, as follows:

$$u = \frac{\overline{s}}{\overline{h}}$$

$$= \frac{60 \text{ m/veh}}{2.5 \text{ s/veh}}$$

$$= \underline{\underline{24 \text{ m/s} \ (86.4 \text{ km/h})}}$$

## 5.3    BASIC TRAFFIC STREAM MODELS

While the preceding definitions and relationships provide the basis for the measurement and calculation of traffic stream parameters, it is also essential to understand the interaction of the individual macroscopic measures in order to fully analyze the operational performance of the traffic stream. The models that describe these interactions are discussed in the following sections, and it will be shown that Eq. 5.14 serves the important function of linking specific models of traffic into a consistent, generalized model.

### 5.3.1    Speed-Density Model

The most intuitive starting point for developing a consistent, generalized traffic model is to focus on the relationship between speed and density. To begin, consider a section of highway with only a single vehicle on it. Under these conditions, the density (veh/km) will be very low and the driver will be able to travel freely at a speed close to the design speed of the highway. This speed is referred to as the free-flow speed because vehicle speed is not inhibited by the presence of other vehicles. As more and more vehicles begin to use a section of highway, the traffic density will increase and the average operating speed of vehicles will decline from the free-flow value as drivers slow to allow for the maneuvers of other vehicles. Eventually, the highway section will become so congested (will have such a high density) that the traffic will come to a stop ($u = 0$), and the density will be determined by the length of the vehicles and the spaces that drivers leave between them. This high-density condition is referred to as the jam density.

One possible representation of the process described above is the linear relationship shown in Fig. 5.1. Mathematically, such a relationship can be expressed as

$$u = u_f \left( 1 - \frac{k}{k_j} \right)$$

(5.15)

*where*

$u$ = space-mean speed in km/h,
$u_f$ = free-flow speed in km/h,
$k$ = density in veh/km, and
$k_j$ = jam density in veh/km.

The advantage of using a linear representation of the speed-density relationship is that it provides a basic insight into the relationships among traffic flow, speed, and density interactions without clouding these insights by the additional complexity that a nonlinear speed-density relationship introduces. However, it is important to note that field studies have shown that the speed-density relationship tends to be nonlinear at low densities and high densities (those that approach the jam density). In fact, the overall speed-density relationship is better represented by three relationships: (1) a nonlinear relationship at low densities that has speed slowly declining from the free-flow value, (2) a linear relationship over the large medium-density region (speed declining linearly with density as shown in Eq. 5.15), and (3) a nonlinear relationship near the jam density as the speed asymptotically approaches zero with increasing density. For the purposes of exposition, we present only traffic stream models that are based on the assumption of a linear speed-density relationship. Examples of nonlinear speed-density relationships are provided elsewhere [Pipes 1967; Drew 1965].

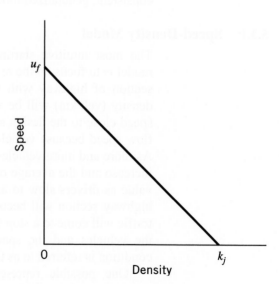

**Figure 5.1** Illustration of a typical linear speed-density relationship.

### 5.3.2 Flow-Density Model

Using the assumption of a linear speed-density relationship as shown in Eq. 5.15, a parabolic flow-density model can be obtained by substituting Eq. 5.15 into Eq. 5.14:

$$q = u_f \left( k - \frac{k^2}{k_j} \right) \qquad (5.16)$$

where all terms are as defined previously.

The general form of Eq. 5.16 is shown in Fig. 5.2. Note in this figure that the maximum flow rate, $q_{cap}$, represents the highest rate of traffic flow that the highway is capable of handling. This is referred to as the traffic flow at capacity, or simply the capacity of the roadway. The traffic density that corresponds to this capacity flow rate is $k_{cap}$, and the corresponding speed is $u_{cap}$. Equations for $q_{cap}$, $k_{cap}$, and $u_{cap}$ can be derived by differentiating Eq. 5.16, because at maximum flow

$$\frac{dq}{dk} = u_f \left( 1 - \frac{2k}{k_j} \right) = 0 \qquad (5.17)$$

and because the free-flow speed ($u_f$) is not equal to zero,

$$k_{cap} = \frac{k_j}{2} \qquad (5.18)$$

Substituting Eq. 5.18 into Eq. 5.15 gives

$$u_{cap} = u_f \left( 1 - \frac{k_j}{2k_j} \right) \qquad (5.19)$$

$$= \frac{u_f}{2}$$

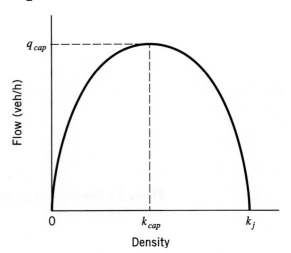

**Figure 5.2** Illustration of the parabolic flow-density relationship.

and using Eq. 5.18 and Eq. 5.19 in Eq. 5.14 gives

$$q_{cap} = u_{cap}k_{cap}$$
$$= \frac{u_f k_j}{4}$$

(5.20)

### 5.3.3 Speed-Flow Model

Again returning to the linear speed-density model (Eq. 5.15), a corresponding speed-flow model can be developed by rearranging Eq. 5.15 to

$$k = k_j\left(1 - \frac{u}{u_f}\right)$$

(5.21)

and by substituting Eq. 5.21 into Eq. 5.14,

$$q = k_j\left(u - \frac{u^2}{u_f}\right)$$

(5.22)

The speed-flow model defined by Eq. 5.22 again gives a parabolic function, as shown in Fig. 5.3. Note that Fig. 5.3 shows that two speeds are possible for flows, $q$, up to the highway's capacity, $q_{cap}$ (this follows from the two densities possible for given flows as shown in Fig. 5.2). It is desirable, for any given flow, to keep the average space-mean speed on the upper portion of the speed-flow curve (above $u_{cap}$). When speeds drop below $u_{cap}$, traffic is in a highly congested and unstable condition.

All of the flow, speed, and density relationships and their interactions are graphically represented in Fig. 5.4.

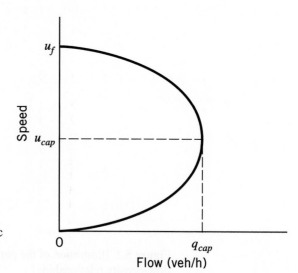

**Figure 5.3** Illustration of the parabolic speed-flow relationship.

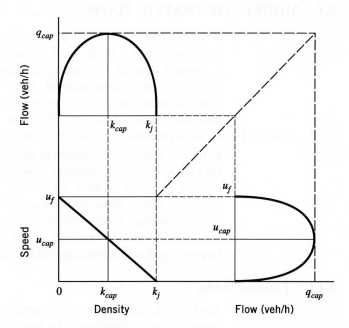

**Figure 5.4** Flow-density, speed-density, and speed-flow relationships (assuming a linear speed-density model).

## EXAMPLE 5.3   APPLICATION OF SPEED-FLOW-DENSITY RELATIONSHIPS

A section of highway is known to have a free-flow speed of 90 km/h and a capacity of 3300 veh/h. In a given hour, 2100 vehicles were counted at a specified point along this highway section. If the linear speed-density relationship shown in Eq. 5.15 applies, what would you estimate the space-mean speed of these 2100 vehicles to be?

### SOLUTION

The jam density is first determined from Eq. 5.20 as

$$k_j = \frac{4q_{cap}}{u_f}$$

$$= \frac{4 \times 3300}{90}$$

$$= 146.7 \ \text{veh/km}$$

Rearranging Eq. 5.22 to solve for $u$,

$$\frac{k_j}{u_f}u^2 - k_j u + q = 0$$

Substituting,

$$\frac{146.7}{90}u^2 - 146.7u + 2100 = 0$$

which gives $u = \underline{72.14 \ \text{km/h}}$ or $\underline{17.86 \ \text{km/h}}$. Both of these speeds are feasible, as shown in Fig. 5.3.

## 5.4    MODELS OF TRAFFIC FLOW

With the basic relationships among traffic flow, speed, and density formalized, attention can now be directed toward a more microscopic view of traffic flow. That is, instead of simply modeling the number of vehicles passing a specified point on a highway in some time interval, there is considerable analytic value in modeling the time between the arrivals of successive vehicles (the concept of vehicle time headway presented earlier). The most simplistic approach to vehicle arrival modeling is to assume that all vehicles are equally or uniformly spaced. This results in what is termed a deterministic, uniform arrival pattern. Under this assumption, if the traffic flow is 360 veh/h, the number of vehicles arriving in any 5-minute time interval is 30 and the headway between all vehicles is 10 seconds (because $h$ will equal $3600/q$). However, actual observations show that such uniformity of traffic flow is not always realistic because some 5-minute intervals are likely to have more or less traffic flow than other 5-minute intervals. Thus a representation of vehicle arrivals that goes beyond the deterministic, uniform assumption is often warranted.

### 5.4.1    Poisson Model

Models that account for the nonuniformity of flow are derived by assuming that the pattern of vehicle arrivals (at a specified point) corresponds to some random process. The problem then becomes one of selecting a probability distribution that is a reasonable representation of observed traffic arrival patterns. An example of such a distribution is the Poisson distribution (the limitations of which will be discussed later), which is expressed as

$$P(n) = \frac{(\lambda t)^n e^{-\lambda t}}{n!}$$

(5.23)

*where*

$P(n)$ = probability of having $n$ vehicles arrive in time $t$,
$\lambda$ = average vehicle flow or arrival rate in vehicles per unit time,
$t$ = duration of the time interval over which vehicles are counted, and
$e$ = base of the natural logarithm ($e = 2.718$).

---

### EXAMPLE 5.4    VEHICLE ARRIVALS AS A POISSON PROCESS

An observer counts 360 veh/h at a specific highway location. Assuming that the arrival of vehicles at this highway location is Poisson distributed, estimate the probabilities of having 0, 1, 2, 3, 4, and 5 or more vehicles arriving over a 20-second time interval.

**SOLUTION**

The average arrival rate, $\lambda$, is 360 veh/h, or 0.1 vehicles per second (veh/s). Using this in Eq. 5.23 with $t = 20$ seconds, the probabilities of having exactly 0, 1, 2, 3, and 4 vehicles arrive are

$$P(0) = \frac{(0.1 \times 20)^0 \, e^{-0.1(20)}}{0!} = \underline{\underline{0.135}}$$

$$P(1) = \frac{(0.1 \times 20)^1 \, e^{-0.1(20)}}{1!} = \underline{\underline{0.271}}$$

$$P(2) = \frac{(0.1 \times 20)^2 \, e^{-0.1(20)}}{2!} = \underline{\underline{0.271}}$$

$$P(3) = \frac{(0.1 \times 20)^3 \, e^{-0.1(20)}}{3!} = \underline{\underline{0.180}}$$

$$P(4) = \frac{(0.1 \times 20)^4 \, e^{-0.1(20)}}{4!} = \underline{\underline{0.090}}$$

For five or more vehicles,

$$P(n \geq 5) = 1 - P(n < 5)$$
$$= 1 - 0.135 - 0.271 - 0.271 - 0.180 - 0.090$$
$$= \underline{\underline{0.053}}$$

A histogram of these probabilities is shown in Fig. 5.5.

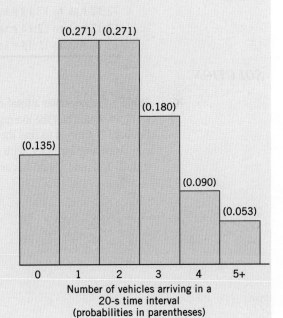

**Figure 5.5** Histogram of the Poisson distribution for $\lambda = 0.1$ vehicles per second.

Number of vehicles arriving in a 20-s time interval (probabilities in parentheses)

**EXAMPLE 5.5** **VEHICLE ARRIVALS AS A POISSON PROCESS WITH DETAILED VEHICLE-ARRIVAL DATA**

Traffic data are collected in 60-second intervals at a specific highway location as shown in Table 5.1. Assuming the traffic arrivals are Poisson distributed and continue at the same rate as that observed in the 15 time periods shown, what is the probability that six or more vehicles will arrive in each of the next three 60-second time intervals (12:15 P.M. to 12:16 P.M., 12:16 P.M. to 12:17 P.M., and 12:17 P.M. to 12:18 P.M.)?

**Table 5.1** Observed Traffic Data for Example 5.5

| Time period | Observed number of vehicles |
|---|---|
| 12:00 P.M. to 12:01 P.M. | 3 |
| 12:01 P.M. to 12:02 P.M. | 5 |
| 12:02 P.M. to 12:03 P.M. | 4 |
| 12:03 P.M. to 12:04 P.M. | 10 |
| 12:04 P.M. to 12:05 P.M. | 7 |
| 12:05 P.M. to 12:06 P.M. | 4 |
| 12:06 P.M. to 12:07 P.M. | 8 |
| 12:07 P.M. to 12:08 P.M. | 11 |
| 12:08 P.M. to 12:09 P.M. | 9 |
| 12:09 P.M. to 12:10 P.M. | 5 |
| 12:10 P.M. to 12:11 P.M. | 3 |
| 12:11 P.M. to 12:12 P.M. | 10 |
| 12:12 P.M. to 12:13 P.M. | 9 |
| 12:13 P.M. to 12:14 P.M. | 7 |
| 12:14 P.M. to 12:15 P.M. | 6 |

*SOLUTION*

Table 5.1 shows that a total of 101 vehicles arrive in the 15-minute period from 12:00 P.M. to 12:15 P.M. Thus the average arrival rate, $\lambda$, is 0.112 veh/s (101/900). As in Example 5.4, Eq. 5.23 is applied to find the probabilities of exactly 0, 1, 2, 3, 4, and 5 vehicles arriving.

Applying Eq. 5.23, with $\lambda = 0.112$ veh/s and $t = 60$ seconds, the probabilities of having 0, 1, 2, 3, 4, and 5 vehicles arriving in a 60-second time interval are (using $\lambda t = 6.733$)

$$P(0) = \frac{(6.733)^0 \, e^{-6.733}}{0!} = \underline{\underline{0.0012}}$$

$$P(1) = \frac{(6.733)^1 \, e^{-6.733}}{1!} = \underline{\underline{0.008}}$$

$$P(2) = \frac{(6.733)^2 \, e^{-6.733}}{2!} = \underline{\underline{0.027}}$$

$$P(3) = \frac{(6.733)^3 \, e^{-6.733}}{3!} = \underline{\underline{0.0606}}$$

$$P(4) = \frac{(6.733)^4 e^{-6.733}}{4!} = \underline{\underline{0.102}}$$

$$P(5) = \frac{(6.733)^5 e^{-6.733}}{5!} = \underline{\underline{0.137}}$$

The summation of these probabilities is the probability that 0 to 5 vehicles will arrive in any given 60-second time interval, which is

$$P(n \le 5) = \sum_{i=0}^{5} P(n)$$

$$= 0.0012 + 0.008 + 0.027 + 0.0606 + 0.102 + 0.137$$

$$= 0.3358$$

So 1 minus $P(n \le 5)$ is the probability that 6 or more vehicles will arrive in any 60-second time interval, which is

$$P(n \ge 6) = 1 - P(n \le 5)$$

$$= 1 - 0.3358$$

$$= 0.6642$$

The probability that 6 or more vehicles will arrive in three successive time intervals ($t_1$, $t_2$, and $t_3$) is simply the product of probabilities, which is

$$P(n \ge 6) \text{ for three successive time periods} = \prod_{t_i=1}^{3} P(n \ge 6)$$

$$= (0.6642)^3$$

$$= \underline{\underline{0.293}}$$

The assumption of Poisson vehicle arrivals also implies a distribution of the time intervals between the arrivals of successive vehicles (time headway). To show this, note that the average arrival rate is

$$\lambda = \frac{q}{3600} \tag{5.24}$$

*where*

$\lambda$ = average vehicle arrival rate in veh/s,

$q$ = flow in veh/h, and

3600 = number of seconds per hour.

Substituting Eq. 5.24 into Eq. 5.23 gives

$$P(n) = \frac{(qt/3600)^n e^{-qt/3600}}{n!} \tag{5.25}$$

Note that the probability of having no vehicles arrive in a time interval of length $t$, $P(0)$, is equivalent to the probability of a vehicle headway, $h$, being greater than or equal to the time interval $t$. So from Eq. 5.25,

$$P(0) = P(h \geq t)$$
$$= e^{-qt/3600}$$

(5.26)

This distribution of vehicle headways is known as the negative exponential distribution and is often simply referred to as the exponential distribution.

**EXAMPLE 5.6    HEADWAYS AND THE NEGATIVE EXPONENTIAL DISTRIBUTION**

Consider the traffic situation in Example 5.4 (360 veh/h). Again assume that the vehicle arrivals are Poisson distributed. What is the probability that the headway between successive vehicles will be less than 8 seconds, and what is the probability that the headway between successive vehicles will be between 8 and 10 seconds?

**SOLUTION**

By definition, $P(h < t) = 1 - P(h \geq t)$. This expression gives the probability that the headway will be less than 8 seconds as

$$P(h < t) = 1 - e^{-qt/3600}$$
$$= 1 - e^{-360(8)/3600}$$
$$= \underline{0.551}$$

To determine the probability that the headway will be between 8 and 10 seconds, compute the probability that the headway will be greater than or equal to 10 seconds:

$$P(h \geq t) = e^{-qt/3600}$$
$$= e^{-360(10)/3600}$$
$$= \underline{0.368}$$

So the probability that the headway will be between 8 and 10 seconds is $\underline{0.081}$ ($1 - 0.551 - 0.368$).

To help in visualizing the shape of the exponential distribution, Fig. 5.6 shows the probability distribution implied by Eq. 5.26, with the flow, $q$, equal to 360 veh/h as in Example 5.4.

## 5.4.2    Limitations of the Poisson Model

Empirical observations have shown that the assumption of Poisson-distributed traffic arrivals is most realistic in lightly congested traffic conditions. As traffic flows become heavily congested or when traffic signals cause cyclical traffic stream disturbances, other distributions of traffic flow become more appropriate. The primary limitation of the Poisson model of vehicle arrivals is the constraint imposed

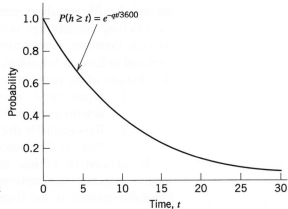

**Figure 5.6** Exponentially distributed probabilities of headways greater than or equal to $t$, with $q = 360$ veh/h.

by the Poisson distribution that the mean of period observations equals the variance. For example, the mean of period-observed traffic in Example 5.5 is 6.733 and the corresponding variance, $\sigma^2$, is 7.210. Because these two values are close, the Poisson model was appropriate for this example. If the variance is significantly greater than the mean, the data are said to be overdispersed, and if the variance is significantly less than the mean, the data are said to be underdispersed. In either case the Poisson distribution is no longer appropriate, and another distribution should be used. Such distributions are discussed in detail in more specialized sources [Transportation Research Board 1975; Poch and Mannering 1996; Lord and Mannering 2010].

## 5.5   QUEUING THEORY AND TRAFFIC FLOW ANALYSIS

The formation of traffic queues during congested periods is a source of considerable delay and results in a loss of highway performance. Under extreme conditions, queuing delay can account for 90% or more of a motorist's total trip travel time. Given this, it is essential in traffic analysis to develop a clear understanding of the characteristics of queue formation and dissipation along with mathematical formulations that can predict queuing-related elements.

As is well known, the problem of queuing is not unique to traffic analysis. Many non-transportation fields, such as the design and operation of industrial plants, retail stores, service-oriented industries, and computer networks, must also give serious consideration to the problem of queuing. The impact of queues on performance and productivity in manufacturing, retailing, and other fields has led to numerous theories of queuing behavior (the process by which queues form and dissipate). As will be shown, the models of traffic flow presented earlier (uniform, deterministic arrivals and Poisson arrivals) will form the basis for studying traffic queues within the more general context of queuing theory.

### 5.5.1   Dimensions of Queuing Models

The purpose of traffic queuing models is to provide a means to estimate important measures of highway performance, including vehicle delay and traffic queue lengths. Such estimates are critical to roadway design (the required length of left-turn bays and the number of lanes at intersections) and traffic operations control, including the timing of traffic signals at intersections.

Queuing models are derived from underlying assumptions regarding arrival patterns, departure characteristics, and queue disciplines. Traffic arrival patterns were explored in Section 5.4, where, given an average vehicle arrival rate ($\lambda$), two possible distributions of the time between the arrival of successive vehicles were considered:

1. Equal time intervals (derived from the assumption of uniform, deterministic arrivals)
2. Exponentially distributed time intervals (derived from the assumption of Poisson-distributed arrivals)

In addition to vehicle arrival assumptions, the derivation of traffic queuing models requires assumptions relating to vehicle departure characteristics. Of particular interest is the distribution of the amount of time it takes a vehicle to depart—for example, the time to pass through an intersection at the beginning of a green signal, the time required to pay a toll at a toll booth, or the time a driver takes before deciding to proceed after stopping at a stop sign. As was the case for arrival patterns, given an average vehicle departure rate (denoted as $\mu$, in vehicles per unit time), the assumption of a deterministic or exponential distribution of departure times is appropriate.

Another important aspect of queuing models is the number of available departure channels. For most traffic applications only one departure channel will exist, such as a highway lane or group of lanes passing through an intersection. However, multiple departure channels are encountered in some traffic applications, such as at toll booths on turnpikes and at entrances to bridges.

The final necessary assumption relates to the queue discipline. In this regard, two options have been popularized in the development of queuing models: first-in, first-out (FIFO), indicating that the first vehicle to arrive is the first to depart, and last-in, first-out (LIFO), indicating that the last vehicle to arrive is the first to depart. For virtually all traffic-oriented queues, the FIFO queuing discipline is the more appropriate of the two.

Queuing models are often identified by three alphanumeric values. The first value indicates the arrival rate assumption, the second value gives the departure rate assumption, and the third value indicates the number of departure channels. For traffic arrival and departure assumptions, the uniform, deterministic distribution is denoted $D$ and the exponential distribution is denoted $M$. Thus a $D/D/1$ queuing model assumes deterministic arrivals and departures with one departure channel. Similarly, an $M/D/1$ queuing model assumes exponentially distributed arrival times, deterministic departure times, and one departure channel.

### 5.5.2 *D/D/*1 Queuing

The case of deterministic arrivals and departures with one departure channel ($D/D/1$ queue) is an excellent starting point in understanding queuing models because of its simplicity. The $D/D/1$ queue lends itself to an intuitive graphical or mathematical solution that is best illustrated by an example.

**EXAMPLE 5.7**   *D/D/1* **QUEUING WITH CONSTANT ARRIVAL AND DEPARTURE RATES**

Vehicles arrive at an entrance to a recreational park. There is a single gate (at which all vehicles must stop), where a park attendant distributes a free brochure. The park opens at 8:00 A.M., at which time vehicles begin to arrive at a rate of 480 veh/h. After 20 minutes the arrival flow rate declines to 120 veh/h, and it continues at that level for the remainder of the day. If the time required to distribute the brochure is 15 seconds, and assuming *D/D/1* queuing, describe the operational characteristics of the queue.

*SOLUTION*

Begin by putting arrival and departure rates into common units of vehicles per minute:

$$\lambda = \frac{480 \text{ veh/h}}{60 \text{ min/h}} = 8 \text{ veh/min} \quad \text{for } t \le 20 \text{ min}$$

$$\lambda = \frac{120 \text{ veh/h}}{60 \text{ min/h}} = 2 \text{ veh/min} \quad \text{for } t > 20 \text{ min}$$

$$\mu = \frac{60 \text{ s/min}}{15 \text{ s/veh}} = 4 \text{ veh/min} \quad \text{for all } t$$

Equations for the total number of vehicles that have arrived and departed up to a specified time, *t*, can now be written. Define *t* as the number of minutes after the start of the queuing process (in this case the number of minutes after 8:00 A.M.). The total number of vehicle arrivals at time *t* is equal to

$$8t \quad \text{for } t \le 20 \text{ min}$$

and

$$160 + 2(t - 20) \quad \text{for } t > 20 \text{ min}$$

Similarly, the number of vehicle departures is

$$4t \quad \text{for all } t$$

The preceding equations can be illustrated graphically as shown in Fig. 5.7. When the arrival curve is above the departure curve, a queue condition exists. The point at which the arrival curve meets the departure curve is the moment when the queue dissipates (no more queue exists). In this example, the point of queue dissipation can be determined graphically by inspection of Fig. 5.7, or analytically by equating appropriate arrival and departure equations, that is,

$$160 + 2(t - 20) = 4t$$

Solving for *t* gives *t* = 60 minutes. Thus the queue that began to form at 8:00 A.M. will dissipate 60 minutes later (9:00 A.M.), at which time 240 vehicles will have arrived and departed (4 veh/min × 60 min).

Another aspect of interest is individual vehicle delay. Under the assumption of a FIFO queuing discipline, the delay of an individual vehicle is given by the horizontal distance between arrival and departure curves starting from the time of the vehicle's arrival in the queue. So, by inspection of Fig. 5.7, the 160th vehicle to arrive will have the longest delay, 20 minutes (the longest horizontal distance between arrival and departure curves), and vehicles arriving after the 239th vehicle will encounter no queue delay because the queue will have dissipated and the departure rate will continue to exceed the arrival rate. It follows that with the LIFO queuing discipline, the first vehicle to arrive would have to wait until the entire queue clears (60 minutes of delay).

The total length of the queue at a specified time, expressed as the number of vehicles, is given by the vertical distance between arrival and departure curves at that time. For example, at 10 minutes after the start of the queuing process (8:10 A.M.) the queue is 40 vehicles long, and the longest queue (longest vertical distance between arrival and departure curves) will occur at $t = 20$ minutes and is 80 vehicles long (see Fig. 5.7).

Total vehicle delay, defined as the summation of the delays for the individual vehicles, is given by the total area between the arrival and departure curves (see Fig. 5.7) and, in this case, is in units of vehicle-minutes. In this example, the area between the arrival and departure curves can be determined by summing triangular areas, giving total delay, $D_t$, as

$$D_t = \frac{1}{2}(80 \times 20) + \frac{1}{2}(80 \times 40)$$

$$= \underline{2400 \text{ veh-min}}$$

Finally, because 240 vehicles encounter queuing delay (as previously determined), the average delay per vehicle is 10 minutes (2400 veh-min/240 veh), and the average queue length is 40 vehicles (2400 veh-min/60 min).

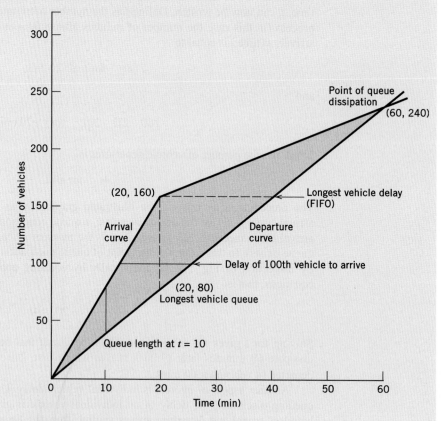

**Figure 5.7** *D/D/*1 queuing diagram for Example 5.7.

**EXAMPLE 5.8**   **$D/D/1$ QUEUING WITH TIME-VARYING ARRIVAL RATE AND CONSTANT DEPARTURE RATE**

Vehicles start arriving at an entrance to a national park at 5:45 A.M. and a park booth collecting entrance fees opens at 6:00 A.M. and process vehicles at a deterministic rate of 20 veh/min. It is estimated that 20% of the arriving vehicles have priority passes that allow them to access a separate processing booth that also opens at 6:00 A.M. but processes vehicles at a slower deterministic rate of 15 veh/min. The deterministic arrival rate of vehicles is a function of time such that $\lambda(t) = 17.2 - 0.2t$ where $t$ is in minutes after 5:45 A.M. Determine the average delay per vehicle for the non-priority-pass vehicles and the priority-pass vehicles (starting with their arrival at 6:45 A.M.) until their respective queues clear, assuming $D/D/1$ queuing.

**SOLUTION**

This problem has time-dependent deterministic arrivals and departure rates that do not vary over time. Begin by computing the arrivals of non-priority-pass vehicles. Because 20% of all arrivals are priority-pass vehicles, the non-priority-pass vehicle arrivals will be (starting at 5:45 A.M.):

$$= \int_0^t 17.2 - 0.2t \ dt$$

$$= 17.2t - 0.1t^2 - 0.2\left(17.2t - 0.1t^2\right)$$

$$= 17.2t - 0.1t^2 - 3.44t + 0.02t^2$$

$$= 13.76t - 0.08t^2$$

The time required to clear the queue of the non-priority-pass vehicles (with a departure rate of 20 veh/min) is determined as:

$$13.76t - 0.08t^2 = 20(t-15)$$

$$-0.08t^2 - 6.24t + 300 = 0$$

$$t = 33.60 \min$$

At $t = 33.60$ minutes, the total number of non-priority-pass vehicle arrivals is

$$= 13.76t - 0.08t^2$$

$$= 13.76(33.60) - 0.08(33.60)^2$$

$$= 372.02 \text{ veh}$$

So the total delay for the non-priority-pass vehicles, $D_{np}$ will be the area under the arrival function minus the area under the departure function, which will be a simple triangle with a height of 372.02 vehicles and a base of 18.06 min (33.60 − 15):

$$D_{np} = \int_0^{33.60} 13.76t - 0.08t^2 \ dt - \frac{1}{2}(33.60 - 15)(372.02)$$

$$= 6.88(33.60)^2 - 0.0267(33.60)^3 - 3459.786$$

$$= 3294.65 \text{ veh-min}$$

So the average delay per vehicle for the non-priority-pass vehicles is <u>8.86 min</u> (3294.65/372.02).

For the priority-pass vehicles, the arrival function is 20% of the total vehicle arrival function or,

$$= 0.2 \int_0^t 17.2 - 0.2t \ dt$$

$$= 0.2 \left( 17.2t - 0.1t^2 \right)$$

$$= 3.44t - 0.02t^2$$

The time required to clear the queue of the priority-pass vehicles (with a departure rate of 15 veh/min) is:

$$3.44t - 0.02t^2 = 15(t - 15)$$

$$-0.02t^2 - 11.56t + 225 = 0$$

$$t = 18.84 \text{ min}$$

At $t = 18.84$ minutes, the total number of priority-pass vehicle arrivals is

$$= 3.44t - 0.02t^2$$

$$= 3.44(18.84) - 0.02(18.84)^2$$

$$= 57.71 \text{ veh}$$

So the total delay for the priority-pass vehicles, $D_p$ will again be the area under the arrival function minus the area under the departure function, which will be a simple triangle with a height of 57.71 vehicles and a base of 3.84 min (18.84 − 15):

$$D_{np} = \int_0^{18.84} 3.44t - 0.02t^2 \ dt - \frac{1}{2}(18.84 - 15)(57.71)$$

$$= 1.72(18.84)^2 - 0.0067(18.84)^3 - 110.803$$

$$= 454.90 \text{ veh-min}$$

So the average delay per vehicle for the priority-pass vehicles is <u>7.88 min</u> (454.9/57.71). Thus the priority pass saves an average of 0.98 min, or about 59 s.

**EXAMPLE 5.9**    *D/D/1 QUEUING WITH TIME-VARYING ARRIVAL AND DEPARTURE RATES*

After observing arrivals and departures at a highway toll booth over a 60-minute time period, an observer notes that the arrival and departure rates (or service rates) are deterministic, but instead of being uniform, they change over time according to a known function. The arrival rate is given by the function $\lambda(t) = 2.2 + 0.17t - 0.0032t^2$, and the departure rate is given by $\mu(t) = 1.2 + 0.07t$, where $t$ is in minutes after the beginning of the observation period and $\lambda(t)$ and $\mu(t)$ are in vehicles per minute. Determine the total vehicle delay at the toll booth and the longest queue, assuming $D/D/1$ queuing.

**SOLUTION**

Note that this problem is an example of a time-dependent deterministic queue because the deterministic arrival and departure rates change over time. Begin by computing the time to queue dissipation by equating vehicle arrivals and departures:

$$\int_0^t 2.2 + 0.17t - 0.0032t^2 \, dt = \int_0^t 1.2 + 0.07t \, dt$$

$$2.2t + 0.085t^2 - 0.00107t^3 = 1.2t + 0.035t^2$$

$$-0.00107t^3 + 0.05t^2 + t = 0$$

This gives $t = 61.8$ minutes. Therefore, the total vehicle delay (the area between the arrival and departure functions) is

$$D_t = \int_0^{61.8} 2.2t + 0.085t^2 - 0.00107t^3 \, dt - \int_0^{61.8} 1.2t + 0.035t^2 \, dt$$

$$= 1.1t^2 + 0.0283t^3 - 0.0002675t^4 - 0.6t^2 - 0.0117t^3 \, \big|_0^{61.8}$$

$$= -0.0002675(61.8)^4 + 0.0166(61.8)^3 + 0.5(61.8)^2$$

$$= \underline{1925.8 \text{ veh-min}}$$

The queue length (in vehicles) at any time $t$ is given by the function

$$Q(t) = \int_0^t 2.2 + 0.17t - 0.0032t^2 \, dt - \int_0^t 1.2 + 0.07t \, dt$$

$$= -0.00107t^3 + 0.05t^2 + t$$

Solving for the time at which the maximum queue length occurs yields

$$\frac{dQ(t)}{dt} = -0.00321t^2 + 0.1t + 1 = 0$$

$$t = \underline{39.12 \text{ min}}$$

Substituting with $t = 39.12$ minutes gives the maximum queue length:

$$Q(39.12) = -0.00107t^3 + 0.05t^2 + t \big|_0^{39.12}$$
$$= -0.00107(39.12)^3 + 0.05(39.12)^2 + 39.12$$
$$= \underline{\underline{51.58 \text{ veh}}}$$

**EXAMPLE 5.10    DETERMINING A REQUIRED DEPARTURE RATE**

A parking garage has a single processing booth where cars pay for parking. The garage opens at 6:00 A.M. and vehicles start arriving at 6:00 A.M. at a deterministic rate of $\lambda(t) = 6.1 - 0.22t$ where $\lambda(t)$ is in vehicles per minute and $t$ is in minutes after 6:00 A.M. What is the minimum constant departure rate (from 6:00 A.M. on) needed to ensure that the queue length does not exceed 10 vehicles?

*SOLUTION*

Let $\mu$ be the unknown departure rate (in veh/min) so that the queue length (in vehicles) at any time $t$ is:

$$Q(t) = \int_0^t 6.1 - 0.22t \, dt - \int_0^t \mu \, dt$$
$$= 6.1t - 0.11t^2 - \mu t$$

Using this and solving for the time at which the maximum queue length occurs gives

$$\frac{dQ(t)}{dt} = 6.1 - 0.22t - \mu = 0$$
$$t = \frac{6.1 - \mu}{0.22}$$

Substituting this value of $t$ into the queue-length equation with a maximum $Q(t) = 10$ vehicles as specified in the problem gives

$$Q(t) = 10 = 6.1 \left( \frac{6.1 - \mu}{0.22} \right) - 0.11 \left( \frac{6.1 - \mu}{0.22} \right)^2 - \mu \left( \frac{6.1 - \mu}{0.22} \right)$$

which gives $2.271\mu^2 - 27.727 + 74.566 = 0$. Solving for $\mu$ gives possible solutions as 8.21 veh/min and 4 veh/min, so the minimum departure rate would be <u>4 veh/min</u>.

### 5.5.3   *M/D/1 Queuing*

The assumption of exponentially distributed times between the arrivals of successive vehicles (Poisson arrivals) will, in some cases, give a more realistic representation of traffic flow than the assumption of uniformly distributed arrival times. Therefore, the *M/D/1* queue (exponentially distributed arrivals, deterministic departures, and one departure channel) has some important applications within the traffic analysis field. Although a graphical solution to an *M/D/1* queue is difficult, a mathematical solution is straightforward. Defining a new term (traffic intensity) for the ratio of average arrival to departure rates as

$$\rho = \frac{\lambda}{\mu} \tag{5.27}$$

*where*

$\rho$ = traffic intensity, unitless,
$\lambda$ = average arrival rate in vehicles per unit time, and
$\mu$ = average departure rate in vehicles per unit time,

and assuming that $\rho$ is less than 1, it can be shown that for an *M/D/1* queue the following queuing performance equations apply:

$$\overline{Q} = \frac{\rho^2}{2(1-\rho)} \tag{5.28}$$

$$\overline{w} = \frac{\rho}{2\mu(1-\rho)} \tag{5.29}$$

$$\overline{t} = \frac{2-\rho}{2\mu(1-\rho)} \tag{5.30}$$

*where*

$\overline{Q}$ = average length of queue in vehicles,
$\overline{w}$ = average waiting time in the queue, in unit time per vehicle,
$\overline{t}$ = average time spent in the system (the summation of average waiting time in the queue and average departure time), in unit time per vehicle, and
Other terms are as defined previously.

It is important to note that under the assumption that the traffic intensity is less than 1 ($\lambda < \mu$), the *D/D/1* queue will predict no queue formation. However, a queuing model that is derived based on random arrivals or departures, such as the *M/D/1* queuing model, will predict queue formations under such conditions. Also, note that the *M/D/1* queuing model presented here is based on steady-state conditions (constant average arrival and departure rates), with randomness arising from the assumed probability distribution of arrivals. This contrasts with the time-varying deterministic queuing case, presented in Example 5.9, in which arrival and departure rates changed over time but randomness was not present.

## EXAMPLE 5.11 M/D/1 QUEUING: PARK-ENTRANCE APPLICATION

Consider the entrance to the recreational park described in Example 5.7. However, let the average arrival rate be 180 veh/h and Poisson distributed (exponential times between arrivals) over the entire period from park opening time (8:00 A.M.) until closing at dusk. Compute the average length of queue (in vehicles), average waiting time in the queue, and average time spent in the system, assuming $M/D/1$ queuing.

**SOLUTION**

Putting arrival and departure rates into common units of vehicles per minute gives

$$\lambda = \frac{180 \text{ veh/h}}{60 \text{ min/h}} = 3 \text{ veh/min} \quad \text{for all } t$$

$$\mu = \frac{60 \text{ s/min}}{15 \text{ s/veh}} = 4 \text{ veh/min} \quad \text{for all } t$$

and

$$\rho = \frac{\lambda}{\mu} = \frac{3 \text{ veh/min}}{4 \text{ veh/min}} = 0.75$$

For the average length of queue (in vehicles), Eq. 5.28 is applied:

$$\overline{Q} = \frac{0.75^2}{2(1-0.75)}$$

$$= 1.125 \text{ veh}$$

For average waiting time in the queue, Eq. 5.29 gives

$$\overline{w} = \frac{0.75}{2(4)(1-0.75)}$$

$$= 0.375 \text{ min/veh}$$

For average time spent in the system [queue time plus departure (service) time], Eq. 5.30 is used:

$$\overline{t} = \frac{2-0.75}{2(4)(1-0.75)}$$

$$= 0.625 \text{ min/veh}$$

or, alternatively, because the departure (service) time is $1/\mu$ (the 0.25 minutes it takes the park attendant to distribute the brochure),

$$\bar{t} = \bar{w} + \frac{1}{\mu}$$

$$= 0.375 + \frac{1}{4}$$

$$= \underline{\underline{0.625 \text{ min/veh}}}$$

### 5.5.4 *M/M/*1 Queuing

A queuing model that assumes one departure channel and exponentially distributed departure times in addition to exponentially distributed arrival times (an *M/M/*1 queue) is applicable in some traffic applications. For example, exponentially distributed departure patterns might be a reasonable assumption at a toll booth, where some arriving drivers have the correct toll and can be processed quickly, and others do not have the correct toll, producing a distribution of departures about some mean departure rate. Under standard *M/M/*1 assumptions, it can be shown that the following queuing performance equations apply (again assuming that $\rho$ is less than 1):

$$\bar{Q} = \frac{\rho^2}{1 - \rho} \tag{5.31}$$

$$\bar{w} = \frac{\lambda}{\mu(\mu - \lambda)} \tag{5.32}$$

$$\bar{t} = \frac{1}{\mu - \lambda} \tag{5.33}$$

*where*

$\bar{Q}$ = average length of queue in vehicles,

$\bar{w}$ = average waiting time in the queue, in unit time per vehicle,

$\bar{t}$ = average time spent in the system ($\bar{w} + 1/\mu$), in unit time per vehicle, and

Other terms are as defined previously.

---

**EXAMPLE 5.12    M/M/1 QUEUING: PARKING-LOT APPLICATION**

Assume that the park attendant in Examples 5.7 and 5.11 takes an average of 15 seconds to distribute brochures, but the distribution time varies depending on whether park patrons have questions relating to park operating policies. Given an average arrival rate of 180 veh/h as in Example 5.11, compute the average length of queue (in vehicles), average waiting time in the queue, and average time spent in the system, assuming *M/M/*1 queuing.

**SOLUTION**

Using the average arrival rate, departure rate, and traffic intensity as determined in Example 5.11, the average length of queue is (from Eq. 5.31)

$$\overline{Q} = \frac{0.75^2}{1 - 0.75}$$
$$= 2.25 \text{ veh}$$

the average waiting time in the queue is (from Eq. 5.32)

$$\overline{w} = \frac{3}{4(4-3)}$$
$$= 0.75 \text{ min/veh}$$

and the average time spent in the system is (from Eq. 5.33)

$$\overline{t} = \frac{1}{4-3}$$
$$= 1 \text{ min/veh}$$

### 5.5.5   *M/M/N* Queuing

A more general formulation of the $M/M/1$ queue is the $M/M/N$ queue, where $N$ is the total number of departure channels. $M/M/N$ queuing is a reasonable assumption at toll booths on turnpikes or at toll bridges, where there is often more than one departure channel available (more than one toll booth open). A parking lot is another example, with $N$ being the number of parking stalls in the lot and the departure rate, $\mu$, being the exponentially distributed times of parking duration. $M/M/N$ queuing is also frequently encountered in non-transportation applications such as checkout lines at retail stores, security checks at airports, and so on.

The following equations describe the operational characteristics of $M/M/N$ queuing. Note that unlike the equations for $M/D/1$ and $M/M/1$, which require that the traffic intensity, $\rho$, be less than 1, the following equations allow $\rho$ to be greater than 1 but apply only when $\rho/N$ (which is called the utilization factor) is less than 1.

$$P_0 = \frac{1}{\displaystyle\sum_{n_c=0}^{N-1} \frac{\rho^{n_c}}{n_c!} + \frac{\rho^N}{N!(1 - \rho/N)}} \tag{5.34}$$

$$P_n = \frac{\rho^n P_0}{n!} \qquad \text{for } n \le N \tag{5.35}$$

$$P_n = \frac{\rho^n P_0}{N^{n-N} N!} \quad \text{for } n \geq N \tag{5.36}$$

$$P_{n>N} = \frac{P_0 \, \rho^{N+1}}{N!N(1-\rho/N)} \tag{5.37}$$

*where*

$P_0$ = probability of having no vehicles in the system,

$P_n$ = probability of having $n$ vehicles in the system,

$P_{n>N}$ = probability of waiting in a queue (the probability that the number of vehicles in the system is greater than the number of departure channels),

$n$ = number of vehicles in the system,

$N$ = number of departure channels,

$n_c$ = departure channel number, and

$\rho$ = traffic intensity ($\lambda/\mu$).

$$\overline{Q} = \frac{P_0 \, \rho^{N+1}}{N!N} \left[ \frac{1}{(1-\rho/N)^2} \right] \tag{5.38}$$

$$\overline{w} = \frac{\rho + \overline{Q}}{\lambda} - \frac{1}{\mu} \tag{5.39}$$

$$\overline{t} = \frac{\rho + \overline{Q}}{\lambda} \tag{5.40}$$

*where*

$\overline{Q}$ = average length of queue (in vehicles),

$\overline{w}$ = average waiting time in the queue, in unit time per vehicle,

$\overline{t}$ = average time spent in the system, in unit time per vehicle, and

Other terms are as defined previously.

## EXAMPLE 5.13   *M/M/N* QUEUING: TOLL-BRIDGE APPLICATION

At an entrance to a toll bridge, four toll booths are open. Vehicles arrive at the bridge at an average rate of 1200 veh/h, and at the booths, drivers take an average of 10 seconds to pay their tolls. Both the arrival and departure rates can be assumed to be exponentially distributed. How would the average queue length, time in the system, and probability of waiting in a queue change if a fifth toll booth were opened?

## SOLUTION

Using the equations for *M/M/N* queuing, we first compute the four-booth case. Note that $\mu$ = 6 veh/min and $\lambda$ = 20 veh/min, and therefore $\rho$ = 3.333. Also, because $\rho/N$ = 0.833 (which is less than 1), Eqs. 5.34 to 5.40 can be used. The probability of having no vehicles in the system with four booths open (using Eq. 5.34) is

$$P_0 = \cfrac{1}{1 + \cfrac{3.333}{1!} + \cfrac{3.333^2}{2!} + \cfrac{3.333^3}{3!} + \cfrac{3.333^4}{4!(0.1667)}}$$

$$= 0.0213$$

The average queue length is (from Eq. 5.38)

$$\overline{Q} = \frac{0.0213(3.333)^5}{4!4}\left[\frac{1}{(0.1667)^2}\right]$$

$$= 3.287 \text{ veh}$$

The average time spent in the system is (from Eq. 5.40)

$$\overline{t} = \frac{3.333 + 3.287}{20}$$

$$= 0.331 \text{ min/veh}$$

And the probability of having to wait in a queue is (from Eq. 5.37)

$$P_{n>N} = \frac{0.0213(3.333)^5}{4!4(0.1667)}$$

$$= 0.548$$

With a fifth booth open, the probability of having no vehicles in the system is (from Eq. 5.34)

$$P_0 = \cfrac{1}{1 + \cfrac{3.333}{1!} + \cfrac{3.333^2}{2!} + \cfrac{3.333^3}{3!} + \cfrac{3.333^4}{4!} + \cfrac{3.333^5}{5!(0.3333)}}$$

$$= 0.0318$$

The average queue length is (from Eq. 5.38)

$$\overline{Q} = \frac{0.0318(3.333)^6}{5!5}\left[\frac{1}{(0.3333)^2}\right]$$

$$= 0.654 \text{ veh}$$

The average time spent in the system is (from Eq. 5.40)

$$\overline{t} = \frac{3.333 + 0.654}{20}$$

$$= 0.199 \text{ min/veh}$$

And the probability of having to wait in a queue is (from Eq. 5.37)

$$P_{n>N} = \frac{0.0318(3.333)^6}{5!5(0.3333)}$$

$$= 0.218$$

So opening a fifth booth would reduce the average queue length by 2.633 veh (3.287 − 0.654), the average time in the system by 0.132 min/veh (0.331 − 0.199), and the probability of waiting in a queue by 0.330 (0.548 − 0.218).

**EXAMPLE 5.14    *M/M/N* QUEUING: PARKING-LOT APPLICATION**

A convenience store has four available parking spaces. The owner predicts that the duration of customer shopping (the time that a customer's vehicle will occupy a parking space) is exponentially distributed with a mean of 6 minutes. The owner knows that in the busiest hour customer arrivals are exponentially distributed with a mean arrival rate of 20 customers per hour. What is the probability that a customer will not find an open parking space when arriving at the store?

**SOLUTION**

Putting mean arrival and departure rates in common units gives $\mu = 10$ veh/h and $\lambda = 20$ veh/h. So $\rho = 2.0$, and because $\rho/N = 0.5$ (which is less than 1), Eqs. 5.34 to 5.40 can be used. The probability of having no vehicles in the system with four parking spaces available (using Eq. 5.34) is

$$P_0 = \frac{1}{1 + \dfrac{2}{1!} + \dfrac{2^2}{2!} + \dfrac{2^3}{3!} + \dfrac{2^4}{4!(0.5)}}$$

$$= 0.1304$$

Thus the probability of not finding an open parking space upon arrival is (from Eq. 5.37)

$$P_{n>N} = \frac{0.1304(2)^5}{4!4(0.5)}$$

$$= \underline{0.087}$$

## 5.6    TRAFFIC ANALYSIS AT HIGHWAY BOTTLENECKS

Some of the most severe congestion problems occur at highway bottlenecks, which are defined as a portion of highway with a lower capacity ($q_{cap}$) than the incoming section of highway. This reduction in capacity can originate from a number of sources, including a decrease in the number of highway lanes and reduced shoulder widths (which tend to cause drivers to slow and thus effectively reduce highway capacity, as will be discussed in Chapter 6). There are two general types of traffic bottlenecks—those that are recurring and those that are incident induced. Recurring bottlenecks exist where the highway itself limits capacity—for example, by a physical reduction in the number of lanes. Traffic congestion at such bottlenecks results from recurring traffic flows that exceed the vehicle capacity of the highway in the bottleneck area. In contrast, incident-induced bottlenecks occur as a result of vehicle breakdowns or accidents that effectively reduce highway capacity by restricting the through movement of traffic. Because incident-induced bottlenecks are unanticipated and temporary in nature, they have features that distinguish them from recurring bottlenecks, such as the possibility that the capacity resulting from an incident-induced bottleneck may change over time. For example, an accident may initially stop traffic flow completely, but as the wreckage is cleared, partial capacity (one lane open) may be provided for a period of time before full capacity is eventually restored. A feature shared by recurring and incident-induced bottlenecks is the adjustment in traffic flow that may occur as travelers choose other routes and/or different trip departure times, to avoid the bottleneck area, in response to visual information or traffic advisories.

The analysis of traffic flow at bottlenecks can be undertaken using the queuing models discussed in Section 5.5. The most intuitive approach to analyzing traffic congestion at bottlenecks is to assume $D/D/1$ queuing.

**EXAMPLE 5.15    $D/D/1$ QUEUING: HIGHWAY BOTTLENECK APPLICATION**

An incident occurs on a freeway that has a capacity in the northbound direction, before the incident, of 4000 veh/h and a constant flow of 2900 veh/h during the morning commute (no adjustments to traffic flow result from the incident). At 8:00 A.M. a traffic accident closes the freeway to all traffic. At 8:12 A.M. the freeway is partially opened with a capacity of 2000 veh/h. Finally, the wreckage is removed, and the freeway is restored to full capacity (4000 veh/h) at 8:31 A.M. Assume $D/D/1$ queuing to determine time of queue dissipation, longest queue length, total delay, average delay per vehicle, and longest wait of any vehicle (assuming FIFO).

**SOLUTION**

Let $\mu$ be the full-capacity departure rate and $\mu_r$ be the restrictive partial-capacity departure rate. Putting arrival and departure rates in common units of vehicles per minute, we have

$$\mu = \frac{4000 \text{ veh/h}}{60 \text{ min/h}} = 66.67 \text{ veh/min}$$

$$\mu_r = \frac{2000 \text{ veh/h}}{60 \text{ min/h}} = 33.33 \text{ veh/min}$$

$$\lambda = \frac{2900 \text{ veh/h}}{60 \text{ min/h}} = 48.33 \text{ veh/min}$$

The arrival rate is constant over the entire time period, and the total number of vehicles is equal to $\lambda t$, where $t$ is the number of minutes after 8:00 A.M. The total number of departing vehicles is

$$
\begin{array}{ll}
0 & \text{for } t \le 12 \text{ min} \\
\mu_r(t-12) & \text{for } 12 \text{ min} < t \le 31 \text{ min} \\
633.33 + \mu(t-31) & \text{for } t > 31 \text{ min}
\end{array}
$$

Note that the value of 633.33 in the departure rate function for $t > 31$ is based on the preceding departure rate function [33⅓ (31 − 12)]. These arrival and departure rates can be represented graphically as shown in Fig. 5.8. As discussed in Section 5.5, for $D/D/1$ queuing, the queue will dissipate at the intersection point of the arrival and departure curves, which can be determined as

$$\lambda t = 633.33 + \mu(t - 31) \qquad \text{or} \qquad t = \underline{78.16 \text{ min}} \text{ (just after 9:18 A.M.)}$$

At this time a total of 3777.5 vehicles (48.33 × 78.16) will have arrived and departed (for the sake of clarity, fractions of vehicles are used). The longest queue (longest vertical distance between arrival and departure curves) occurs at 8:31 a.m. and is

$$\begin{aligned}
Q_{max} &= \lambda t - \mu_r \left(t - 12\right) \\
&= 48.33(31) - 33.33(19) \\
&= \underline{865 \text{ veh}}
\end{aligned}$$

Total vehicle delay is (using equations for triangular and trapezoidal areas to calculate the total area between the arrival and departure curves)

$$\begin{aligned}
D_t &= \frac{1}{2}(12)(580) + \frac{1}{2}(580 + 1498.33)(19) - \frac{1}{2}(19)(633.33) \\
&\quad + \frac{1}{2}(1498.33 - 633.33)(78.16 - 31) \\
&= \underline{37,604.2 \text{ veh-min}}
\end{aligned}$$

**Figure 5.8** $D/D/1$ queuing diagram for Example 5.15.

The average delay per vehicle is <u>9.95 min</u> (37,604.2/3777.5). The longest wait of any vehicle (the longest horizontal distance between the arrival and departure curves), assuming a FIFO queuing discipline, will be the delay time of the 633.33rd vehicle to arrive. This vehicle will arrive 13.1 minutes (633.33/48.33) after 8:00 A.M. and will depart at 8:31 A.M., being delayed a total of <u>17.9 min</u>.

## NOMENCLATURE FOR CHAPTER 5

| | | | |
|---|---|---|---|
| $D$ | deterministic arrivals or departures | $Q_{max}$ | maximum length of queue |
| $D_t$ | total vehicle delay | $s$ | vehicle spacing |
| $h$ | vehicle time headway | $t$ | time |
| $k$ | traffic density | $\bar{t}$ | average time spent in the system |
| $k_j$ | traffic jam density | $u$ | space-mean speed (also denoted $\bar{u}_s$) |
| $k_{cap}$ | traffic density at capacity | $u_i$ | spot speed of vehicle $i$ |
| $l$ | roadway length | $u_f$ | free-flow speed |
| $M$ | exponentially distributed arrivals or departures | $u_{cap}$ | speed at capacity |
| $n$ | number of vehicles | $\bar{u}_s$ | space-mean speed (also denoted simply as $u$) |
| $n_c$ | departure channel number | $\bar{u}_t$ | time-mean speed |
| $N$ | total number of departure channels | $\bar{w}$ | average time waiting in the queue |
| $q$ | traffic flow | $\lambda$ | arrival rate |
| $q_{cap}$ | traffic flow at capacity (maximum traffic flow) | $\mu$ | departure rate |
| $Q$ | length of queue | $\rho$ | traffic intensity |
| $\bar{Q}$ | average length of queue | | |

## REFERENCES

Drew, D. R. "Deterministic Aspects of Freeway Operations and Control." *Highway Research Record*, 99, 1965.

Lord, D., and F. Mannering. "The Statistical Analysis of Crash-Frequency Data: A Review and Assessment of Methodological Alternatives," *Transportation Research Part A*, vol. 44, no. 5, 2010.

Pipes, L. A. "Car Following Models and the Fundamental Diagram of Road Traffic." *Transportation Research*, vol. 1, no. 1, 1967.

Poch, M., and F. Mannering. "Negative Binomial Analysis of Intersection-Accident Frequencies," *Journal of Transportation Engineering*, vol. 122, no. 2, March/April 1996.

Transportation Research Board. *Traffic Flow Theory: A Monograph.* Special Report 165. Washington, DC: National Research Council, 1975.

Transportation Research Board. *Highway Capacity Manual.* Washington, DC: National Research Council, 2010.

## PROBLEMS

**5.1** Assume you are observing traffic in a single lane of a highway at a specific location. You measure the average headway and average spacing of passing vehicles as 3.0 seconds and 45 m, respectively. Calculate the flow, average speed, and density of the traffic stream in this lane.

**5.2** Assume you are an observer standing at a point along a three-lane roadway. All vehicles in lane 1 are traveling at 50 km/h, all vehicles in lane 2 are traveling at 75 km/h, and all vehicles in lane 3 are traveling at 100 km/h. There is also a constant spacing of 0.8 kilometer between vehicles. If you collect spot speed data for all vehicles as they cross your observation point, for 30 minutes, what will be the time-mean speed and space-mean speed for this traffic stream?

**5.3** Four race cars are traveling on a 4.0-kilometer tri-oval track. The four cars are traveling at constant speeds of

315 km/h, 305 km/h, 300 km/h, and 290 km/h, respectively. Assume you are an observer standing at a point on the track for a period of 30 minutes and are recording the instantaneous speed of each vehicle as it crosses your point. What is the time-mean speed and space-mean speed for these vehicles for this time period? (Note: Be careful with rounding.)

**5.4** For Problem 5.3, calculate the space-mean speed assuming you were given only an aerial photo of the circling race cars and the constant travel speed of each of the vehicles.

**5.5** On a specific westbound section of highway, studies show that the speed-density relationship is

$$u = u_f \left[ 1 - \left( \frac{k}{k_j} \right)^{3.5} \right]$$

It is known that the capacity is 3800 veh/h and the jam density is 140 veh/km. What is the space-mean speed of the traffic at capacity, and what is the free-flow speed?

**5.6** A section of highway has a speed-flow relationship of the form

$$q = au^2 + bu$$

It is known that at capacity (which is 2900 veh/h) the space-mean speed of traffic is 50 km/h. Determine the speed when the flow is 1400 veh/h and the free-flow speed.

**5.7** A section of highway has the following flow-density relationship:

$$q = 80k - 0.4k^2$$

What is the capacity of the highway section, the speed at capacity, and the density when the highway is at one-quarter of its capacity?

**5.8** An observer has determined that the time headways between successive vehicles on a section of highway are exponentially distributed and that 65% of the headways between vehicles are 9 seconds or greater. If the observer decides to count traffic in 30-second time intervals, estimate the probability of the observer counting exactly four vehicles in an interval.

**5.9** At a specified point on a highway, vehicles are known to arrive according to a Poisson process. Vehicles are counted in 20-second intervals, and vehicle counts are taken in 120 of these time intervals. It is noted that no cars arrive in 18 of these 120 intervals. Approximate the number of these 120 intervals in which exactly three cars arrive.

**5.10** For the data collected in Problem 5.9, estimate the percentage of time headways that will be 10 seconds or greater and those that will be less than 6 seconds.

**5.11** A vehicle pulls out onto a single-lane highway that has a flow rate of 300 veh/h (Poisson distributed). The driver of the vehicle does not look for oncoming traffic. Road conditions and vehicle speeds on the highway are such that it takes 1.7 seconds for an oncoming vehicle to stop once the brakes are applied. Assuming a standard driver reaction time of 2.5 seconds, what is the probability that the vehicle pulling out will get in an accident with oncoming traffic?

**5.12** Consider the conditions in Problem 5.11. How short would the driver reaction times of oncoming vehicles have to be for the probability of an accident to equal 0.20?

**5.13** Vehicles arrive at a single toll booth beginning at 8:00 A.M. They arrive and depart according to a uniform deterministic distribution. However, the toll booth does not open until 8:10 A.M. The average arrival rate is 8 veh/min and the average departure rate is 10 veh/min. Assuming $D/D/1$ queuing, when does the initial queue clear and what are the total delay, the average delay per vehicle, longest queue length (in vehicles), and the wait time of the 100th vehicle to arrive (assuming first-in-first-out)?

**5.14** Vehicles begin to arrive at a park entrance at 7:45 A.M. at a constant rate of six per minute and at a constant rate of four vehicles per minute from 8:00 A.M. on. The park opens at 8:00 A.M. and the manager wants to set the departure rate so that the average delay per vehicle is no greater than 9 minutes (measured from the time of the first arrival until the total queue clears). Assuming $D/D/1$ queuing, what is the minimum departure rate needed to achieve this?

**5.15** A toll booth on a turnpike is open from 8:00 A.M. to 12 midnight. Vehicles start arriving at 7:45 A.M. at a uniform deterministic rate of six per minute until 8:15 A.M. and from then on at two per minute. If vehicles are processed at a uniform deterministic rate of six per minute, determine when the queue will dissipate, the total delay, the maximum queue length (in vehicles), the longest vehicle delay under FIFO, and the longest vehicle delay under LIFO.

**5.16** Vehicles begin to arrive at a parking lot at 6:00 A.M. at a rate of eight per minute. Due to an accident on the access highway, no vehicles arrive from 6:20 to 6:30 A.M. From 6:30 A.M. on, vehicles arrive at a rate of two per minute. The parking lot attendant processes incoming vehicles (collects parking fees) at a rate of four per minute throughout the day. Assuming $D/D/1$ queuing, determine total vehicle delay.

**5.17** Vehicles begin to arrive at a toll booth at eight vehicles per minute from 9 A.M. to 10 A.M. The booth opens at 9:10 A.M. and services at a rate of ten vehicles per minute until 9:40 A.M. From 9:40 A.M. until 10 A.M. the service rate is six vehicles per minute. Assuming $D/D/1$ queuing, what is the total vehicle delay from 9 A.M. to 10 A.M. assuming $D/D/1$ queuing?

**5.18** The arrival rate at a parking lot is 6 veh/min. Vehicles start arriving at 6:00 P.M., and when the queue reaches 36 vehicles, service begins. If company policy is that total vehicle delay should be equal to 500 veh-min, what is the departure rate? (Assume $D/D/1$ queuing and a constant service rate.)

**5.19** At 8:00 A.M. there are 10 vehicles in a queue at a toll booth and vehicles are arriving at a rate of $\lambda(t) = 6.9$

$- 0.2t$. Beginning at 8 A.M., vehicles are being serviced at a rate of $\mu(t) = 2.1 + 0.3t$ ($\lambda(t)$ and $\mu(t)$ are in vehicles per minute and $t$ is in minutes after 8:00 A.M.). Assuming $D/D/1$ queuing, what is the maximum queue length, and what would the total delay be from 8:00 A.M. until the queue clears?

**5.20** At the end of a sporting event, vehicles begin leaving a parking lot at $\lambda(t) = 12 - 0.25t$ and vehicles are processed at $\mu(t) = 2.5 + 0.5t$ ($t$ is in minutes and $\lambda(t)$ and $\mu(t)$ are in vehicles per minute). Assuming $D/D/1$ queuing, determine the total vehicle delay, longest queue, and the wait time of the 50th vehicle to arrive assuming first-in-first out and $D/D/1$ queuing.

**5.21** Vehicles arrive at a single park-entrance booth where a brochure is distributed. At 8 A.M. there are 20 vehicles in the queue and vehicles continue to arrive at the deterministic rate of $\lambda(t) = 4.2 - 0.1t$, where $\lambda(t)$ is in vehicles per minute and $t$ is in minutes after 8:00 A.M. From 8 A.M. until 8:10 A.M., vehicles are served at a constant deterministic rate of three per minute. Starting at 8:10 A.M., another brochure-distributing person is added and the brochure-service rate increases to six per minute (still at a single booth). Assuming $D/D/1$ queuing, determine the longest queue, the total delay from 8 A.M. until the queue dissipates; and the wait time of the 40th vehicle to arrive.

**5.22** Vehicles arrive at a single toll booth beginning at 7:00 A.M. at a rate of 8 veh/min. Service also starts at 7:00 A.M. at a rate of $\mu(t) = 6 + 0.2t$ where $\mu(t)$ is in vehicles per minute and $t$ is in minutes after 7:00 A.M. Assuming $D/D/1$ queuing, determine when the queue will clear, the total delay, and the maximum queue length in vehicles.

**5.23** Vehicles begin arriving at a single toll-road booth at 8:00am at a time-dependent deterministic rate of $\lambda(t) = 2 + 0.1t$ (with $\lambda(t)$ in veh/min and $t$ in minutes). At 8:07 A.M. the toll booth opens and vehicles are serviced at a constant deterministic rate of 6 veh/min. Assuming $D/D/1$ queuing, what is the average delay per vehicle from 8:00 A.M. until the initial queue clears and what is the delay of the 20th vehicle to arrive?

**5.24** Vehicles begin to arrive at a toll booth at 8:50 A.M. with an arrival rate of $\lambda(t) = 4.1 + 0.01t$ [with $t$ in minutes and $\lambda(t)$ in vehicles per minute]. The toll booth opens at 9:00 A.M. and processes vehicles at a rate of 12 per minute throughout the day. Assuming $D/D/1$ queuing, when will the queue dissipate and what will be the total vehicle delay?

**5.25** Vehicles begin to arrive at a toll booth at 7:50 A.M. with an arrival rate of $\lambda(t) = 5.2 - 0.01t$ (with $t$ in minutes after 7:50 A.M. and $\lambda$ in vehicles per minute). The toll booth opens at 8:00 A.M. and serves vehicles at a rate of $\mu(t) = 3.3 + 2.4t$ (with $t$ in minutes after 8:00 A.M. and $\mu$ in vehicles per minute). Once the service rate reaches 10 veh/min, it stays at that level for the rest of the day. If queuing is $D/D/1$, when will the queue that formed at 7:50 A.M. be cleared?

**5.26** Vehicles arrive at a freeway on-ramp meter at a constant rate of six per minute starting at 6:00 A.M. Service begins at 6:00 A.M. such that $\mu(t) = 2 + 0.5t$, where $\mu(t)$ is in veh/min and $t$ is in minutes after 6:00 A.M. What is the total delay and the maximum queue length (in vehicles)?

**5.27** Vehicles arrive at a toll booth according to the function $\lambda(t) = 5.2 - 0.20t$, where $\lambda(t)$ is in vehicles per minute and $t$ is in minutes. The toll booth operator processes one vehicle every 20 seconds. Determine total delay, maximum queue length, and the time that the 20th vehicle to arrive waits from its arrival to its departure.

**5.28** There are 10 vehicles in a queue when an attendant opens a toll booth. Vehicles arrive at the booth at a rate of 4 per minute. The attendant opens the booth and improves the service rate over time following the function $\mu(t) = 1.1 + 0.30t$, where $\mu(t)$ is in vehicles per minute and $t$ is in minutes. When will the queue clear, what is the total delay, and what is the maximum queue length?

**5.29** Vehicles begin to arrive at a parking lot at 6:00 A.M. with an arrival rate function (in vehicles per minute) of $\lambda(t) = 1.2 + 0.3t$, where $t$ is in minutes. At 6:10 A.M. the parking lot opens and processes vehicles at a rate of 12 per minute. What is the total delay and the maximum queue length?

**5.30** At a parking lot, vehicles arrive according to a Poisson process and are processed (parking fee collected) at a uniform deterministic rate at a single station. The mean arrival rate is 4.2 veh/min and the processing rate is 5 veh/min. Determine the average length of queue, the average time spent in the system, and the average waiting time in the queue.

**5.31** Consider the parking lot and conditions described in Problem 5.30. If the rate at which vehicles are processed became exponentially distributed (instead of deterministic) with a mean processing rate of 5 veh/min, what would be the average length of queue, the average time spent in the system, and the average waiting time in the queue?

**5.32** Vehicles arrive at a toll booth with a mean arrival rate of 3 veh/min (the time between arrivals is exponentially distributed). The toll booth operator processes vehicles (collects tolls) at a uniform deterministic rate of one every 15 seconds. What is the average length of queue, the average time spent in the system, and the average waiting time in the queue?

**5.33** A business owner decides to pass out free transistor radios (along with a promotional brochure) at a booth in a parking lot. The owner begins giving the radios away at 9:15 A.M. and continues until 10:00 A.M. Vehicles start arriving for the radios at 8:45 A.M. at a uniform deterministic rate of 4 per minute and continue to arrive at this rate until 9:15 A.M. From 9:15 to 10:00 A.M. the arrival rate becomes 8 per minute. The radios and brochures are distributed at a uniform deterministic rate of 11 cars per minute over the 45-minute time period. Determine total delay, maximum queue length, and longest vehicle delay assuming FIFO and LIFO.

**5.34** Consider the conditions described in Problem 5.33. Suppose the owner decides to accelerate the radio-brochure distribution rate (in veh/min) so that the queue that forms will be cleared by 9:45 A.M. What would this new distribution rate be?

**5.35** A ferryboat queuing lane holds 40 vehicles. If vehicles are processed (tolls collected) at a uniform deterministic rate of 5 vehicles per minute and processing begins when the lane reaches capacity, what is the uniform deterministic arrival rate if the vehicle queue is cleared 35 minutes after vehicles begin to arrive?

**5.36** At a toll booth, vehicles arrive and are processed (tolls collected) at uniform deterministic rates $\lambda$ and $\mu$, respectively. The arrival rate is 3 veh/min. Processing begins 15 minutes after the arrival of the first vehicle, and the queue dissipates $t$ minutes after the arrival of the first vehicle. Letting the number of vehicles that must actually wait in a queue be $x$, develop an expression for determining processing rates in terms of $x$.

**5.37** Vehicles arrive at a recreational park booth at a uniform deterministic rate of 5 veh/min. If uniform deterministic processing of vehicles (collecting of fees) begins 20 minutes after the first arrival and the total delay is 3200 veh-min, how long after the arrival of the first vehicle will it take for the queue to be cleared?

**5.38** Trucks begin to arrive at a truck weigh station (with a single scale) at 6:00 A.M. at a deterministic but time-varying rate of $\lambda(t) = 4.3 - 0.22t$ [$\lambda(t)$ is in veh/min and $t$ is in minutes]. The departure rate is a constant 2 veh/min (time to weigh a truck is 30 seconds). When will the queue that forms be cleared, what will be the total delay, and what will be the maximum queue length?

**5.39** Commercial trucks begin to arrive at a seaport entry plaza at 7:50 a.m., at the rate of $\lambda(t) = 6.3 - 0.25t$ veh/min, with $t$ in minutes. The plaza opens at 8:00 a.m. For the first 10 minutes, one processing booth is open. After the first 10 minutes until the queue clears, two processing booths are open. Each booth processes trucks at a uniform rate of 2 per minute. What is the average delay per vehicle, the maximum queue length, and the average queue length?

**5.40** Vehicles begin to arrive at a remote parking lot after the start of a major sporting event. They are arriving at a deterministic but time-varying rate of $\lambda(t) = 3.3 - 0.1t$ [$\lambda(t)$ is in veh/min and $t$ is in minutes]. The parking lot attendant processes vehicles (assigns spaces and collects fees) at a deterministic rate at a single station. A queue exceeding four vehicles will back up onto a congested street, and is to be avoided. How many vehicles per minute must the attendant process to ensure that the queue does not exceed four vehicles?

**5.41** A truck weighing station has a single scale. The time between truck arrivals at the station is exponentially distributed with a mean arrival rate of 1.6 veh/min. The time it takes vehicles to be weighed is exponentially distributed with a mean rate of 2.1 veh/min. When more than 5 trucks are in the system, the queue backs up onto the highway and interferes with through traffic. What is the probability that the number of trucks in the system will exceed 5?

**5.42** Consider the convenience store described in Example 5.14. The owner is concerned about customers not finding an available parking space when they arrive during the busiest hour. How many spaces must be provided for there to be less than a 1% chance of an arriving customer not finding an open parking space?

**5.43** Vehicles arrive at a toll bridge at a rate of 420 veh/h (the time between arrivals is exponentially distributed). Two toll booths are open and each can process arrivals (collect tolls) at a mean rate of 12 seconds per vehicle (the processing time is also exponentially distributed). What is the total time spent in the system by all vehicles in a 1-hour period?

**5.44** Vehicles leave an airport parking facility (arrive at parking fee collection booths) at a rate of 500 veh/h (the time between arrivals is exponentially distributed). The parking facility has a policy that the average time a patron spends in a queue waiting to pay for parking is not to exceed 5 seconds. If the time required to pay for parking is exponentially distributed with a mean of 15 seconds, what is the smallest number of payment processing booths that must be open to keep the average time spent in a queue below 5 seconds?

**5.45** A freeway with two northbound lanes is shut down because of an accident. At the time of the accident, the traffic flow rate is 1200 vehicles per hour per lane and the flow remains at this level. The capacity of the freeway is 2200 vehicles per hour per lane when not impacted by an accident. The freeway is shut down completely for 20 minutes after the accident and then one lane is open for 20 minutes and finally both lanes are opened (40 minutes after the accident). What is the average delay per vehicle resulting from the accident (assuming $D/D/1$ queuing)?

**5.46** A four-lane highway has a normal capacity of 1800 vehicles per hour per lane. In the southbound direction, a vehicle disablement on the roadway shoulder occurs at 4:30 p.m. Due to rubbernecking, the capacity in the southbound direction is reduced to 1200 veh/h/lane at this time. At 4:45 p.m., the disabled vehicle is removed from the shoulder and the capacity increases to 1500 veh/h/lane. At 5:00 p.m. the roadway capacity returns to its full value of 1800 veh/h/lane. From 4:30 p.m. until the queue clears the traffic flow rate in the southbound direction is 1600 veh/h/lane. What is the average delay per vehicle, the maximum queue length, and the average queue length in the southbound direction resulting from the incident (assuming $D/D/1$ queuing)?

**5.47** Five minivans and three trucks are traveling on a 4.8 kilometer circular track and complete a full lap in 98.0, 108.0, 113.0, 108.0, 102.0, 101.0, 85.0, and 95 seconds, respectively. Assuming all the vehicles are traveling at constant speeds, what is the time-mean speed of the minivans? Pay attention to rounding.

   a)   163.732 km/h
   b)   171.867 km/h
   c)   163.327 km/h
   d)   163.600 km/h

**5.48** Vehicles arrive at an intersection at a rate of 400 veh/h according to a Poisson distribution. What is the probability that more than five vehicles will arrive in a one-minute interval?

   a)   0.7944
   b)   0.6560
   c)   0.6547
   d)   0.1552

**5.49** In studying of traffic flow at a highway toll booth over a course of 60 minutes, it is determined that the arrival and departure rates are deterministic, but not uniform. The arrival rate is found to vary according to the function $\lambda(t) = 1.8 + 0.25t - 0.0030t^2$. The departure rate function is $\mu(t) = 1.4 + 0.11t$. In both of these functions, $t$ is in minutes after the beginning of the observation and $\lambda(t)$ and $\mu(t)$ are in vehicles per minute. At what time does the maximum queue length occur?

   a)   49.4 min
   b)   2.7 min
   c)   19.4 min
   d)   60.0 min

**5.50** A theme park has a single entrance gate where visitors must stop and pay for parking. The average arrival rate during the peak hour is 150 veh/h and is Poisson distributed. It takes, on average, 20 seconds per vehicle (exponentially distributed) to pay for parking. What is the average waiting time for this queuing system?

   a)   4.167 min/veh
   b)   2.0 min/veh
   c)   1.667 min/veh
   d)   0.833 min/veh

**5.51** At an impaired driver checkpoint, the time required to conduct the impairment test varies (according to an exponential distribution) depending on the compliance of the driver, but takes 60 seconds on average. If an average of 30 vehicles per hour arrive (according to a Poisson distribution) at the checkpoint, determine the average time spent in the system.

   a)   0.033 min/veh
   b)   1.5 min/veh
   c)   1.0 min/veh
   d)   2.0 min/veh

**5.52** A toll road with three toll booths has an average arrival rate of 850 veh/h and drivers take an average of 12 seconds to pay their tolls. If the arrival and departure times are determined to be exponentially distributed, how would the probability of waiting in a queue change if a fourth toll both were opened?

   a)   0.088
   b)   0.534
   c)   0.313
   d)   0.847

# Chapter 6

# Highway Capacity and Level-of-Service Analysis

## 6.1 INTRODUCTION

The underlying objective of traffic analysis is to quantify a roadway's performance with regard to specified traffic volumes. This performance can be measured in terms of travel delay (as the roadway becomes increasingly congested) as well as other factors. The comparative performance of various roadway segments (which is determined from an analysis of traffic) is important because it can be used as a basis to allocate limited roadway construction and improvement funds. The purpose of this chapter is to apply the elements of uninterrupted traffic flow theory covered in Chapter 5 to the practical field analysis of traffic flow and capacity on freeways, multilane highways, and two-lane highways.

The main challenge of such a process is to adapt the theoretical formulations to the wide range of conditions that occur in the field. These diverse field conditions must be taken into account in a traffic analysis methodology, yet the methodology must remain theoretically consistent. For example, in Chapter 5, capacity ($q_{cap}$) is simply defined as the highest traffic flow rate that the roadway is capable of supporting. For applied traffic analysis, a consistent and reasonably precise method of determining capacity must be developed within this definition. Because it can readily be shown that the capacity of a roadway segment is a function of factors such as roadway type (freeway, multilane highway, or two-lane highway), free-flow speed, number of lanes, and widths of lanes and shoulders, the method of capacity determination clearly must account for a wide variety of physical and operational roadway characteristics.

Additionally, recall that Chapter 5 defines traffic flow on the basis of units of vehicles per hour. Two practical issues arise concerning this unit of measure. First, in many cases vehicular traffic consists of a variety of vehicle types with substantially different performance characteristics. These performance differentials are likely to be magnified by changing roadway geometries, such as upgrades or downgrades, which have a differential effect on the acceleration and deceleration capabilities of the various types of vehicles; for example, grades have a greater impact on the performance of large trucks than on automobiles. As a result, traffic must be defined not only in terms of vehicles per unit time but also in terms of vehicle composition, because it is clear that a 1500-veh/h traffic flow consisting of 100% automobiles will differ significantly with regard to operating speed and traffic density from a 1500-veh/h traffic flow that consists of 50% automobiles and 50% heavy trucks.

171

The other flow-related concern is the temporal distribution of traffic. In practice, the analysis of roadway traffic usually focuses on the most critical condition, which is the most congested hour within a 24-hour daily period (the temporal distribution of traffic is discussed in more detail in Section 6.7). However, within this most congested peak hour, traffic flow is likely to be nonuniform. It is therefore necessary to arrive at some method of defining and measuring the nonuniformity of flow within the peak hour.

To summarize, the objective of applied traffic analysis is to provide a practical method of quantifying the degree of traffic congestion and to relate this to the overall traffic-related performance of the roadway. The following sections of this chapter discuss and demonstrate accepted standards for applied traffic analysis for the three major types of uninterrupted-flow roadways: freeways, multilane highways, and two-lane highways (one lane in each direction).

## 6.2   LEVEL-OF-SERVICE CONCEPT

The *Highway Capacity Manual* (HCM), produced by the Transportation Research Board [2010], is a synthesis of the state of the art in methodologies for quantifying traffic operational performance and capacity utilization (congestion level) for a variety of transportation facilities. One of the foundations of the HCM is the concept of level of service (LOS). The level of service represents a qualitative ranking of the traffic operational conditions experienced by users of a facility under specified roadway, traffic, and traffic control (if present) conditions. Current practice designates six levels of service ranging from A to F, with level of service A representing the best operating conditions and level of service F the worst.

A number of operational performance measures, such as speed, flow, and density, can be measured or calculated for any transportation facility. To apply the level-of-service concept to traffic analysis, it is necessary to select a performance measure that is representative of how motorists actually perceive the quality of service they are receiving on a facility. Motorists tend to evaluate their received quality of service in terms of factors such as speed and travel time, freedom to maneuver, traffic interruptions, and comfort and convenience. Thus, it is important to select a measure that encompasses some or all of these factors. The performance measure that is selected for level-of-service (LOS) analysis for a particular transportation facility is referred to as the service measure.

The HCM [Transportation Research Board 2010] defines the LOS categories for freeways and multilane highways as follows:

**Level of service A.** LOS A represents free-flow conditions (traffic operating at free-flow speeds, as defined in Chapter 5). Individual users are virtually unaffected by the presence of others in the traffic stream. Freedom to select speeds and to maneuver within the traffic stream is extremely high. The general level of comfort and convenience provided to drivers is excellent.

**Level of service B.** LOS B also allows speeds at or near free-flow speeds, but the presence of other users in the traffic stream begins to be noticeable. Freedom to select speeds is relatively unaffected, but there is a slight decline in the freedom to maneuver within the traffic stream relative to LOS A.

**Level of service C.** LOS C has speeds at or near free-flow speeds, but the freedom to maneuver is noticeably restricted (lane changes require careful attention on the part of drivers). The general level of comfort and convenience declines significantly at this level. Disruptions in the traffic stream, such as an incident (for example, vehicular accident or disablement), can result in significant queue formation and vehicular delay. In contrast, the effects of incidents at LOS A or LOS B are minimal, with only minor delay in the immediate vicinity of the event.

**Level of service D.** LOS D represents the conditions where speeds begin to decline slightly with increasing flow. The freedom to maneuver becomes more restricted, and drivers experience reductions in physical and psychological comfort. Incidents can generate lengthy queues because the higher density associated with this LOS provides little space to absorb disruptions in the traffic flow.

**Level of service E.** LOS E represents operating conditions at or near the roadway's capacity. Even minor disruptions to the traffic stream, such as vehicles entering from a ramp or vehicles changing lanes, can cause delays as other vehicles give way to allow such maneuvers. In general, maneuverability is extremely limited, and drivers experience considerable physical and psychological discomfort.

**Level of service F.** LOS F describes a breakdown in vehicular flow. Queues form quickly behind points in the roadway where the arrival flow rate temporarily exceeds the departure rate, as determined by the roadway's capacity (see Chapter 5). Such points occur at incidents and on- and off-ramps, where incoming traffic results in capacity being exceeded. Vehicles typically operate at low speeds under these conditions and are often required to come to a complete stop, usually in a cyclic fashion. The cyclic formation and dissipation of queues is a key characterization of LOS F.

A visual perspective of the level-of-service definitions for freeways is provided in Fig. 6.1. In dealing with level of service it is important to remember that when the traffic volume is at or near the roadway capacity (which will be shown as a function of the prevailing traffic and physical characteristics of the roadway), the roadway is operating at LOS E. This, however, is not a desirable condition because under LOS E conditions there is considerable driver discomfort, which could increase the likelihood of vehicular crashes and overall delay. In roadway design, the possibility of degradation of level of service to LOS E should be avoided, although this is not always possible due to financial and environmental constraints that may limit the design speed, number of lanes, and other factors affecting roadway capacity.

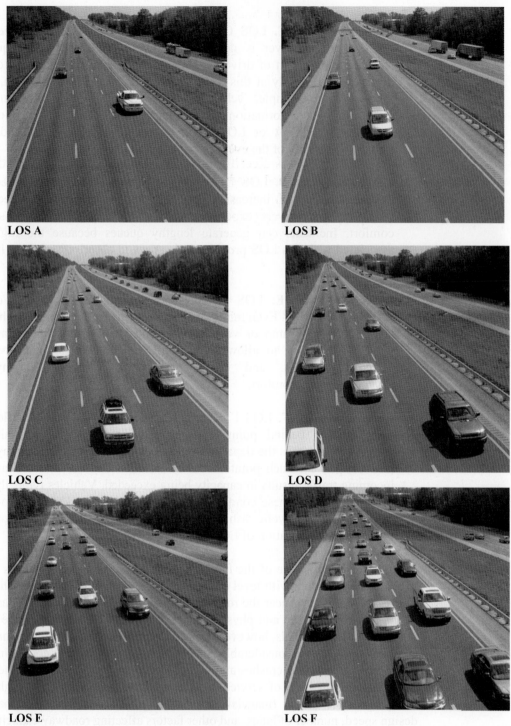

**Figure 6.1** Illustration of freeway level of service (A to F).

## 6.3   LEVEL-OF-SERVICE DETERMINATION

There are several steps in a basic level-of-service determination for an uninterrupted-flow facility. The remainder of this section describes the general details of each step, as applicable to uninterrupted-flow facility analyses. Facility-specific details of these steps are described in the sections that follow.

### 6.3.1   Base Conditions and Capacity

The determination of a roadway's level of service begins with the specification of base roadway conditions. Recall that in the introduction to this chapter, the effects of vehicle performance and roadway design characteristics on traffic flow were discussed qualitatively. In practice, the effects of such factors on traffic flow are measured quantitatively, relative to the base traffic and roadway design conditions. For uninterrupted-flow roadways, base conditions can be categorized as those relating to roadway conditions, such as lane widths, lateral clearances, access frequency, and terrain; and traffic stream conditions such as the effects of heavy vehicles (large trucks, buses, and RVs) and driver population characteristics. Base conditions are defined as those conditions that represent unrestrictive geometric and traffic conditions. Additionally, base conditions are assumed to consist of favorable environmental conditions (such as dry roadways).

The capacity of a particular roadway segment will be greatest when all roadway and traffic conditions meet or exceed their base values. Empirical studies have identified the values of these base conditions for which the capacity of a roadway segment is maximized. Values in excess of the base conditions will not increase the capacity of the roadway, but values more restrictive than the base conditions will result in a lower capacity. For example, studies have identified a base lane width of 3.6 m. That is, lane widths in excess of 3.6 m will not result in increased capacity; however, lane widths less than 3.6 m will result in a reduction in capacity. Capacity values for base conditions have been determined for all uninterrupted-flow facility types from field studies. It should be noted that for purposes of level-of-service analysis, capacity is defined not as the absolute maximum flow rate ever observed for a particular facility type, but rather as the maximum flow rate that can be reasonably expected on a recurring basis.

Because all base conditions for a particular roadway type are seldom realized in practice, methods for converting the measured flow rate into an equivalent analysis flow rate in terms of passenger cars for the given traffic conditions and estimating the actual free-flow speed for the given roadway conditions are needed. The following sections describe the procedures for arriving at flow and speed values for given roadway and traffic conditions.

### 6.3.2   Determine Free-Flow Speed

Free-flow speed (*FFS*) is a term that was introduced in Chapter 5 as the speed of traffic as the traffic density approaches zero. In practice, *FFS* is governed by roadway design characteristics (horizontal and vertical curves, lane and shoulder widths, and median design), the frequency of access points, the complexity of the driving environment (possible distractions from roadway signs and the like), and posted speed limits.

The free-flow speed must be determined given the characteristics of the roadway segment. *FFS* is the mean speed of traffic as measured when flow rates are low to moderate (specific values are given under the individual sections for each roadway type). Ideally, *FFS* should be measured directly in the field at the site of interest. However, if this is not possible or feasible, an alternative method can be employed to arrive at an estimate of *FFS* under the prevailing conditions. This method makes adjustments to a base *FFS* (*BFFS*) depending on the physical characteristics of the roadway segment, such as lane width, shoulder width, and access frequency. This method has the same basic structure for the various roadway types, but contains adjustment factors and values appropriate for each roadway type.

### 6.3.3 Determine Analysis Flow Rate

One of the fundamental inputs to a traffic analysis is the actual traffic volume on the roadway, in vehicles per hour, which is given the symbol *V*. Generally, the highest volume in a 24-hour period (the peak-hour volume) is used for *V* in traffic analysis computations. However, this hourly volume needs to be adjusted to reflect the temporal variation of traffic demand within the analysis hour, the impacts due to heavy vehicles, and, in the case of freeway and multilane roadways, the characteristics of the driving population. To account for these effects, the hourly volume is divided by adjustment factors to obtain an equivalent flow rate in terms of passenger cars per hour (pc/h). Additionally, the flow rate is expressed on a per-lane basis (pc/h/ln) by dividing by the number of lanes in the analysis segment.

### 6.3.4 Calculate Service Measure(s) and Determine LOS

Once the previous steps have been completed, all that remains is to calculate the value of the service measure and then determine the LOS from the service measure value. For freeways and multilane highways, this is a relatively straightforward task. However, for two-lane highways, there are actually two service measures, and the calculation of these and the subsequent LOS determination are more involved.

## 6.4 BASIC FREEWAY SEGMENTS

A basic freeway segment is defined as a section of a divided roadway having two or more lanes in each direction, full access control, and traffic that is unaffected by merging or diverging movements near ramps. It is important to note that capacity analysis for divided roadways focuses on the traffic flow in one direction only. This is reasonable because the objective is to measure the highest level of congestion. Due to directional imbalance of traffic flows — for example, morning rush hours having higher volumes going toward the central city and evening rush hours having higher volumes going away from the central city — consideration of traffic volumes in both directions is likely to seriously understate the true level of traffic congestion.

Table 6.1 provides the level-of-service criteria corresponding to traffic density, speed, volume-to-capacity ratio, and maximum flow rate. A graphical representation of this table is provided in Fig. 6.2. The maximum service flow rate is simply the maximum flow rate, under base conditions, that can be sustained for a given level of service. This value is related to speed and density as discussed in Chapter 5. This speed-flow-density relationship is central to the analysis of basic freeway segments, as will be outlined in the remainder of this section.

### 6.4.1    Base Conditions and Capacity

The base conditions for a basic freeway segment are defined as [Transportation Research Board 2010]

- 3.6-m minimum lane widths
- 1.8-m minimum right-shoulder clearance between the edge of the travel lane and objects (utility poles, retaining walls, etc.) that influence driver behavior
- 0.6-m minimum median lateral clearance
- Only passenger cars in the traffic stream
- Five or more lanes in each travel direction (urban areas only)
- 3.2-km or greater interchange spacing
- Level terrain (no grades greater than 2%)
- A driver population of mostly familiar roadway users

These conditions represent a high operating level, with a free-flow speed of 110 km/h or higher.

The capacity, $c$, for basic freeway segments, in passenger cars per hour per lane (pc/h/ln), is given in Table 6.2. From Table 6.1, note that, by definition, the upper boundary of LOS E corresponds to the value of capacity and a $v/c$ of 1.0. Other values of $v/c$ for a specific level of service are obtained by simply dividing the maximum flow rate for that level of service by capacity (the maximum flow rate at LOS E).

### 6.4.2    Service Measure

The service measure for basic freeway segments is density. Density, as discussed in Chapter 5, is typically measured in terms of passenger cars per kilometer per lane (pc/km/ln) and therefore provides a good measure of the relative mobility of individual vehicles in the traffic stream. A low traffic stream density gives individual vehicles the ability to change lanes and speeds with relative ease, while a high density makes it very difficult for individual vehicles to maneuver within the traffic stream. Thus, traffic density is the primary determinant of freeway level of service.

Recall Eq. 5.14 from Chapter 5:

$$q = uk \tag{6.1}$$

*where*

$q$ = flow in veh/h,
$u$ = speed in km/h, and
$k$ = density in veh/km.

Density is therefore calculated as flow divided by speed. The following sections will describe how to arrive at flow and speed values for the given roadway and traffic conditions. Once the flow and speed values have been determined according to the given conditions, a density can be calculated and then referenced in Table 6.1 or Fig. 6.2 to arrive at a level of service for the freeway segment.

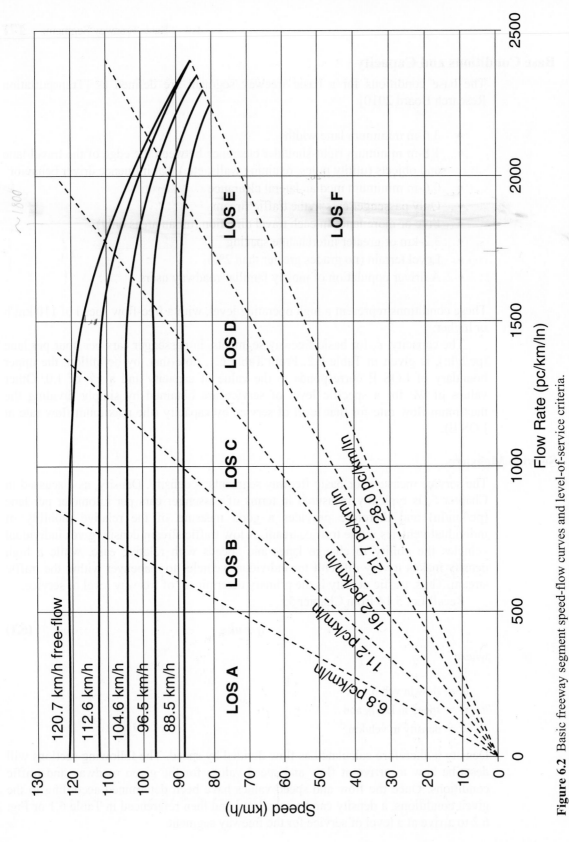

**Figure 6.2** Basic freeway segment speed-flow curves and level-of-service criteria.

Adapted from Exhibit 11-6, p. 11-8 with permission of the Transportation Research Board, *Highway Capacity Manual* 2010, Copyright, National Academy of Sciences, Washington, D.C. Figure has been converted to SI units for use in this edition. Values from the source material are in U.S. Customary units.

**Table 6.1** LOS Criteria for Basic Freeway Segments

| Criterion | LOS | | | | |
|---|---|---|---|---|---|
| | A | B | C | D | E |
| *FFS* = 120.7 km/h | | | | | |
| Maximum density (pc/km/ln) | 6.8 | 11.2 | 16.2 | 21.7 | 28.0 |
| Average speed (km/h) | 120.7 | 118.7 | 109.9 | 98.0 | 85.7 |
| Maximum v/c | 0.34 | 0.55 | 0.74 | 0.89 | 1.00 |
| Maximum flow rate (pc/h/ln) | 820 | 1330 | 1780 | 2125 | 2400 |
| *FFS* = 112.6 km/h | | | | | |
| Maximum density (pc/km/ln) | 6.8 | 11.2 | 16.2 | 21.7 | 28.0 |
| Average speed (km/h) | 112.6 | 112.6 | 107.2 | 97.2 | 85.7 |
| Maximum v/c | 0.32 | 0.53 | 0.72 | 0.88 | 1.00 |
| Maximum flow rate (pc/h/ln) | 765 | 1260 | 1735 | 2110 | 2400 |
| *FFS* = 104.6 km/h | | | | | |
| Maximum density (pc/km/ln) | 6.8 | 11.2 | 16.2 | 21.7 | 28.0 |
| Average speed (km/h) | 104.6 | 104.6 | 102.9 | 94.8 | 84.0 |
| Maximum v/c | 0.30 | 0.50 | 0.71 | 0.87 | 1.00 |
| Maximum flow rate (pc/h/ln) | 710 | 1170 | 1670 | 2055 | 2350 |
| *FFS* = 96.5 km/h | | | | | |
| Maximum density (pc/km/ln) | 6.8 | 11.2 | 16.2 | 21.7 | 28.0 |
| Average speed (km/h) | 96.5 | 96.5 | 96.5 | 92.0 | 82.2 |
| Maximum v/c | 0.29 | 0.47 | 0.68 | 0.87 | 1.00 |
| Maximum flow rate (pc/h/ln) | 655 | 1080 | 1565 | 1995 | 2300 |
| *FFS* = 88.5 km/h | | | | | |
| Maximum density (pc/km/ln) | 6.8 | 11.2 | 16.2 | 21.7 | 28.0 |
| Average speed (km/h) | 88.5 | 88.5 | 88.5 | 88.0 | 80.4 |
| Maximum v/c | 0.27 | 0.44 | 0.64 | 0.85 | 1.00 |
| Maximum flow rate (pc/h/ln) | 600 | 990 | 1435 | 1910 | 2250 |

*Note:* Density is the primary determinant of LOS. Maximum flow rate values are rounded to the nearest 5 passenger cars.

**Table 6.2** Relationship Between Free-Flow Speed and Capacity on Basic Freeway Segments

*Source*: Transportation Research Board, *Highway Capacity Manual* 2010. Washington, D.C. National Academy of Sciences. Table has been converted to SI units for use in this edition. Values from the source material are in U.S. Customary units.

| Free-flow speed (km/h) | Capacity (pc/h/ln) |
|---|---|
| 120.7 | 2400 |
| 112.6 | 2400 |
| 104.6 | 2350 |
| 96.5 | 2300 |
| 88.5 | 2250 |

## 6.4.3 Determine Free-Flow Speed

For basic freeway segments, *FFS* is the mean speed of passenger cars operating in flow rates up to 1300 passenger cars per hour per lane (pc/h/ln). If *FFS* is to be estimated rather than measured, the following equation can be used. It accounts for the roadway characteristics of lane width, right-shoulder lateral clearance, and ramp density.

$$FFS = 121.3 - f_{LW} - f_{LC} - 7.96TRD^{0.84} \qquad (6.2)$$

*where*

$$
\begin{aligned}
FFS &= \text{estimated free-flow speed in km/h,} \\
f_{LW} &= \text{adjustment for lane width in km/h,} \\
f_{LC} &= \text{adjustment for lateral clearance in km/h,} \\
7.96TRD^{0.84} &= \text{adjustment for total ramp density in km/h (with } TRD \text{ in ramps/km).}
\end{aligned}
$$

The constant value of 121.3 in Eq. 6.2 is considered to be the base free-flow speed (*BFFS*) and applies to freeways in urban and rural areas. The HCM [Transportation Research Board 2010] recommends that the calculated free-flow speed be rounded to the nearest *FFS* values shown in Figure 6.2 and Table 6.1. The following sections describe the procedures for estimating the adjustment factor values.

### Lane Width Adjustment

When lane widths are narrower than the base 3.6 m, the adjustment factor $f_{LW}$ is used to reflect the impact on free-flow speed. Such an adjustment is needed because narrow lanes cause traffic to slow as a result of reduced psychological comfort and limits on driver maneuvering and accident avoidance options. Thus, *FFS* under these conditions is less than the value that would be observed if base lane widths were provided. The adjustment factors used in current practice are presented in Table 6.3.

**Table 6.3** Adjustment for Lane Width

| Lane width (m) | Reduction in free-flow speed, $f_{LW}$ (km/h) |
| --- | --- |
| 3.6 | 0.0 |
| 3.3 | 3.1 |
| 3.0 | 10.6 |

Reproduced with permission of the Transportation Research Board, *Highway Capacity Manual* 2010, Copyright, National Academy of Sciences, Washington, D.C. Exhibit 11-8, p. 11-11. Table has been converted to SI units for use in this edition. Values from the source material are in U.S. Customary units.

### Lateral Clearance Adjustment

When obstructions are closer than 1.8 m (at the roadside) from the traveled pavement, the adjustment factor $f_{LC}$ is used to reflect the impact on *FFS*. Again, these conditions lead to reduced psychological comfort for the driver and consequently reduced speeds. An obstruction is a right-side object that can either be continuous (such as a retaining wall or barrier) or periodic (such as light posts or utility poles). Table 6.4 provides corrections for obstructions on the right side of the roadway.

**Table 6.4**  Adjustment for Right-Shoulder Lateral Clearance

| Right-shoulder lateral clearance (m) | Reduction in free-flow speed, $f_{LC}$ (km/h), lanes in one direction | | | |
|:---:|:---:|:---:|:---:|:---:|
| | 2 | 3 | 4 | ≥5 |
| ≥ 1.8 | 0.0 | 0.0 | 0.0 | 0.0 |
| 1.5 | 1.0 | 0.7 | 0.3 | 0.2 |
| 1.2 | 1.9 | 1.3 | 0.7 | 0.4 |
| 0.9 | 2.9 | 1.9 | 1.0 | 0.6 |
| 0.6 | 3.9 | 2.6 | 1.3 | 0.8 |
| 0.3 | 4.8 | 3.2 | 1.6 | 1.1 |
| 0 | 5.8 | 3.9 | 1.9 | 1.3 |

Reproduced with permission of the Transportation Research Board, *Highway Capacity Manual* 2010, Copyright, National Academy of Sciences, Washington, D.C. Adapted from Exhibit 11-9, p. 11-12. Table has been converted to SI units for use in this edition. Values from the source material are in U.S. Customary units.

**Total Ramp Density Adjustment**

Ramp density provides a measure of the impact of merging and diverging traffic on free-flow speed. Total ramp density is the number of on- and off-ramps (in one direction) within a distance of 5 kilometers upstream and 5 kilometers downstream of the midpoint of the analysis segment, divided by 10 kilometers.

### 6.4.4   Determine Analysis Flow Rate

The analysis flow rate is calculated using the following equation:

$$v_p = \frac{V}{PHF \times N \times f_{HV} \times f_p} \tag{6.3}$$

*where*

$v_p$ = 15-min passenger car equivalent flow rate (pc/h/ln),
$V$ = hourly volume (veh/h),
$PHF$ = peak-hour factor,
$N$ = number of lanes,
$f_{HV}$ = heavy-vehicle adjustment factor, and
$f_p$ = driver population factor.

The adjustment factors $PHF$, $f_{HV}$, and $f_p$ are described next.

**Peak-Hour Factor**

As previously mentioned, vehicle arrivals during the period of analysis [typically the highest hourly volume within a 24-h period (peak hour)] will likely be nonuniform. To account for this varying arrival rate, the peak 15-min vehicle arrival rate within the analysis hour is usually used for practical traffic analysis purposes. The peak-hour factor has been developed for this purpose, and is defined as the ratio of the hourly volume to the maximum 15-min flow rate expanded to an hourly volume, as follows:

$$PHF = \frac{V}{V_{15} \times 4}$$   (6.4)

*where*

$PHF$ = peak-hour factor,
$V$ = hourly volume for hour of analysis,
$V_{15}$ = maximum 15-min volume within hour of analysis, and
4 = number of 15-min periods per hour.

Equation 6.4 indicates that the further the *PHF* is from unity, the more *peaked* or nonuniform the traffic flow is during the hour. For example, consider two roads both of which have a peak-hour volume, $V$, of 1800 veh/h. The first road has 600 vehicles arriving in the highest 15-min interval, and the second road has 500 vehicles arriving in the highest 15-min interval. The first road has a more nonuniform flow, as indicated by its *PHF* of 0.75 [1800/(600 × 4)], which is further from unity than the second road's *PHF* of 0.90 [1800/(500 × 4)].

**Heavy-Vehicle Adjustment**

Large trucks, buses, and recreational vehicles have performance characteristics (slow acceleration and inferior braking) and dimensions (length, height, and width) that have an adverse effect on roadway capacity. Recall that base conditions stipulate that no heavy vehicles are present in the traffic stream, and when prevailing conditions indicate the presence of such vehicles, the adjustment factor $f_{HV}$ is used to translate the traffic stream from base to prevailing conditions. The $f_{HV}$ correction term is found using a two-step process. The first step is to determine the passenger car equivalent (PCE) for each large truck, bus, and recreational vehicle in the traffic stream. These values represent the number of passenger cars that would consume the same amount of roadway capacity as a single large truck, bus, or recreational vehicle. These passenger car equivalents are denoted $E_T$ for large trucks and buses and $E_R$ for recreational vehicles, and are a function of roadway grades because steep grades will tend to magnify the poor performance of heavy vehicles as well as the sight distance problems caused by their larger dimensions (the visibility afforded to drivers in vehicles following heavy vehicles). For segments of freeway that contain a mix of grades, an extended segment analysis can be used as long as no single grade is steep enough or long enough to significantly impact the overall operations of the segment. As a guideline, an extended segment analysis can be used for freeway segments where no single grade that is less than 3% is more than 0.8 km long, or no single grade that is 3% or greater is longer than 0.4 km. If an extended segment analysis is used, the terrain must be generally classified according to the following definitions [Transportation Research Board 2010]:

**Level terrain.** Any combination of horizontal and vertical alignment permitting heavy vehicles to maintain approximately the same speed as passenger cars. This generally includes short grades of no more than 2%.

**Rolling terrain.** Any combination of horizontal and vertical alignment that causes heavy vehicles to reduce their speed substantially below those of passenger cars but does not cause heavy vehicles to operate at their limiting speed $[F_{net}(V) \neq 0]$ for the given terrain for any significant length of time or at frequent intervals due to high grade resistance, as illustrated in Fig. 2.6.

**Mountainous terrain.** Any combination of horizontal and vertical alignment that causes heavy vehicles to operate at their limiting speed for significant distances or at frequent intervals.

The passenger car equivalency factors for an extended segment analysis can be obtained from Table 6.5.

**Table 6.5** Passenger Car Equivalents (PCEs) for Extended Freeway Segments

| Factor | Type of terrain | | |
| --- | --- | --- | --- |
| | Level | Rolling | Mountainous |
| $E_T$ (trucks and buses) | 1.5 | 2.5 | 4.5 |
| $E_R$ (RVs) | 1.2 | 2.0 | 4.0 |

Reproduced with permission of the Transportation Research Board, *Highway Capacity Manual 2010*, Copyright, National Academy of Sciences, Washington, D.C. Adapted from Exhibit 11-9, p.11-12.

Any grade that does not meet the conditions for an extended segment analysis must be analyzed as a separate segment because of its significant impact on traffic operations. In these cases, grade-specific PCE values must be used. Tables 6.6 and 6.7 provide these values for positive grades (upgrades). These tables assume typical large trucks (with average weight-to-power ratios between 75 and 90 kg/kW) and recreational vehicles (with average weight-to-power ratios between 20 and 40 kg/kW). Note that the equivalency factors presented in these tables increase with increasing grade and length of grade, but decrease with increasing heavy vehicle percentage. This decrease with increasing percentage is due to the fact that heavy vehicles tend to group together as their percentages increase on steep, extended grades, thus decreasing their adverse impact on the traffic stream.

Sometimes it is necessary to determine the cumulative effect on traffic operations of several significant grades in succession. For this situation, a distance-weighted average may be used if all grades are less than 4% or the total combined length of the grades is less than 1220 m. For example, a 2% upgrade for 305 m followed immediately by a 3% upgrade for 610 m would use the equivalency factor for a 2.67% upgrade $[(2 \times 305 + 3 \times 610)/915]$ for 915 m or 0.915 km. For information on additional analysis situations involving composite grades, refer to the *Highway Capacity Manual* [Transportation Research Board 2010]. These situations include combining two or more successive grades when the grades exceed 4% or the combined length is greater than 1220 m, determining the length of a grade that starts or ends on a vertical curve, and determining the point of greatest traffic impact in a series of grades (for example, if a long 5% grade were immediately followed by a 2% grade, the end of the 5% grade would be used, as this would be the point of minimum vehicle speed).

**Table 6.6** Passenger Car Equivalents ($E_T$) for Trucks and Buses on Specific Upgrades

| Upgrade (%) | Length (km) | Percentage of trucks and buses | | | | | | | | |
|---|---|---|---|---|---|---|---|---|---|---|
| | | 2 | 4 | 5 | 6 | 8 | 10 | 15 | 20 | 25 |
| < 2 | All | 1.5 | 1.5 | 1.5 | 1.5 | 1.5 | 1.5 | 1.5 | 1.5 | 1.5 |
| ≥ 2–3 | 0.0–0.4 | 1.5 | 1.5 | 1.5 | 1.5 | 1.5 | 1.5 | 1.5 | 1.5 | 1.5 |
| | > 0.4–0.8 | 1.5 | 1.5 | 1.5 | 1.5 | 1.5 | 1.5 | 1.5 | 1.5 | 1.5 |
| | > 0.8–1.2 | 1.5 | 1.5 | 1.5 | 1.5 | 1.5 | 1.5 | 1.5 | 1.5 | 1.5 |
| | > 1.2–1.6 | 2.0 | 2.0 | 2.0 | 2.0 | 1.5 | 1.5 | 1.5 | 1.5 | 1.5 |
| | > 1.6–2.4 | 2.5 | 2.5 | 2.5 | 2.5 | 2.0 | 2.0 | 2.0 | 2.0 | 2.0 |
| | > 2.4 | 3.0 | 3.0 | 2.5 | 2.5 | 2.0 | 2.0 | 2.0 | 2.0 | 2.0 |
| > 3–4 | 0.0–0.4 | 1.5 | 1.5 | 1.5 | 1.5 | 1.5 | 1.5 | 1.5 | 1.5 | 1.5 |
| | > 0.4–0.8 | 2.0 | 2.0 | 2.0 | 2.0 | 2.0 | 2.0 | 1.5 | 1.5 | 1.5 |
| | > 0.8–1.2 | 2.5 | 2.5 | 2.0 | 2.0 | 2.0 | 2.0 | 2.0 | 2.0 | 2.0 |
| | > 1.2–1.6 | 3.0 | 3.0 | 2.5 | 2.5 | 2.5 | 2.5 | 2.0 | 2.0 | 2.0 |
| | > 1.6–2.4 | 3.5 | 3.5 | 3.0 | 3.0 | 3.0 | 3.0 | 2.5 | 2.5 | 2.5 |
| | > 2.4 | 4.0 | 3.5 | 3.0 | 3.0 | 3.0 | 3.0 | 2.5 | 2.5 | 2.5 |
| > 4–5 | 0.0–0.4 | 1.5 | 1.5 | 1.5 | 1.5 | 1.5 | 1.5 | 1.5 | 1.5 | 1.5 |
| | > 0.4–0.8 | 3.0 | 2.5 | 2.5 | 2.5 | 2.0 | 2.0 | 2.0 | 2.0 | 2.0 |
| | > 0.8–1.2 | 3.5 | 3.0 | 3.0 | 3.0 | 2.5 | 2.5 | 2.5 | 2.5 | 2.5 |
| | > 1.2–1.6 | 4.0 | 3.5 | 3.5 | 3.5 | 3.0 | 3.0 | 3.0 | 3.0 | 3.0 |
| | > 1.6 | 5.0 | 4.0 | 4.0 | 4.0 | 3.5 | 3.5 | 3.0 | 3.0 | 3.0 |
| > 5–6 | 0.0–0.4 | 2.0 | 2.0 | 1.5 | 1.5 | 1.5 | 1.5 | 1.5 | 1.5 | 1.5 |
| | > 0.4–0.5 | 4.0 | 3.0 | 2.5 | 2.5 | 2.0 | 2.0 | 2.0 | 2.0 | 2.0 |
| | > 0.5–0.8 | 4.5 | 4.0 | 3.5 | 3.0 | 2.5 | 2.5 | 2.5 | 2.5 | 2.5 |
| | > 0.8–1.2 | 5.0 | 4.5 | 4.0 | 3.5 | 3.0 | 3.0 | 3.0 | 3.0 | 3.0 |
| | > 1.2–1.5 | 5.5 | 5.0 | 4.5 | 4.0 | 3.0 | 3.0 | 3.0 | 3.0 | 3.0 |
| | > 1.6 | 6.0 | 5.0 | 5.0 | 4.5 | 3.5 | 3.5 | 3.5 | 3.5 | 3.5 |
| > 6 | 0.0–0.4 | 4.0 | 3.0 | 2.5 | 2.5 | 2.5 | 2.5 | 2.0 | 2.0 | 2.0 |
| | > 0.4–0.5 | 4.5 | 4.0 | 3.5 | 3.5 | 3.5 | 3.0 | 2.5 | 2.5 | 2.5 |
| | > 0.5–0.8 | 5.0 | 4.5 | 4.0 | 4.0 | 3.5 | 3.0 | 2.5 | 2.5 | 2.5 |
| | > 0.8–1.2 | 5.5 | 5.0 | 4.5 | 4.5 | 4.0 | 3.5 | 3.0 | 3.0 | 3.0 |
| | > 1.2–1.6 | 6.0 | 5.5 | 5.0 | 5.0 | 4.5 | 4.0 | 3.5 | 3.5 | 3.5 |
| | > 1.6 | 7.0 | 6.0 | 5.5 | 5.5 | 5.0 | 4.5 | 4.0 | 4.0 | 4.0 |

Reproduced with permission of the Transportation Research Board, *Highway Capacity Manual* 2010, Copyright, National Academy of Sciences, Washington, D.C. Adapted from Exhibit 11-11, p. 11-16. Table has been converted to SI units for use in this edition. Values from the source material are in U.S. Customary units.

Negative grades (downgrades) also have an impact on equivalency factors because the comparatively poor braking characteristics of heavy vehicles have a more deleterious effect on the traffic stream than the level-terrain case. Table 6.8 gives the passenger car equivalents for trucks and buses on downgrades. It is assumed that recreational vehicles are not significantly impacted by downgrades, and therefore downgrade values for $E_R$ are drawn from the level-terrain column in Table 6.5.

**Table 6.7** Passenger Car Equivalents ($E_R$) for RVs on Specific Upgrades

| Upgrade (%) | Length (km) | Percentage of RVs | | | | | | | | |
|---|---|---|---|---|---|---|---|---|---|---|
| | | 2 | 4 | 5 | 6 | 8 | 10 | 15 | 20 | 25 |
| ≤ 2 | All | 1.2 | 1.2 | 1.2 | 1.2 | 1.2 | 1.2 | 1.2 | 1.2 | 1.2 |
| > 2–3 | 0.0–0.8 | 1.2 | 1.2 | 1.2 | 1.2 | 1.2 | 1.2 | 1.2 | 1.2 | 1.2 |
| | > 0.8 | 3.0 | 1.5 | 1.5 | 1.5 | 1.5 | 1.5 | 1.2 | 1.2 | 1.2 |
| > 3–4 | 0.0–0.4 | 1.2 | 1.2 | 1.2 | 1.2 | 1.2 | 1.2 | 1.2 | 1.2 | 1.2 |
| | > 0.4–0.8 | 2.5 | 2.5 | 2.0 | 2.0 | 2.0 | 2.0 | 1.5 | 1.5 | 1.5 |
| | > 0.8 | 3.0 | 2.5 | 2.5 | 2.5 | 2.0 | 2.0 | 2.0 | 1.5 | 1.5 |
| 4–5 | 0.0–0.4 | 2.5 | 2.0 | 2.0 | 2.0 | 1.5 | 1.5 | 1.5 | 1.5 | 1.5 |
| | > 0.4–0.8 | 4.0 | 3.0 | 3.0 | 3.0 | 2.5 | 2.5 | 2.0 | 2.0 | 2.0 |
| | > 0.8 | 4.5 | 3.5 | 3.0 | 3.0 | 3.0 | 2.5 | 2.5 | 2.0 | 2.0 |
| > 5 | 0.0–0.4 | 4.0 | 3.0 | 2.5 | 2.5 | 2.5 | 2.0 | 2.0 | 2.0 | 1.5 |
| | > 0.4–0.8 | 6.0 | 4.0 | 4.0 | 3.5 | 3.0 | 3.0 | 2.5 | 2.5 | 2.0 |
| | > 0.8 | 6.0 | 4.5 | 4.0 | 4.5 | 3.5 | 3.0 | 3.0 | 2.5 | 2.0 |

Reproduced with permission of the Transportation Research Board, *Highway Capacity Manual* 2010, Copyright, National Academy of Sciences, Washington, D.C. Adapted from Exhibit 11-12, p. 11-17. Table has been converted to SI units for use in this edition. Values from the source material are in U.S. Customary units.

**Table 6.8** Passenger Car Equivalents ($E_T$) for Trucks and Buses on Specific Downgrades

| Downgrade (%) | Length (km) | Percentage of trucks | | | |
|---|---|---|---|---|---|
| | | 5 | 10 | 15 | 20 |
| < 4 | All | 1.5 | 1.5 | 1.5 | 1.5 |
| 4–5 | ≤ 6.4 | 1.5 | 1.5 | 1.5 | 1.5 |
| 4–5 | > 6.4 | 2.0 | 2.0 | 2.0 | 1.5 |
| > 5–6 | ≤ 6.4 | 1.5 | 1.5 | 1.5 | 1.5 |
| > 5–6 | > 6.4 | 5.5 | 4.0 | 4.0 | 3.0 |
| > 6 | ≤ 6.4 | 1.5 | 1.5 | 1.5 | 1.5 |
| > 6 | > 6.4 | 7.5 | 6.0 | 5.5 | 4.5 |

Reproduced with permission of the Transportation Research Board, *Highway Capacity Manual* 2010, Copyright, National Academy of Sciences, Washington, D.C. Adapted from Exhibit 11-13, p. 11-17. Table has been converted to SI units for use in this edition. Values from the source material are in U.S. Customary units.

Once the appropriate equivalency factors have been obtained, the following equation is applied to arrive at the heavy-vehicle adjustment factor $f_{HV}$:

$$f_{HV} = \frac{1}{1 + P_T\left(E_T - 1\right) + P_R\left(E_R - 1\right)} \qquad (6.5)$$

Table 6.5 or 6.7

Prop of RVs in Traffic Stream

*where*

$f_{HV}$ = heavy-vehicle adjustment factor,

$P_T$ = proportion of trucks and buses in the traffic stream,

$P_R$ = proportion of recreational vehicles in the traffic stream,

$E_T$ = passenger car equivalent for trucks and buses, from Table 6.5, 6.6, or 6.8, and

$E_R$ = passenger car equivalent for recreational vehicles, from Table 6.5 or 6.7.

As an example of how the heavy-vehicle adjustment factor is computed, consider a freeway with a 1.6-km 4% upgrade with a traffic stream having 8% trucks, 2% buses, and 2% recreational vehicles. Tables 6.6 and 6.7 must be used because the grade is too steep and long for Table 6.5 to apply. The corresponding equivalency factors for this roadway are $E_T = 2.5$ (for a combined truck and bus percentage of 10) and $E_R = 3.0$, as obtained from Tables 6.6 and 6.7, respectively. Also, from the given percentages of heavy vehicles in the traffic stream, $P_T = 0.1$ and $P_R = 0.02$. Substituting these values into Eq. 6.5 gives $f_{HV} = 0.84$, or a 16% reduction in effective roadway capacity relative to the base condition of no heavy vehicles in the traffic stream.

**✳ Driver Population Adjustment**

Under base conditions, the traffic stream is assumed to consist of regular weekday drivers and commuters. Such drivers have a high familiarity with the roadway and generally maneuver and respond to the maneuvers of other drivers in a safe and predictable fashion. There are times, however, when the traffic stream has a driver population that is less familiar with the roadway in question (such as weekend drivers or recreational drivers). Such drivers can cause a significant reduction in roadway capacity relative to the base condition of having only familiar drivers.

To account for the composition of the driver population, the adjustment factor $f_p$ is used, and its recommended range is 0.85–1.00. Normally, the analyst should select a value of 1.00 for primarily commuter (or familiar-driver) traffic streams. But for other driver populations (for example, a large percentage of tourists), the loss in roadway capacity can vary from 1% to 15%. The exact value of the driver population adjustment factor is dependent on local conditions such as roadway characteristics and the surrounding environment (possible driver distractions such as scenic views and the like). When the driver population consists of a significant percentage of unfamiliar users, judgment is necessary to determine the exact value of this factor. This usually involves collection of data on local conditions (for further information, see [Transportation Research Board 2010]).

### 6.4.5   Calculate Density and Determine LOS

With all the terms in the previous equations defined, these equations can now be applied to determine freeway level of service and freeway capacity. The final step before level of service can be determined is to calculate the density of the traffic stream. The alternative notation to Eq. 6.1 is shown in Eq. 6.6, which will be used in subsequent example problems (for consistency with the *Highway Capacity Manual*).:

$$D = \frac{v_p}{S} \tag{6.6}$$

*where*

$D =$ density in pc/km/ln,
$v_p =$ flow rate in pc/h/ln, and
$S =$ average passenger car speed in km/h.

The average passenger car speed is found by reading it from the $y$-axis of Fig. 6.2 for the corresponding flow rate ($v_p$) and free-flow speed. Once the density value is calculated, the level of service can be read from Table 6.1 or Fig. 6.2.

Application of the process for determining basic freeway segment capacity and level of service will now be demonstrated by example.

**EXAMPLE 6.1**   **BASIC FREEWAY SEGMENT LOS WITH GENERAL TERRAIN CLASSIFICATION**

A six-lane urban freeway (three lanes in each direction) is on rolling terrain with 3.3-m lanes, obstructions 0.6 m from the right edge of the traveled pavement, and nine ramps within 5 kilometers upstream and 5 kilometers downstream of the midpoint of the analysis segment. The traffic stream consists primarily of commuters. A directional weekday peak-hour volume of 2300 vehicles is observed, with 700 vehicles arriving in the most congested 15-min period. If the traffic stream has 15% large trucks and buses and no recreational vehicles, determine the level of service.

**SOLUTION**

Determine the free-flow speed according to Eq. 6.2.

$$FFS = 121.3 - f_{LW} - f_{LC} - 7.96TRD^{0.84}$$

*with*

$f_{LW} = 3.1$ km/h (Table 6.3),
$f_{LC} = 2.6$ km/h (Table 6.4), and
$TRD = \dfrac{9}{10} = 0.9$ ramps/km

$$FFS = 121.3 - 3.1 - 2.6 - 7.96(0.9)^{0.84} = 108.3 \text{ km/h}$$

Rounding this *FFS* value to the nearest *FFS* in Table 6.2 gives a *FFS* of 104.6 km/h. Determine the flow rate according to Eq. 6.3:

$$v_p = \frac{V}{PHF \times N \times f_{HV} \times f_p}$$

*with*

$PHF = \dfrac{2300}{700 \times 4} = 0.821$
$N = 3$ (given),
$f_p = 1.0$ (commuters), and
$E_T = 2.5$ (rolling terrain, Table 6.5).

From Eq. 6.5 we obtain:

$$f_{HV} = \frac{1}{1 + 0.15(2.5 - 1)} = 0.816$$

So,

$$v_p = \frac{2300}{0.821 \times 3 \times 0.816 \times 1.0} = 1144.4 \rightarrow 1145 \text{ pc/h/ln}$$

Obtaining average passenger car speed from Fig. 6.2 for a flow rate of 1145 and a *FFS* of 104.6 km/h yields an *S* of 104.6 km/h. In this case, the average speed is still the same as the *FFS* because the flow rate is low enough such that it is still on the linear/flat part of the speed-flow curve.

Now, density can be calculated with Eq. 6.6:

$$D = \frac{1145}{104.6} = 10.9 \text{ pc/km/ln}$$

From Table 6.1, it can be seen that this corresponds to <u>LOS B</u> (6.8 [max density for LOS A] < 10.9 < 11.2 [max density for LOS B]). Thus, this freeway segment operates at level of service B.

This problem can also be solved graphically by applying Fig. 6.2. Using this figure, draw a vertical line up from 1145 pc/h/ln (on the figure's *x*-axis) and find that this line intersects the 104.6 km/h free-flow speed curve in the LOS B density region (the dashed diagonal lines).

## EXAMPLE 6.2    BASIC FREEWAY SEGMENT LOS WITH A SPECIFIC GRADE

Consider the freeway and traffic conditions in Example 6.1. At some point further along the roadway there is a 6% upgrade that is 2.4 km long. All other characteristics are the same as in Example 6.1. What is the level of service of this portion of the roadway, and how many vehicles can be added before the roadway reaches capacity (assuming that the proportion of vehicle types and the peak-hour factor remain constant)?

## SOLUTION

To determine the LOS of this segment of the freeway, we note that all adjustment factors are the same as those in Example 6.1 except $f_{HV}$, which must now be determined using an equivalency factor, $E_T$, drawn from the specific-upgrade tables (in this case Table 6.6). From Table 6.6, $E_T = 3.5$, which gives

$$f_{HV} = \frac{1}{1 + 0.15(3.5 - 1)} = 0.727$$

So,

$$v_p = \frac{2300}{0.821 \times 3 \times 0.727 \times 1.0} = 1284.5 \rightarrow 1285 \text{ pc/h/ln}$$

From Fig. 6.2, the average passenger car speed (*S*) is still 104.6 km/h; thus

$$D = \frac{1285}{104.6} = 12.3 \ \text{pc/km/ln}$$

which gives <u>LOS C</u> from Table 6.1 (11.2 [max density for LOS B] < 12.3 < 16.2 [max density for LOS C]).

To determine how many vehicles can be added before capacity is reached, the hourly volume at capacity must be computed. Recall that capacity corresponds to a volume-to-capacity ratio of 1.0 (the threshold between LOS E and LOS F). For a free-flow speed of 104.6 km/h, the capacity is 2350 pc/h/ln. Equation 6.3 is rearranged and used to solve for the hourly volume based upon this capacity:

$$v_p = \frac{V}{PHF \times N \times f_{HV} \times f_p} \ \Rightarrow \ 2350 = \frac{V}{0.821 \times 3 \times 0.727 \times 1.0}$$

which gives $V = 4208$ veh/h. This means that about <u>1908 vehicles</u> (4208 − 2300) can be added during the peak hour before capacity is reached. It should be noted that the assumption that the peak-hour factor will remain constant as the roadway approaches capacity is not very realistic. In practice it is observed that as a roadway approaches capacity, *PHF* gets closer to 1. This implies that the flow rate over the peak hour becomes more uniform. This uniformity is the result of, among other factors, motorists adjusting their departure and arrival times to avoid congested periods within the peak hour.

## 6.5  MULTILANE HIGHWAYS

Multilane highways are similar to freeways in most respects, except for a few key differences:

- Vehicles may enter or leave the roadway at at-grade intersections and driveways (multilane highways do not have full access control).
- Multilane highways may or may not be divided (by a barrier or median separating opposing directions of flow), whereas freeways are always divided.
- Traffic signals may be present.
- Design standards (such as design speeds) are sometimes lower than those for freeways.
- The visual setting and development along multilane highways are usually more distracting to drivers than in the freeway case.

Multilane highways usually have four or six lanes (both directions), have posted speed limits between 60 and 100 km/h, and can have physical medians, medians that are two-way left-turn lanes (TWLTLs), or opposing directional volumes that may not be divided by a median at all. Two examples of multilane highways are shown in Fig. 6.3.

The determination of level of service on multilane highways closely mirrors the procedure for freeways. The main differences lie in some of the adjustment factors and their values. The procedure we present is valid only for sections of highway that are not significantly influenced by large queue formations and dissipations resulting from traffic signals (this is generally taken as having traffic signals spaced 3.2 km apart or more), do not have significant on-street parking, do not have bus stops with high usage, and do not have significant pedestrian activity.

Table 6.9 provides the level-of-service criteria corresponding to traffic density, speed, volume-to-capacity ratio, and the maximum flow rates for multilane highways. A graphical representation of this table is provided in Fig. 6.4.

**Figure 6.3** Examples of multilane highways.

**Table 6.9** LOS Criteria for Multilane Highways

| Criterion | LOS | | | | |
|---|---|---|---|---|---|
| | A | B | C | D | E |
| *FFS* = 96.5 km/h | | | | | |
| Maximum density (pc/km/ln) | 6.8 | 11.2 | 16.2 | 21.7 | 24.9 |
| Average speed (km/h) | 96.5 | 96.5 | 95.6 | 91.3 | 88.5 |
| Maximum *v/c* | 0.30 | 0.49 | 0.70 | 0.90 | 1.00 |
| Maximum flow rate (pc/h/ln) | 655 | 1080 | 1550 | 1980 | 2200 |
| *FFS* = 88.5 km/h | | | | | |
| Maximum density (pc/km/ln) | 6.8 | 11.2 | 16.2 | 21.7 | 25.5 |
| Average speed (km/h) | 55.0 | 55.0 | 54.9 | 52.9 | 51.2 |
| Maximum *v/c* | 0.29 | 0.47 | 0.68 | 0.88 | 1.00 |
| Maximum flow rate (pc/h/ln) | 600 | 990 | 1430 | 1850 | 2100 |
| *FFS* = 80.5 km/h | | | | | |
| Maximum density (pc/km/ln) | 6.8 | 11.2 | 16.2 | 21.7 | 26.7 |
| Average speed (km/h) | 50.0 | 50.0 | 50.0 | 48.9 | 47.5 |
| Maximum *v/c* | 0.27 | 0.45 | 0.65 | 0.86 | 1.00 |
| Maximum flow rate (pc/h/ln) | 545 | 900 | 1300 | 1710 | 2000 |
| *FFS* = 72.4 km/h | | | | | |
| Maximum density (pc/km/ln) | 6.8 | 11.2 | 16.2 | 21.7 | 28.0 |
| Average speed (km/h) | 45.0 | 45.0 | 45.0 | 44.4 | 42.2 |
| Maximum *v/c* | 0.26 | 0.43 | 0.62 | 0.82 | 1.00 |
| Maximum flow rate (pc/h/ln) | 490 | 810 | 1170 | 1550 | 1900 |

*Note:* Density is the primary determinant of LOS. Maximum flow rate values are rounded to the nearest 5 passenger cars.

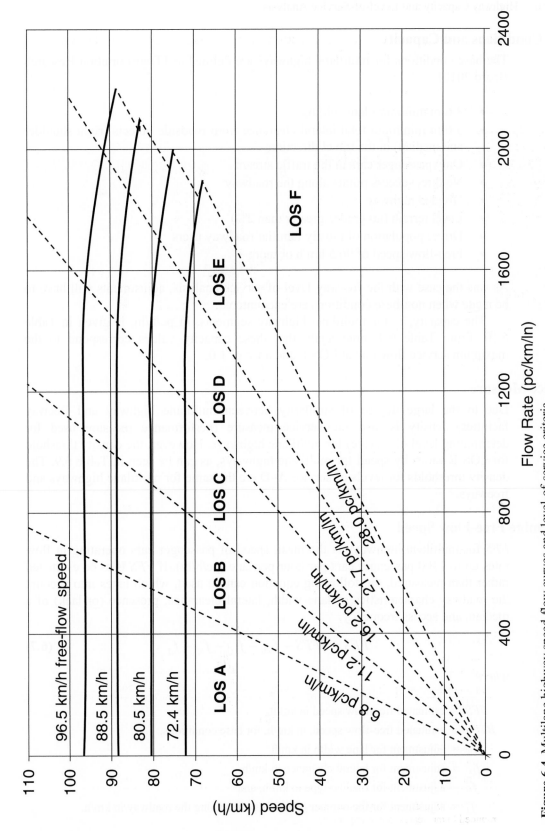

**Figure 6.4** Multilane highway speed-flow curves and level-of-service criteria.
Reproduced with permission of the Transportation Research Board, *Highway Capacity Manual* 2010, Copyright, National Academy of Sciences, Washington, D.C. Adapted from Exhibit 14-5, p. 14-5. Figure has been converted to SI units for use in this edition. Value from the source material are in U.S. Customary units.

191

### 6.5.1    Base Conditions and Capacity

The base conditions for multilane highways are defined as [Transportation Research Board 2010]

- 3.6-m minimum lane widths
- 3.6-m minimum total lateral clearance from roadside objects (right shoulder and median) in the travel direction
- Only passenger cars in the traffic stream
- No direct access points along the roadway
- Divided highway
- Level terrain (no grades greater than 2%)
- Driver population of mostly familiar roadway users
- Free-flow speed of 96.5 km/h or more

As was the case with the freeway level-of-service analysis, adjustments will have to be made when non-base conditions are encountered.

The capacity, $c$, for multilane highway segments, in pc/h/ln, is given in Table 6.10. From Table 6.9, note again that these capacity values correspond to the maximum service flow rate at LOS E and a $v/c$ of 1.0.

### 6.5.2    Service Measure

Due to the large degree of similarity between multilane highway and freeway facilities, density is also the service measure (performance measure used for determining level of service) for multilane highways. However, the density threshold for LOS E varies by speed for multilane highways, as can be seen in Table 6.9. The density thresholds for levels of service A–D are the same for multilane highways and freeways.

### 6.5.3    Determine Free-Flow Speed

*FFS* for multilane highways is the mean speed of passenger cars operating in flow rates up to 1400 passenger cars per hour per lane (pc/h/ln). If *FFS* is to be estimated rather than measured, the following equation can be used, which takes into account the roadway characteristics of lane width, lateral clearance, presence (or lack) of a median, and access frequency:

$$FFS = BFFS - f_{LW} - f_{LC} - f_M - f_A \qquad (6.7)$$

*where*

$FFS$ = estimated free-flow speed in km/h,

$BFFS$ = estimated free-flow speed, in km/h, for base conditions,

$f_{LW}$ = adjustment for lane width in km/h,

$f_{LC}$ = adjustment for lateral clearance in km/h,

$f_M$ = adjustment for median type in km/h, and

$f_A$ = adjustment for the number of access points along the roadway in km/h.

As can be seen, this equation closely resembles Eq. 6.2 in the freeway section. Both include adjustments for lane width and lateral clearance, and the access frequency adjustment is similar to the ramp density adjustment. The main difference is that Eq. 6.7 also includes an adjustment for median type. The presence of a physical barrier or wide separation between opposing flows (such as a TWLTL) will lead to higher free-flow speeds than if there is no separation or physical barrier between opposing flows. This adjustment is not included for freeways since, by definition, all freeways are divided. As was the case for freeways, the HCM [Transportation Research Board 2010] recommends that the calculated free-flow speed be rounded to the nearest $FFS$ values shown in Figure 6.4 and Table 6.9.

| | Free-flow speed | Capacity |
|---|---|---|
| **Table 6.10** Relationship Between Free-Flow Speed and Capacity on Multilane Highway Segments | (km/h) | (pc/h/ln) |
| *Source*: Transportation Research Board, *Highway Capacity Manual* 2010. Washington, D.C. National Academy of Sciences. Table has been converted to SI units for use in this edition. Values from the source material are in U.S. Customary units. | 96.5 | 2200 |
| | 88.5 | 2100 |
| | 80.5 | 2000 |
| | 72.4 | 1900 |

As for $BFFS$, many factors can influence the free-flow speed, with the posted speed limit often being a significant one. For multilane highways, research has found that free-flow speeds, under base conditions, are about 11 km/h higher than the speed limit for 65- and 70-km/h posted-speed-limit roadways, and about 8 km/h higher for 80-km/h and higher posted-speed-limit roadways. The following sections describe the procedures for estimating the adjustment factor values.

### Lane Width Adjustment

The same lane width adjustment factor values are used for multilane highways as are used for freeways. Thus, Table 6.3 should be used for multilane highways as well.

### Lateral Clearance Adjustment

The adjustment factor for potentially restrictive lateral clearances ($f_{LC}$) is determined first by computing the total lateral clearance, which is defined as

$$TLC = LC_R + LC_L \tag{6.8}$$

*where*

> $TLC$ = total lateral clearance in m,
>
> $LC_R$ = lateral clearance on the right side of the travel lanes to obstructions (retaining walls, utility poles, signs, trees, etc.), and
>
> $LC_L$ = lateral clearance on the left side of the travel lanes to obstructions.

For undivided highways, there is no adjustment for left-side lateral clearance because this is already taken into account in the $f_M$ term (thus $LC_L = 1.8$ m in Eq. 6.8). If an individual lateral clearance (either left or right side) exceeds 1.8 m, 1.8 m is used in

Eq. 6.8. Finally, highways with TWLTLs are considered to have $LC_L$ equal to 1.8 m. Once Eq. 6.8 is applied, the value for $f_{LC}$ can be determined directly from Table 6.11.

**Table 6.11** Adjustment for Lateral Clearance

*Total lateral clearance is the sum of the lateral clearances of the median (if greater than 1.8 m, use 1.8 m) and shoulder (if greater than 1.8 m, use 1.8 m). Therefore, for purposes of analysis, total lateral clearance cannot exceed 3.6 m.

Reproduced with permission of the Transportation Research Board, *Highway Capacity Manual* 2010, Copyright, National Academy of Sciences, Washington, D.C. Adapted from Exhibit 14-9, p. 14-12. Table has been converted to SI units for use in this edition. Values from the source material are in U.S. Customary units.

| Total lateral clearance* (m) | Reduction in free-flow speed (km/h) | |
|---|---|---|
| | Four-lane highways | Six-lane highways |
| 3.6 | 0.0 | 0.0 |
| 3.0 | 0.6 | 0.6 |
| 2.4 | 1.5 | 1.5 |
| 1.8 | 2.1 | 2.1 |
| 1.2 | 3.0 | 2.7 |
| 0.6 | 5.8 | 4.5 |
| 0.0 | 8.7 | 6.3 |

### Median Adjustment

Values for the adjustment factor for median type, $f_M$, are provided in Table 6.12. This table shows that undivided highways have a free-flow speed that is 2.6 km/h lower than divided highways (which include those with two-way left-turn lanes).

**Table 6.12** Adjustment for Median Type

Reproduced with permission of the Transportation Research Board, *Highway Capacity Manual* 2010, Copyright, National Academy of Sciences, Washington, D.C. Adapted from Exhibit 14-10, p. 14-12. Table has been converted to SI units for use in this edition. Values from the source material are in U.S. Customary units.

| Median type | Reduction in free-flow speed (km/h) |
|---|---|
| Undivided highways | 2.6 |
| Divided highways (including TWLTLs) | 0.0 |

### Access Frequency Adjustment

The final adjustment factor in Eq. 6.7 is for the number of access points per kilometer, $f_A$. Access points are defined to include intersections and driveways (on the right side of the highway in the direction being considered) that significantly influence traffic flow, and thus do not generally include driveways to individual residences or service driveways at commercial sites. Adjustment values for access point frequency are provided in Table 6.13.

**Table 6.13** Adjustment for Access-Point Frequency

Reproduced with permission of the Transportation Research Board, *Highway Capacity Manual* 2010, Copyright, National Academy of Sciences, Washington, D.C. Adapted from Exhibit 14-11, p. 14-12. Table has been converted to SI units for use in this edition. Values from the source material are in U.S. Customary units.

| Access points/ kilometer | Reduction in free-flow speed (km/h) |
|---|---|
| 0 | 0.0 |
| 6 | 4.0 |
| 12 | 8.0 |
| 18 | 12.0 |
| ≥ 24 | 16.0 |

### 6.5.4 Determine Analysis Flow Rate

The analysis flow rate for multilane highways is determined in the same manner as for freeways, using Eq. 6.3 and the remainder of the procedure outlined in Section 6.4.4. There is one minor difference for multilane highways—the guidelines for an extended segment analysis. An extended segment (general terrain type) analysis can be used for multilane highway segments if grades of 3% or less do not extend for more than 1.6 km or any grades greater than 3% do not extend for more than 0.8 km.

### 6.5.5 Calculate Density and Determine LOS

The procedure for calculating density and determining LOS for multilane highways is essentially the same as for freeways (see Section 6.4.5). Equation 6.6 is applied to arrive at a density. However, slightly different speed-flow curves and level-of-service criteria are used for multilane highways. Table 6.9 shows the level-of-service criteria for multilane highways, and Fig. 6.4 shows the corresponding speed-flow curves for multilane highways.

The average passenger car speed is found by reading it from the $y$-axis of Fig. 6.4 for the corresponding analysis flow rate ($v_p$) and free-flow speed. Once the density value is calculated, the level of service can be read from Table 6.9 or Fig. 6.4.

---

**EXAMPLE 6.3    MULTILANE HIGHWAY FREE-FLOW SPEED**

A four-lane undivided highway (two lanes in each direction) has 3.3-m lanes, with 1.2-m shoulders on the right side. There are four access points per kilometer, and the posted speed limit is 80 km/h. What is the estimated free-flow speed?

*SOLUTION*

This problem can be solved by direct application of Eq. 6.7 to arrive at an estimated free-flow speed:

$$FFS = BFFS - f_{LW} - f_{LC} - f_M - f_A$$

*with*

$BFFS$ = 88 km/h (assume $FFS$ = posted speed + 8 km/h),

$f_{LW}$ = 3.1 km/h (Table 6.3),

$f_{LC}$ = 0.6 km/h (Table 6.11, with $TLC = 1.2 + 1.8 = 3.0$ from Eq. 6.8, with $LC_L = $ 1.8 m because the highway is undivided),

$f_M$ = 2.6 km/h (Table 6.12), and

$f_A$ = 2.67 km/h (Table 6.13, by interpolation).

Substitution gives

$$FFS = 88 - 3.1 - 0.6 - 2.6 - 2.67 = \underline{\underline{79.03 \text{ km/h}}}$$

which means that the more restrictive roadway characteristics relative to the base conditions result in a reduction in free-flow speed of 8.97 km/h. Note that for further analysis, this *FFS* value should be rounded to 80.5 km/h.

**EXAMPLE 6.4**   **MULTILANE HIGHWAY LOS**

A six-lane divided highway (three lanes in each direction) is on rolling terrain with 1.2 access points per kilometer and has 3.0-m lanes, with a 1.5-m shoulder on the right side and a 0.9-m shoulder on the left side. The peak-hour factor is 0.80, and the directional peak-hour volume is 3000 vehicles per hour. There are 6% large trucks, 2% buses, and 2% recreational vehicles. A significant percentage of nonfamiliar roadway users are in the traffic stream (the driver population adjustment factor is estimated as 0.95). No speed studies are available, but the posted speed limit is 90 km/h. Determine the level of service.

**SOLUTION**

We begin by determining *FFS* by applying Eq. 6.7:

$$FFS = BFFS - f_{LW} - f_{LC} - f_M - f_A$$

with

$BFFS$ = 98 km/h (assume $FFS$ = posted speed + 8 km/h),
$f_{LW}$ = 10.6 km/h (Table 6.3),
$f_{LC}$ = 1.5 km/h (Table 6.11, with $TLC$ = 1.5 + 0.9 = 2.4 from Eq. 6.8,
$f_M$ = 0.0 km/h (Table 6.12), and
$f_A$ = 0.8 km/h (Table 6.13, by interpolation).

Substitution gives

$$FFS = 98.0 - 10.6 - 1.5 - 0.0 - 0.8 = 85.1 \text{ km/h}$$

Rounding this *FFS* value to the nearest 8 km/h gives a *FFS* of 88.5 km/h. Next we determine the analysis flow rate using Eq. 6.3:

$$v_p = \frac{V}{PHF \times N \times f_{HV} \times f_p}$$

with

$V$ = 3000 veh/h (given),
$PHF$ = 0.80 (given),
$N$ = 3 (given),
$f_p$ = 0.95 (given),
$E_T$ = 2.5 (Table 6.5), and
$E_R$ = 2.0 (Table 6.5).

From Eq. 6.5, we find

$$f_{HV} = \frac{1}{1 + 0.08(2.5 - 1) + 0.02(2 - 1)} = 0.877$$

Substitution gives

$$v_p = \frac{3000}{0.8 \times 3 \times 0.877 \times 0.95} = 1500.3 \text{ pc/h/ln}$$

Using Fig. 6.4, for $FFS$ = 88.5 km/h, note that the 1500.3-pc/h/ln flow rate intersects this curve in the LOS D density region. Therefore, this highway is operating at <u>LOS D</u>.

**EXAMPLE 6.5    MULTILANE HIGHWAY CAPACITY**

A local manufacturer wishes to open a factory near the segment of highway described in Example 6.4. How many large trucks can be added to the peak-hour directional volume before capacity is reached? (Add only trucks and assume that the $PHF$ remains constant.)

*SOLUTION*

Note that $FFS$ will remain unchanged at 88.5 km/h. Table 6.9 shows that capacity for $FFS =$ 88.5 km/h is 2100 pc/h/ln. The current number of large trucks and buses in the peak-hour traffic stream is 240 ($0.08 \times 3000$) and the current number of recreational vehicles is 60 ($0.02 \times 3000$). Let us denote the number of new trucks added as $V_{nt}$, so the combination of Eqs. 6.3 and 6.5 gives

$$v_p = \frac{V + V_{nt}}{(PHF)(N)\left[\cfrac{1}{1 + \left(\cfrac{240 + V_{nt}}{V + V_{nt}}\right)(E_T - 1) + \left(\cfrac{60}{V + V_{nt}}\right)(E_R - 1)}\right](f_p)}$$

*with*

$$\begin{aligned}
v_p &= 2100 \text{ pc/h/ln}, \\
V &= 3000 \text{ veh/h (Example 6.4)}, \\
PHF &= 0.80 \text{ (Example 6.4)}, \\
N &= 3 \text{ (Example 6.4)}, \\
f_p &= 0.95 \text{ (Example 6.4)}, \\
E_T &= 2.5 \text{ (Example 6.4), and} \\
E_R &= 2.0 \text{ (Example 6.4)}.
\end{aligned}$$

$$2100 = \frac{3000 + V_{nt}}{(0.80)(3)\left[\cfrac{1}{1 + \left(\cfrac{240 + V_{nt}}{3000 + V_{nt}}\right)(2.5 - 1) + \left(\cfrac{60}{3000 + V_{nt}}\right)(2 - 1)}\right](0.95)}$$

which gives $V_{nt} = 547$, which is the number of trucks that can be added to the peak-hour volume before capacity is reached.

## 6.6   TWO-LANE HIGHWAYS

Two-lane highways are defined as roadways with one lane available in each direction. For level-of-service determination, a key distinction between two-lane highways and the freeways and multilane highways previously discussed is that traffic in both directions must now be considered (previously we considered traffic in one direction only). This is because traffic in an opposing direction has a strong influence on level of service. For example, a high opposing traffic volume limits the opportunity to pass slow-moving vehicles (because such a pass requires the passing vehicle to occupy the opposing lane) and thus forces a lower traffic speed—and, as a consequence, a lower level of service. It also follows that any geometric features that restrict passing sight distance (such as sight distance on horizontal and vertical curves) will have an adverse impact on the level of service. Finally, the type of terrain (level, rolling, or mountainous) plays a more critical role in level-of-service calculations, relative to freeways and multilane highways, because of the sometimes limited ability to pass slower-moving vehicles on grades in areas where passing is prohibited due to sight distance restrictions or where opposing traffic does not permit safe passing. Two examples of two-lane highways are shown in Fig. 6.5.

The analysis procedure for two-lane highways in the *Highway Capacity Manual* [Transportation Research Board 2010] provides performance measure values and levels of service that are specific to only one direction of travel.

### 6.6.1   Base Conditions and Capacity

The base conditions for two-lane highways are defined as [Transportation Research Board 2010]

- 3.6-m minimum lane widths
- 1.8-m minimum shoulder widths
- 0% no-passing zones on the highway segment
- Only passenger cars in the traffic stream
- No direct access points along the roadway
- No impediments to through traffic due to traffic control or turning vehicles
- Level terrain (no grades greater than 2%)

The capacity of extended lengths of two-lane highway under base conditions is 1700 passenger cars per hour (pc/h) in one direction, or 3200 pc/h when both directions are considered. Because of interactions between the two directions of traffic flow, the maximum flow rate in the opposing direction is limited to 1500 pc/h when the other direction is at a flow rate of 1700 pc/h.

**Figure 6.5** Examples of two-lane highways.

### 6.6.2 Service Measures

Three service measures have been identified for two-lane highways: (1) percent time spent following, (2) average travel speed, and (3) percent of free-flow speed. Percent time spent following (*PTSF*) is the average percentage of travel time that vehicles must travel behind slower vehicles due to the lack of passing opportunities (because of geometry and/or opposing traffic). *PTSF* is difficult to measure in the field; thus, it is recommended that the percentage of vehicles traveling with headways less than 3 seconds at a representative location be used as a surrogate measure. *PTSF* is generally representative of a driver's freedom to maneuver in the traffic stream. Average travel speed (*ATS*) is simply the length of the analysis segment divided by the average travel time of all vehicles traversing the segment during the analysis period. *ATS* is an indicator of the mobility on a two-lane highway. Percent free-free flow speed (*PFFS*) is the average travel speed of the analysis segment divided by the free-flow speed of the analysis segment. *PFFS* is an indicator of how closely vehicles are able to travel to their desired speed.

The service measure, and corresponding thresholds, that govern the determination of level of service depends on the functional classification of the two-lane highway. The *Highway Capacity Manual* [Transportation Research Board 2010] has defined three classes of two-lane highway:

*Class I*: Two-lane highways on which motorists expect to travel at high speeds, as well as avoid extended following of other vehicles. Class I highways include inter-city routes, primary arterials connecting major traffic generators, daily commuter routes, and primary links in state or national highway networks.

*Class II*: Two-lane highways on which motorists do not necessarily expect to travel at high speeds. Shorter routes and routes that pass through rugged terrain, for which travel speeds will generally be lower than for Class I highways, are typically assigned to Class II. In these situations, motorists primarily want to avoid extended following of other vehicles.

*Class III*: Two-lane highways on which motorists do not expect frequent passing opportunities, or to travel at high speeds. Scenic routes, recreational routes, or routes that pass through moderately developed areas (small towns) are typically assigned to Class III. These routes generally have lower posted speed limits, and in these situations, motorists usually do not mind following other vehicles or traveling at slower speeds, as long they are able to travel at a speed close to the posted speed limit.

Note that the level-of-service criteria for two-lane highways are presented later (in Section 6.6.6).

### 6.6.3 Determine Free-Flow Speed

*FFS* for two-lane highways is the mean speed of all vehicles operating in flow rates up to 200 pc/h total for both directions. Free-flow speeds on two-lane highways typically range from 70 to 105 km/h. If field measurement of *FFS* cannot be made under conditions with a flow rate of 200 pc/h or less, an adjustment can be made with the following equation:

$$FFS = S_{FM} + 0.0125 \left( V_f / f_{HV} \right) \tag{6.9}$$

*where*

$FFS$ = estimated free-flow speed in km/h,
$S_{FM}$ = mean speed of traffic measured in the field in km/h,
$V_f$ = observed flow rate, in veh/h, for the period when field data were obtained, and
$f_{HV}$ = heavy-vehicle adjustment factor as determined by Eq. 6.5.

If *FFS* is to be estimated rather than measured in the field, the following equation can be used, which takes into account roadway characteristics of lane width, shoulder width, and access frequency:

$$FFS = BFFS - f_{LS} - f_A \tag{6.10}$$

*where*

$FFS$ = estimated free-flow speed in km/h,
$BFFS$ = estimated free-flow speed, in km/h, for base conditions,
$f_{LS}$ = adjustment for lane width and shoulder width in km/h, and
$f_A$ = adjustment for the number of access points along the roadway in km/h.

Specific guidance on choosing a value for *BFFS* is not offered, due to the wide range of speed conditions on two-lane highways and the influence of local and regional factors on driver-desired speeds. Speed data and local knowledge of operating conditions on similar facilities can be used in developing an estimate of *BFFS*. The following discussion describes how to determine the adjustment factor values.

## Lane Width and Shoulder Width Adjustment
The adjustment for lane widths and/or shoulder widths that are more restrictive than the base conditions is shown in Table 6.14.

## Access Frequency Adjustment
The adjustment for access frequency is the same as that for multilane highways, and is shown in Table 6.13. Note that the number of access points (unsignalized intersections, driveways) should be counted for both sides of the highway.

**Table 6.14** Adjustment for Lane Width and Shoulder Width

| Lane width (m) | Reduction in free-flow speed (km/h) Shoulder width (m) | | | |
| --- | --- | --- | --- | --- |
| | $\geq 0.0 < 0.6$ | $\geq 0.6 < 1.2$ | $\geq 1.2 < 1.8$ | $\geq 1.8$ |
| 2.7 < 3.0 | 10.3 | 7.7 | 5.6 | 3.5 |
| $\geq 3.0 < 3.3$ | 8.5 | 5.9 | 3.8 | 1.7 |
| $\geq 3.3 < 3.6$ | 7.5 | 4.9 | 2.8 | 0.7 |
| $\geq 3.6$ | 6.8 | 4.2 | 2.1 | 0.0 |

Reproduced with permission of the Transportation Research Board, *Highway Capacity Manual* 2010, Copyright, National Academy of Sciences, Washington, D.C. Exhibit 15-7, p. 15-15. Table has been converted to SI units for use in this edition. Values from the source material are in U.S. Customary units.

### 6.6.4 Determine Analysis Flow Rate

The hourly volume must be adjusted to account for the peak 15-minute flow rate, the terrain, and the presence of heavy vehicles in the traffic stream. The directional analysis flow rate is calculated with the following equation:

$$v_i = \frac{V_i}{PHF \times f_G \times f_{HV}} \tag{6.11}$$

*where*

$v_i$ = 15-min passenger car equivalent flow rate for direction $i$ (pc/h),

$V_i$ = hourly volume for direction $i$ (veh/h),

$i$ = "$d$" for analysis direction, "$o$" for opposing direction

$PHF$ = peak-hour factor,

$f_G$ = grade adjustment factor, and

$f_{HV}$ = heavy-vehicle adjustment factor.

Unlike the analysis flow rate equation (6.3) for freeways and multilane highways, Eq. 6.11 does not contain an adjustment factor for driver population. Although it is reasonable to assume that drivers familiar with the highway will use it more efficiently than recreational or other nonfamiliar users of the facility, studies have yet to identify a significant difference between the two driver populations [Transportation Research Board 2010].

The procedures for determination of $PHF$, $f_G$, and $f_{HV}$ adjustment factor values are described next.

### Peak-Hour Factor

*PHF* for two-lane highways is calculated in a manner consistent with that for freeways and multilane highways. The only distinction is that, because the two-lane highway analysis methodology considers both directions of traffic flow for the analysis of each travel direction, *PHF* should be calculated for both directions of traffic flow combined. While a *PHF* value could be calculated for each individual direction of traffic flow, this could result in an unreasonably high flow rate for the combination of each directional analysis flow rate since the two directions may not peak at the same time.

### Grade Adjustment Factor

The grade adjustment factor accounts for the effect of terrain on the traffic flow. For terrain generally classified as level or rolling, Table 6.15 shows values for the grade adjustment factor for average travel speed and percent time spent following.

### Heavy-Vehicle Adjustment Factor

Just as for freeways and multilane highways, the heavy-vehicle adjustment factor accounts for the effect on traffic flow due to the presence of trucks, buses, and recreational vehicles in the traffic stream. The passenger car equivalency (PCE) values, however, are different from those for freeway and multilane highway segments. The heavy-vehicle PCE values for level and rolling terrain for both *ATS* and *PTSF* are shown in Table 6.16. Two-lane highways in mountainous terrain must be analyzed as specific upgrades and/or downgrades. For details on the procedure used to evaluate two-lane highways on specific grades (for example, a 5% grade 1 km long), the reader is referred to the *Highway Capacity Manual* [Transportation Research Board 2010].

**Table 6.15** Grade Adjustment Factor for Average Travel Speed (*ATS*) and Percent Time Spent Following (*PTSF*)*

| Directional demand flow rate (veh/h) | Average travel speed | | Percent time spent following | |
|---|---|---|---|---|
| | Level terrain | Rolling terrain | Level terrain | Rolling terrain |
| ≤ 100 | 1.00 | 0.67 | 1.00 | 0.73 |
| 200 | 1.00 | 0.75 | 1.00 | 0.80 |
| 300 | 1.00 | 0.83 | 1.00 | 0.85 |
| 400 | 1.00 | 0.90 | 1.00 | 0.90 |
| 500 | 1.00 | 0.95 | 1.00 | 0.96 |
| 600 | 1.00 | 0.97 | 1.00 | 0.97 |
| 700 | 1.00 | 0.98 | 1.00 | 0.99 |
| 800 | 1.00 | 0.99 | 1.00 | 1.00 |
| ≥ 900 | 1.00 | 1.00 | 1.00 | 1.00 |

* Linear interpolation to the nearest 0.01 is recommended.

Reproduced with permission of the Transportation Research Board, *Highway Capacity Manual* 2010, Copyright, National Academy of Sciences, Washington, D.C. Exhibit 15-9, p. 15-17 and Exhibit 15-16, p. 15-23.

**Table 6.16** Passenger Car Equivalents for Heavy Vehicles for Average Travel Speed (*ATS*) and Percent Time Spent Following (*PTSF*)*

| Vehicle type | Directional demand flow rate (veh/h) | Average travel speed | | Percent time spent following | |
|---|---|---|---|---|---|
| | | Level terrain | Rolling terrain | Level terrain | Rolling terrain |
| Trucks and buses, $E_T$ | $\leq 100$ | 1.9 | 2.7 | 1.1 | 1.9 |
| | 200 | 1.5 | 2.3 | 1.1 | 1.8 |
| | 300 | 1.4 | 2.1 | 1.1 | 1.7 |
| | 400 | 1.3 | 2.0 | 1.1 | 1.6 |
| | 500 | 1.2 | 1.8 | 1.0 | 1.4 |
| | 600 | 1.1 | 1.7 | 1.0 | 1.2 |
| | 700 | 1.1 | 1.6 | 1.0 | 1.0 |
| | 800 | 1.1 | 1.4 | 1.0 | 1.0 |
| | $\geq 900$ | 1.0 | 1.3 | 1.0 | 1.0 |
| RVs, $E_R$ | All flows | 1.0 | 1.1 | 1.0 | 1.0 |

* Linear interpolation to the nearest 0.1 is recommended.

Reproduced with permission of the Transportation Research Board, *Highway Capacity Manual* 2010, Copyright, National Academy of Sciences, Washington, D.C. Exhibit 15-11, p. 15-18 and Exhibit 15-18, p. 15-24.

### 6.6.5   Calculate Service Measures

If the highway is Class I, both *ATS* and *PTSF* must be calculated. If the highway is Class II, only *PTSF* needs to be calculated. If the highway is Class III, only *ATS* needs to be calculated. The calculations for these three service measures are described next.

**Average Travel Speed**

The average travel speed depends on the free-flow speed, the analysis flow rate, and an adjustment factor for the percentage of no-passing zones, and is calculated according to Eq. 6.12:

$$ATS_d = FFS - 0.0125(v_d + v_o) - f_{np} \qquad \textbf{(6.12)}$$

*where*

$ATS_d$ = average travel speed in the analysis direction in km/h,

$FFS$ = free-flow speed in km/h, as measured in the field and possibly adjusted by Eq. 6.9 or estimated from Eq. 6.10,

$v_d$ = analysis flow rate for analysis direction in pc/h, as calculated from Eq. 6.11,

$v_o$ = analysis flow rate for opposing direction in pc/h, as calculated from Eq. 6.11, and

$f_{np}$ = adjustment factor for the percentage of no-passing zones, which is determined from Table 6.17.

**Table 6.17** Adjustment for No-Passing Zones on Average Travel Speed*

| Opposing flow rate, $v_o$ (pc/h) | No-passing zones (%) | | | | |
|---|---|---|---|---|---|
| | ≤20 | 40 | 60 | 80 | 100 |
| **FFS ≥ 104.6 km/h** | | | | | |
| ≤100 | 1.8 | 3.5 | 4.5 | 4.8 | 5.0 |
| 200 | 3.5 | 5.3 | 6.3 | 6.4 | 6.8 |
| 400 | 2.6 | 3.7 | 4.3 | 4.5 | 4.7 |
| 600 | 2.3 | 2.4 | 2.7 | 3.1 | 3.2 |
| 800 | 1.1 | 1.6 | 1.9 | 2.3 | 2.4 |
| 1000 | 1.0 | 1.3 | 1.8 | 1.8 | 1.9 |
| 1200 | 1.0 | 1.3 | 1.4 | 1.6 | 1.8 |
| 1400 | 1.0 | 1.1 | 1.4 | 1.4 | 1.4 |
| ≥1600 | 1.0 | 1.1 | 1.1 | 1.1 | 1.3 |
| **FFS = 96.5 km/h** | | | | | |
| ≤100 | 1.1 | 2.7 | 4.0 | 4.5 | 4.7 |
| 200 | 3.1 | 4.7 | 6.0 | 6.4 | 6.8 |
| 400 | 2.3 | 3.2 | 4.0 | 4.3 | 6.3 |
| 600 | 1.8 | 2.1 | 2.6 | 3.1 | 3.2 |
| 800 | 1.0 | 1.4 | 1.8 | 2.1 | 2.3 |
| 1000 | 1.0 | 1.1 | 1.4 | 1.8 | 1.9 |
| 1200 | 0.8 | 1.1 | 1.4 | 1.4 | 1.8 |
| 1400 | 0.8 | 1.0 | 1.3 | 1.3 | 1.4 |
| ≥1600 | 0.8 | 1.0 | 1.1 | 1.1 | 1.1 |
| **FFS = 88.5 km/h** | | | | | |
| ≤100 | 0.8 | 1.9 | 3.5 | 4.2 | 4.3 |
| 200 | 2.4 | 3.9 | 5.6 | 6.3 | 6.6 |
| 400 | 2.1 | 3.1 | 3.9 | 4.3 | 4.5 |
| 600 | 1.4 | 1.8 | 2.6 | 2.9 | 3.1 |
| 800 | 0.8 | 1.1 | 1.8 | 1.9 | 2.3 |
| 1000 | 0.8 | 1.0 | 1.3 | 1.4 | 1.8 |
| 1200 | 0.8 | 1.0 | 1.1 | 1.4 | 1.6 |
| 1400 | 0.8 | 1.0 | 1.1 | 1.1 | 1.4 |
| ≥1600 | 0.8 | 1.0 | 1.0 | 1.0 | 1.1 |

| Opposing flow rate, $v_o$ (pc/h) | No-passing zones (%) | | | | |
|---|---|---|---|---|---|
| | ≤20 | 40 | 60 | 80 | 100 |
| **FFS = 80.5 km/h** | | | | | |
| ≤100 | 0.3 | 1.1 | 3.1 | 3.9 | 4.0 |
| 200 | 1.9 | 3.2 | 5.3 | 6.3 | 6.4 |
| 400 | 1.8 | 2.6 | 3.5 | 4.2 | 4.3 |
| 600 | 1.0 | 1.4 | 2.3 | 2.7 | 3.1 |
| 800 | 0.6 | 1.0 | 1.4 | 1.9 | 2.1 |
| 1000 | 0.6 | 0.6 | 1.1 | 1.4 | 1.8 |
| 1200 | 0.6 | 0.6 | 1.1 | 1.3 | 1.6 |
| 1400 | 0.6 | 0.6 | 1.0 | 1.1 | 1.3 |
| ≥1600 | 0.6 | 0.6 | 0.8 | 0.8 | 0.8 |
| **FFS ≤ 72.4 km/h** | | | | | |
| ≤100 | 0.2 | 0.6 | 2.7 | 3.5 | 3.9 |
| 200 | 1.4 | 2.6 | 5.0 | 6.1 | 6.4 |
| 400 | 1.4 | 0.8 | 3.2 | 4.0 | 4.3 |
| 600 | 0.6 | 0.5 | 2.1 | 2.7 | 2.9 |
| 800 | 0.5 | 0.5 | 1.3 | 1.8 | 1.9 |
| 1000 | 0.5 | 0.5 | 1.0 | 1.3 | 1.8 |
| 1200 | 0.5 | 0.5 | 1.0 | 1.1 | 1.6 |
| 1400 | 0.5 | 0.5 | 1.0 | 1.0 | 1.1 |
| ≥1600 | 0.5 | 0.5 | 0.6 | 0.6 | 1.0 |

* Linear interpolation to the nearest 0.1 is recommended.

Reproduced with permission of the Transportation Research Board, *Highway Capacity Manual* 2010, Copyright, National Academy of Sciences, Washington, D.C. Adapted from Exhibit 15-15, p. 15-22. Table has been converted to SI units for use in this edition. Values from the source material are in U.S. Customary units.

### Percent Time Spent Following

The percent time spent following depends on the analysis flow rate and an adjustment for the combined effect of the percentage of no-passing zones and the directional distribution of traffic, and is calculated according to Eq. 6.13.

$$PTSF_d = BPTSF_d + f_{np}\left(\frac{v_d}{v_d + v_o}\right) \tag{6.13}$$

*where*

$PTSF_d$ = percent time spent following in the analysis direction,

$BPTSF_d$ = base percent time spent following in the analysis direction,

$f_{np}$ = adjustment factor for the percentage of no-passing zones, which is determined from Table 6.18, and

Other terms are as defined previously.

*BPTSF* is calculated according to Eq. 6.14:

$$BPTSF_d = 100\left[1 - \exp\left(av_d^b\right)\right] \tag{6.14}$$

where *a* and *b* are constants determined from Table 6.19.

### Percent Free-Flow Speed

The percent free-flow speed is calculated according to Eq. 6.15:

$$PFFS_d = \frac{ATS_d}{FFS} \tag{6.15}$$

*where*

$PFFS_d$ = percent free-flow speed in the analysis direction,

$ATS_d$ = average travel speed in the analysis direction in km/h, and

$FFS$ = free-flow speed in km/h, as measured in the field and possibly adjusted by Eq. 6.9 or estimated from Eq. 6.10.

**Table 6.18** Adjustment for No-Passing Zones on Percent Time Spent Following*

| Two-way flow rate, $v_d + v_o$ (pc/h) | No-passing zones (%) | | | | | |
|---|---|---|---|---|---|---|
| | 0 | 20 | 40 | 60 | 80 | 100 |
| *Directional split = 50/50* | | | | | | |
| ≤ 200 | 9.0 | 29.2 | 43.4 | 49.4 | 51.0 | 52.6 |
| 400 | 16.2 | 41.0 | 54.2 | 61.6 | 63.8 | 65.8 |
| 600 | 15.8 | 38.2 | 47.8 | 53.2 | 55.2 | 56.8 |
| 800 | 15.8 | 33.8 | 40.4 | 44.0 | 44.8 | 46.6 |
| 1400 | 12.8 | 20.0 | 23.8 | 26.2 | 27.4 | 28.6 |
| 2000 | 10.0 | 13.6 | 15.8 | 17.4 | 18.2 | 18.8 |
| 2600 | 5.5 | 7.7 | 8.7 | 9.5 | 10.1 | 10.3 |
| 3200 | 3.3 | 4.7 | 5.1 | 5.5 | 5.7 | 6.1 |
| *Directional split = 60/40* | | | | | | |
| ≤ 200 | 11.0 | 30.6 | 41.0 | 51.2 | 52.3 | 53.5 |
| 400 | 14.6 | 36.1 | 44.8 | 53.4 | 55.0 | 56.3 |
| 600 | 14.8 | 36.9 | 44.0 | 51.1 | 52.8 | 54.6 |
| 800 | 13.6 | 28.2 | 33.4 | 38.6 | 39.9 | 41.3 |
| 1400 | 11.8 | 18.9 | 22.1 | 25.4 | 26.4 | 27.3 |
| 2000 | 9.1 | 13.5 | 15.6 | 16.0 | 16.8 | 17.3 |
| 2600 | 5.9 | 7.7 | 8.6 | 9.6 | 10.0 | 10.2 |
| *Directional split = 70/30* | | | | | | |
| ≤ 200 | 9.9 | 28.1 | 38.0 | 47.8 | 48.5 | 49.0 |
| 400 | 10.6 | 30.3 | 38.6 | 46.7 | 47.7 | 48.8 |
| 600 | 10.9 | 30.9 | 37.5 | 43.9 | 45.4 | 47.0 |
| 800 | 10.3 | 23.6 | 28.4 | 33.3 | 34.5 | 35.5 |
| 1400 | 8.0 | 14.6 | 17.7 | 20.8 | 21.6 | 22.3 |
| 2000 | 7.3 | 9.7 | 15.7 | 13.3 | 14.0 | 14.5 |
| *Directional split = 80/20* | | | | | | |
| ≤ 200 | 8.9 | 27.1 | 37.1 | 47.0 | 47.4 | 47.9 |
| 400 | 6.6 | 26.1 | 34.5 | 42.7 | 43.5 | 44.1 |
| 600 | 4.0 | 24.5 | 31.3 | 38.1 | 39.1 | 40.0 |
| 800 | 4.8 | 18.5 | 23.5 | 28.4 | 29.1 | 29.9 |
| 1400 | 3.5 | 10.3 | 13.3 | 16.3 | 16.9 | 32.2 |
| 2000 | 3.5 | 7.0 | 8.5 | 10.1 | 10.4 | 10.7 |
| *Directional split = 90/10* | | | | | | |
| ≤ 200 | 4.6 | 24.1 | 33.6 | 43.1 | 43.4 | 43.6 |
| 400 | 0.0 | 20.2 | 28.3 | 36.3 | 36.7 | 37.0 |
| 600 | -3.1 | 16.8 | 23.5 | 30.1 | 30.6 | 31.1 |
| 800 | -2.8 | 10.5 | 15.2 | 19.9 | 20.3 | 20.8 |
| 1400 | -1.2 | 5.5 | 8.3 | 11.0 | 11.5 | 11.9 |

* Linear interpolation to the nearest 0.1 is recommended.

**Table 6.19**  PTSF Coefficients for Use in Eq. 6.14*

| Opposing Flow Rate, $v_o$ (pc/h) | Coefficient $a$ | Coefficient $b$ |
|---|---|---|
| ≤ 200 | −0.0014 | 0.973 |
| 400 | −0.0022 | 0.923 |
| 600 | −0.0033 | 0.870 |
| 800 | −0.0045 | 0.833 |
| 1000 | −0.0049 | 0.829 |
| 1200 | −0.0054 | 0.825 |
| 1400 | −0.0058 | 0.821 |
| ≥ 1600 | −0.0062 | 0.817 |

* Linea.r interpolation of $a$ to the nearest 0.0001 and $b$ to the nearest 0.001 is recommended.

Reproduced with permission of the Transportation Research Board, *Highway Capacity Manual* 2010, Copyright, National Academy of Sciences, Washington, D.C. Exhibit 15-20, p. 15-26.

### 6.6.6   Determine LOS

The first step in the LOS determination is to compare the analysis flow rate, $v_d$, to the directional capacity of 1700 pc/h. If $v_d$ exceeds 1700, the LOS is F, and the analysis ends. In this case, *PTSF* is virtually 100%, and speeds are highly variable and difficult to estimate. If the capacity in the analysis direction is not exceeded, then the combined demand flow rates $(v_d + v_o)$ for both directions must be checked against the two-way capacity of 3200 pc/h.  If the two-capacity is exceeded, refer to the *Highway Capacity Manual* [Transportation Research Board 2010] for further guidance on this situation.

If capacity is not exceeded, the calculated *PTSF*, *ATS*, and/or *PFFS* values are used with Table 6.20 to determine the LOS. For a particular LOS category to apply for Class I highways, the thresholds for both *PTSF* and *ATS* must be met. For example, for LOS B to apply, *PTSF* must be less than or equal to 50% and *ATS* must be greater than 80 km/h. If, for a particular two-lane highway, *PTSF* is 45% and *ATS* is 78 km/h, the LOS would be C.

**Table 6.20**  LOS Criteria for Two-Lane Highways

| LOS | Class I | | Class II | Class III |
|---|---|---|---|---|
| | Percent time spent following (*PTSF*) | Average travel speed (*ATS*) km/h | Percent time spent following (*PTSF*) | Percent free-flow speed (*PFFS*) |
| A | ≤ 35 | > 90 | ≤ 40 | > 91.7 |
| B | ≤ 50 | > 80 | ≤ 55 | > 83.3–91.7 |
| C | ≤ 65 | > 70 | ≤ 70 | > 75.0–83.3 |
| D | ≤ 80 | > 60 | ≤ 85 | > 66.7–75.0 |
| E | > 80 | ≤ 60 | > 85 | ≤ 66.7 |

*Note:* LOS F applies whenever the demand flow rate exceeds the segment capacity.

Reproduced with permission of the Transportation Research Board, *Highway Capacity Manual* 2010, Copyright, National Academy of Sciences, Washington, D.C. Adapted from Exhibit 15-3, p. 15-7. Table has been converted to SI units for use in this edition. Values from the source material are in U.S. Customary units

**EXAMPLE 6.6    TWO-LANE HIGHWAY ANALYSIS FLOW RATES**

One segment of a Class I two-lane highway is on rolling terrain and has an hourly volume of 1000 veh/h (total for both directions), a directional traffic split of 60/40, and $PHF = 0.92$, and the traffic stream contains 5% large trucks, 2% buses, and 6% recreational vehicles. For these conditions, determine the analysis direction [1] and opposing direction [2] flow rates for $ATS$ and $PTSF$.

**SOLUTION**

The first step is to calculate the hourly volume for each direction, which is accomplished by multiplying the two-way volume by the percentage of traffic traveling in each direction, as given by the directional split.

$$V_d = 1000 \times 0.60 = 600 \text{ veh/h}$$
$$V_o = 1000 \times 0.40 = 400 \text{ veh/h}$$

The next step is to calculate the flow rate, in veh/h, that will be used to determine the grade adjustment and PCE values. This is done by dividing the directional hourly volume by the $PHF$.

$$\frac{V_d}{PHF} = \frac{600}{0.92} \cong 652; \quad \frac{V_o}{PHF} = \frac{400}{0.92} \cong 435$$

The values for $ATS$ will be selected first. For rolling terrain,

$f_G$ = 0.98 [1]; 0.92 [2] (Table 6.15)
$E_T$ = 1.6 [1]; 1.9 [2] (Table 6.16)
$E_R$ = 1.1 [1]; 1.1 [2] (Table 6.16)

Substituting the PCE values into Eq. 6.5 gives

$$f_{HV[1]} = \frac{1}{1 + 0.07(1.6-1) + 0.06(1.1-1)} = 0.954$$

$$f_{HV[2]} = \frac{1}{1 + 0.07(1.9-1) + 0.06(1.1-1)} = 0.935$$

Substituting the $f_{HV}$ and $f_G$ values into Eq. 6.11 gives

$$v_d = \frac{600}{0.92 \times 0.98 \times 0.954} = 697.6 \rightarrow 698 \text{ pc/h}$$

$$v_o = \frac{400}{0.92 \times 0.92 \times 0.935} = 505.4 \rightarrow 506 \text{ pc/h}$$

Repeating this process for $PTSF$ results in the following values:

$f_G$ = 0.98 [1]; 0.92 [2] (Table 6.15)
$E_T$ = 1.1 [1]; 1.5 [2] (Table 6.16)
$E_R$ = 1.0 [1]; 1.0 [2] (Table 6.16)

$$f_{HV} = 0.993\ [1];\ 0.966\ [2]\ (Eq.\ 6.5)$$
$$v_d = \underline{671\ pc/h}\ (Eq.\ 6.11)$$
$$v_o = \underline{490\ pc/h}\ (Eq.\ 6.11)$$

## EXAMPLE 6.7  TWO-LANE HIGHWAY LOS

The two-lane highway segment in Example 6.6 has the following additional characteristics: 3.3-m lanes, 0.6-m shoulders, access frequency of six per kilometer, 50% no-passing zones, and a base *FFS* of 90 km/h. Using the analysis flow rates for *ATS* and *PTSF* from Example 6.6, determine the level of service for this two-lane highway segment.

## SOLUTION

We begin by checking whether the highway segment is over capacity. The flow rates for the analysis direction of 698 and 671 for *ATS* and *PTSF*, respectively, are both well below the directional capacity of 1700 pc/h. Furthermore, the combined flow rates $(v_d + v_o)$ of 1204 and 1161 for *ATS* and *PTSF*, respectively, are below the two-way capacity of 3200 pc/h.

Since the facility is not over capacity, we can proceed with the LOS determination. We first estimate the free-flow speed using Eq. 6.10.

$$FFS = BFFS - f_{LS} - f_A$$

*with*

$$BFFS = 90\ km/h\ (given),$$
$$f_{LS} = 4.9\ km/h\ (Table\ 6.14),\ and$$
$$f_A = 4.0\ km/h\ (Table\ 6.13).$$

Substituting these values into Eq. 6.10 gives

$$FFS = 90 - 4.9 - 4.0 = 81.1\ km/h$$

The average travel speed will be calculated first, using Eq. 6.12.

$$ATS = FFS - 0.0125\left(v_d + v_o\right) - f_{np}$$

*with*

$$FFS = 81.1\ km/h\ (from\ previous\ calculation),$$
$$v_d = 698\ pc/h\ (Example\ 6.6),$$
$$v_o = 506\ pc/h\ (Example\ 6.6),\ and$$
$$f_{np} = 2.44\ km/h\ (Table\ 6.17,\ by\ three\text{-}way\ linear\ interpolation\ for\ v_o,\ percent$$
$$no\text{-}passing\ zones,\ and\ free\text{-}flow\ speed).$$

Substituting these values into Eq. 6.12 gives

$$ATS_d = 81.1 - 0.0125\left(698 + 506\right) - 2.44 = 63.6\ km/h$$

The percent time spent following is calculated next, using Eq. 6.13.

$$PTSF_d = BPTSF_d + f_{np}\left(\frac{v_d}{v_d + v_o}\right)$$

*BPTSF* is calculated from Eq. 6.14.

$$BPTSF_d = 100\left[1 - \exp\left(av_d^b\right)\right]$$

*with*

$v_d$ = 698 pc/h (Example 6.6),

$a$ = −0.0028 (Table 6.19, by linear interpolation for $v_o$), and

$b$ = 0.8949 (Table 6.19, by linear interpolation for $v_o$).

Substituting these values into Eq. 6.14 gives

$$BPTSF_d = 100\left[1 - \exp\left(-0.0028\left(698\right)^{0.8949}\right)\right] = 62.5\%$$

$f_{np}$ is found to be 27.75% from Table 6.18, by linear interpolation for two-way flow rate and percent no-passing zones. Note that a three-way linear interpolation is also possible with this table if the directional split does not fall into one of the five predefined categories. Substituting these values into Eq. 6.13 gives

$$PTSF_d = 62.5 + 27.75\left(\frac{671}{671 + 490}\right) = 78.6\%$$

We now determine the LOS for the calculated *ATS* and *PTSF* values from Table 6.20. From this table, for a Class I highway, the <u>LOS is D</u>, as the level of service for both *PTSF* and *ATS* is D.

## 6.7 DESIGN TRAFFIC VOLUMES

In the preceding sections of this chapter, consideration was given to the determination of level of service, given some hourly volume. However, a procedure for selecting an appropriate hourly volume is needed to compute the level of service and to determine the number of lanes that need to be provided in a new roadway design to achieve some specified level of service. The selection of an appropriate hourly volume is complicated by two issues. First, there is considerable variability in traffic volume by time of day, day of week, time of year, and type of roadway. Figure 6.6 shows such variations in traffic volumes by hour of day and day of week for typical intra-city and inter-city routes. Figure 6.7 gives variations by time of year by comparing monthly percentages of the annual average daily traffic, AADT (in units of vehicles per day and computed as the total yearly traffic volume divided by the number of days in the year). The second concern is an outgrowth of the first: Given the temporal variability in traffic flow, what hourly volume should be used for design and/or analysis? To answer this question, consider the example diagram shown in Fig. 6.8. This figure plots hourly volume (as a percentage of AADT) against the cumulative number of hours that exceed this volume, per year. For example, the highest traffic flow in the year, on this sample roadway, would have an hourly volume of $0.148 \times$ AADT (a volume that is exceeded by zero other hours). Sixty hours in the year would have a volume that exceeds $0.11 \times$ AADT.

**Figure 6.6**   Examples of hourly and daily traffic variations for intra-city and inter-city routes.

**Figure 6.7** Example of monthly traffic volume variations for business and recreational access routes.

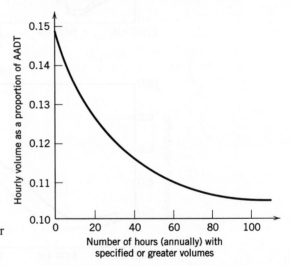

**Figure 6.8** Highest 100 hourly volumes over a one-year period for a typical roadway.

In determining the number of lanes that should be provided on a new or redesigned roadway, it is obvious that using the worst single hour in a year (the hour with the highest traffic flow, which would be $0.148 \times \text{AADT}$ from Fig. 6.8) would be a wasteful use of resources because additional lanes would be provided for a relatively rare occurrence. In contrast, if the 100th highest volume is used for design, the design level of service will be exceeded 100 times a year, which will result in considerable driver delay. Clearly, some compromise between the expense of providing additional capacity (such as additional lanes) and the expense of incurring additional driver delay must be made.

A common practice in the United States is to use a design hour-volume (DHV) that is between the 10th and 50th highest-volume hours of the year, depending on the type and location of the roadway (urban freeway, rural/suburban multilane highway, etc.), local traffic data, and engineering judgment. Perhaps the most common hourly volume used for roadway design is the 30th highest of the year. In practice, the *K*-factor is used to convert annual average daily traffic (AADT) to the 30th highest hourly volume. *K* is defined as

$$K = \frac{\text{DHV}}{\text{AADT}} \tag{6.16}$$

*where*

$K$ = factor used to convert annual average daily traffic to a specified annual hourly volume,

DHV = design hour-volume (typically, the 30th highest annual hourly volume), and

AADT = roadway's annual average daily traffic in veh/day.

For example, Fig. 6.8 shows that the *K*-value corresponding to the 30th highest hourly volume is 0.12. More generally, $K_i$ can be defined as the *K*-factor corresponding to the *i*th highest annual hourly volume. Again, for example, the 20th highest annual hourly volume would have a *K*-value, $K_{20}$, of 0.126, from Fig. 6.8. If *K* is not subscripted, the 30th highest annual hourly volume is assumed ($K = K_{30}$).

Finally, in the design and analysis of some highway types (such as freeways and multilane highways), the concern lies with directional traffic flows. Thus a factor is needed to reflect the proportion of peak-hour traffic volume traveling in the peak direction. This factor is denoted *D* and is used to arrive at the directional design-hour volume (DDHV) by application of

$$\text{DDHV} = K \times D \times \text{AADT} \tag{6.17}$$

*where*

DDHV = directional design-hour volume,

$D$ = directional distribution factor to reflect the proportion of peak-hour traffic volume traveling in the peak direction, and

Other terms are as defined previously.

---

**EXAMPLE 6.8    DETERMINATION OF REQUIRED NUMBER OF FREEWAY LANES**

A freeway is to be designed as a passenger-car–only facility for an AADT of 35,000 vehicles per day. It is estimated that the freeway will have a free-flow speed of 112.6 km/h. The design will be for commuters, and the peak-hour factor is estimated to be 0.85 with 65% of the peak-hour traffic traveling in the peak direction. Assuming that Fig. 6.8 applies, determine the number of lanes required to provide at least LOS C using the highest annual hourly volume and the 30th highest annual hourly volume.

**SOLUTION**

By inspection of Fig. 6.8, the highest annual hourly volume has $K_1 = 0.148$. Application of Eq. 6.16 gives

$$\text{DDHV} = K_1 \times D \times \text{AADT}$$
$$= 0.148 \times 0.65 \times 35,000 = 3367 \text{ veh/h}$$

The next step is to determine the maximum service flow rate that can be accommodated at LOS C for $FFS = 112.6$ km/h. From Table 6.1, we see that this value is 1735 pc/h/ln. Thus, we must provide enough lanes so that the per-lane traffic flow is less than or equal to this value.

We can use Eq. 6.3 to find $v_p$, based on an assumed number of lanes. Comparing the calculated value of $v_p$ to the maximum service flow rate of 1735 will determine whether we have an adequate number of lanes. Assuming a four-lane freeway (two lanes in each direction), Eq. 6.3 gives

$$v_p = \frac{3367}{0.85 \times 2 \times 1.0 \times 1.0} = 1980.6 \text{ pc/h/ln}$$

*with*

$$V = 3367 \text{ (DDHV from above)},$$
$$f_{HV} = 1.0 \text{ (no heavy vehicles), and}$$
$$f_p = 1.0 \text{ (commuters)}.$$

This value is higher than 1735, so we need to provide more lanes. The calculation is repeated, this time with an assumed six-lane freeway (three lanes each direction):

$$v_p = \frac{3367}{0.85 \times 3 \times 1.0 \times 1.0} = 1320.4 \text{ pc/h/ln}$$

Since this value is less than 1735, a six-lane freeway is necessary to provide LOS C operation for the design traffic flow rate.

For the 30th highest hourly annual volume, Fig. 6.8 gives $K_{30} = K = 0.12$, which when used in Eq. 6.16 gives

$$\text{DDHV} = K \times D \times \text{AADT}$$
$$= 0.12 \times 0.65 \times 35,000 = 2730 \text{ veh/h}$$

Again applying Eq. 6.3, with an assumed four-lane freeway, yields

$$v_p = \frac{2730}{0.85 \times 2 \times 1.0 \times 1.0} = 1605.9 \text{ pc/h/ln}$$

This value is less than 1735, so a <u>four-lane freeway</u> (two lanes each direction) is adequate for this design traffic flow rate.

This example demonstrates the impact of the chosen design traffic flow rate on roadway design. Only a four-lane freeway is necessary to provide LOS C for the 30th highest annual hourly volume, as opposed to a six-lane freeway needed to satisfy the level-of-service requirement for the highest annual hourly volume.

# NOMENCLATURE FOR CHAPTER 6

| | |
|---|---|
| AADT | annual average daily traffic |
| $ATS$ | average travel speed (two-lane highways) |
| $BFFS$ | estimated free-flow speed for base conditions |
| $c$ | roadway capacity |
| $D$ | density or factor for directional distribution of traffic |
| DHV | design-hour volume |
| DDHV | directional design-hour volume |
| $E_R$ | passenger car equivalents for recreational vehicles |
| $E_T$ | passenger car equivalents for large trucks and buses |
| FFS | measured or estimated free-flow speed |
| $f_A$ | free-flow speed adjustment factor for access point frequency (multilane and two-lane highways) |
| $f_{d/np}$ | adjustment factor for the combined effect of the directional distribution of traffic and the percentage of no-passing zones (two-lane highways) |
| $f_G$ | grade adjustment factor (two-lane highways) |
| $f_{HV}$ | heavy-vehicle adjustment factor |
| $f_{ID}$ | free-flow speed adjustment factor for interchange density (freeways) |
| $f_{LC}$ | free-flow speed adjustment factor for lateral clearance (freeways and multilane highways) |

| | |
|---|---|
| $f_{LS}$ | free-flow speed adjustment factor for lane and shoulder width(s) (two-lane highways) |
| $f_{LW}$ | free-flow speed adjustment factor for lane width (freeways and multilane highways) |
| $f_M$ | free-flow speed adjustment factor for median type (multilane highways) |
| $f_{np}$ | adjustment factor for the percentage of no-passing zones (two-lane highways) |
| $f_p$ | driver population adjustment factor (freeways and multilane highways) |
| $K_i$ | factor used to convert AADT to $i$th highest annual hourly volume |
| $LC_L$ | left-side lateral clearance (multilane highways) |
| $LC_R$ | right-side lateral clearance (multilane highways) |
| $N$ | number of lanes in one direction |
| PHF | peak-hour factor |
| PTSF | percent time spent following (two-lane highways) |
| PFFS | percent free-flow speed |
| $S_{FM}$ | mean speed of traffic measured in the field (two-lane highways) |
| TLC | total lateral clearance (multilane highways) |
| $v$ | analysis flow rate |
| $V$ | hourly volume |
| $V_{15}$ | highest 15-minute volume |
| $v/c$ | volume-to-capacity ratio |

# REFERENCES

Transportation Research Board. *Highway Capacity Manual*. Washington, DC: National Research Council, 2010.

# PROBLEMS

**6.1** A six-lane freeway (three lanes in each direction) has regular weekday users and currently operates at maximum LOS C conditions. The lanes are 3.3 m wide, the right-side shoulder is 1.2 m wide, and there are two ramps within five kilometers upstream of the segment midpoint and one ramp within five kilometers downstream of the segment midpoint. The highway is on rolling terrain with 10% large trucks and buses (no recreational vehicles), and the peak-hour factor is 0.90. Determine the hourly volume for these conditions.

**6.2** Consider the freeway in Problem 6.1. At one point along this freeway there is a 4% upgrade with a directional hourly traffic volume of 5435 vehicles. If

all other conditions are as described in Problem 6.1, how long can this grade be without the freeway LOS dropping to F?

**6.3** A four-lane freeway (two lanes in each direction) is located on rolling terrain and has 3.6-m lanes, no lateral obstructions within 1.8 m of the pavement edges, and there are two ramps within five kilometers upstream of the segment midpoint and three ramps within five kilometers downstream of the segment midpoint. The traffic stream consists of cars, buses, and large trucks (no recreational vehicles). A weekday directional peak-hour volume of 1800 vehicles (familiar users) is observed, with 700 arriving in the most congested 15-min period. If a level of service no worse than C is desired, determine the maximum number of large trucks and buses that can be present in the peak-hour traffic stream.

**6.4** Consider the freeway and traffic conditions described in Problem 6.3. If 180 of the 1800 vehicles observed in the peak hour were large trucks and buses, what would the level of service of this freeway be on a 5-mi, 6% downgrade?

**6.5** A six-lane freeway (three lanes in each direction) in a scenic area has a measured free-flow speed of 88.5 km/h. The peak-hour factor is 0.80, and there are 8% large trucks and buses and 6% recreational vehicles in the traffic stream. One upgrade is 5% and 0.8 km long. An analyst has determined that the freeway is operating at capacity on this upgrade during the peak hour. If the peak-hour traffic volume is 3900 vehicles, what value for the driver population factor was used?

**6.6** A freeway is being designed for a location in mountainous terrain. The expected free-flow speed is 88.5 km/h. Lane widths will be 3.6 m and shoulder widths will be 1.8 m. During the peak hour, it is expected there will be a directional peak-hour volume of 2700 vehicles, 12% large trucks and buses and 6% recreational vehicles. The *PHF* and driver population adjustment are expected to be 0.88 and 0.90, respectively. If a level of service no worse than D is desired, determine the necessary number of lanes.

**6.7** A segment of four-lane freeway (two lanes in each direction) has a 3% upgrade that is 450 m long followed by a 300-m 4% upgrade. It has 3.6-m lanes and 0.9-m shoulders. The directional hourly traffic flow is 2000 vehicles with 5% large trucks and buses (no recreational vehicles). The total ramp density for this freeway segment is 1.45 ramps per kilometer. If the peak-hour factor is 0.90 and all of the drivers are regular users, what is the level of service of this compound-grade freeway segment?

**6.8** Consider Example 6.2, in which it was determined that 1908 vehicles could be added to the peak hour before capacity is reached. Assuming rolling terrain as in Example 6.1, how many passenger cars could be added to the original traffic mix before peak-hour capacity is reached? (Assume only passenger cars are added and that the number of large trucks and buses originally in the traffic stream remains constant.)

**6.9** An eight-lane freeway (four lanes in each direction) is on rolling terrain and has 3.3-m lanes with a 1.2-m right-side shoulder. The total ramp density is 0.9 ramps per kilometer. The directional

peak-hour traffic volume is 5400 vehicles with 6% large trucks and 5% buses (no recreational vehicles). The traffic stream consists of regular users and the peak-hour factor is 0.95. It has been decided that large trucks will be banned from the freeway during the peak hour. What will the freeway's density and level of service be before and after the ban? (Assume that the trucks are removed and all other traffic attributes are unchanged.)

**6.10** A 5% upgrade on a six-lane freeway (three lanes in each direction) is 0.8 km long. On this segment of freeway, the directional peak-hour volume is 3800 vehicles with 2% large trucks and 4% buses (no recreational vehicles), the peak-hour factor is 0.90, and all drivers are regular users. The lanes are 3.6-m wide, there are no lateral obstructions within 3 m of the roadway, and the total ramp density is 0.6 ramps per kilometer. A bus strike will eliminate all bus traffic, but it is estimated that for each bus removed from the roadway, seven additional passenger cars will be added as travelers seek other means of travel. What are the density, volume-to-capacity ratio, and level of service of the upgrade segment before and after the bus strike?

**6.11** A multilane highway (two lanes in each direction) is on level terrain. The free-flow speed has been measured at 72.4 km/h. The peak-hour directional traffic flow is 1300 vehicles with 6% large trucks and buses and 2% recreational vehicles ($f_p = 0.95$). If the peak-hour factor is 0.85, determine the highway's level of service.

**6.12** Consider the multilane highway in Problem 6.11. If the proportion of vehicle types and peak-hour factor remain constant, how many vehicles can be added to the directional traffic flow before capacity is reached?

**6.13** A six-lane multilane highway (three lanes in each direction) has a peak-hour factor of 0.90, 3.3-m lanes with a 1.2-m right-side shoulder, and a two-way left-turn lane in the median. The directional peak-hour traffic flow is 4000 vehicles with 8% large trucks and buses and 2% recreational vehicles. The driver population factor has been estimated at 0.95. What will the level of service of this highway be on a 4% upgrade that is 2.4-km long if the speed limit is 90 km/h and there are 9 access points per kilometer?

**6.14** A divided multilane highway in a recreational area ($f_p = 0.90$) has four lanes (two lanes in each direction) and is on rolling terrain. The highway has 3-m lanes with a 1.8-m right-side shoulder and a 0.9-m left-side shoulder. The posted speed is 80 km/h.

Previously there were 2.5 access points per kilometer, but recent development has increased the number of access points to 7.5 per kilometer. Before development, the peak-hour factor was 0.95 and the directional hourly volume was 2200 vehicles with 10% large trucks and buses and 3% recreational vehicles. After development, the peak-hour directional flow is 2600 vehicles with the same vehicle percentages and peak-hour factor. What is the level of service before and after the development?

**6.15** A multilane highway has four lanes (two lanes in each direction) and a measured free-flow speed of 88.5 km/h. One upgrade is 5% and is 1.0 km long. Currently trucks are not permitted on the highway, but there are 2% buses (no recreational vehicles) in the directional peak-hour volume of 1900 vehicles (the peak-hour factor is 0.80). Local authorities are considering allowing trucks on this upgrade. If this is done, they estimate that 150 large trucks will use the highway during the peak hour. What would be the level of service before and after the trucks are allowed (assuming the driver population adjustment to be 1.0 before and 0.97 after)?

**6.16** A four-lane undivided multilane highway (two lanes in each direction) has 3.3-m lanes, 1.2-m shoulders, and 6 access points per kilometer. It is determined that the roadway currently operates at capacity with $PHF = 0.80$ and a driver population adjustment of 0.9. If the highway is on level terrain with 8% large trucks and buses (no recreational vehicles) and the speed limit is 90 km/h, what is the directional hourly volume?

**6.17** A new four-lane divided multilane highway (two lanes in each direction) is being planned with 3.6-m lanes, 1.8-m shoulders on both sides, and an 80-km/h speed limit. One 3% downgrade is 7.2 km long, and there will be 2.5 access points per kilometer. The peak-hour directional volume along this grade is estimated to consist of 1800 passenger cars, 140 large trucks, 40 buses, and 10 recreational vehicles. If the peak-hour factor is estimated to be 0.85 and the driver population adjustment factor is expected to be 1.0, what level of service will this segment of highway operate under?

**6.18** A six-lane divided multilane highway (three lanes in each direction) has a measured free-flow speed of 80.5 km/h. It is on mountainous terrain with a traffic stream consisting of 7% large trucks and buses and 3% recreational vehicles. The driver population adjustment is 0.92. One direction of the highway currently operates at maximum LOS C

conditions, and it is known that the highway has $PHF = 0.90$. How many vehicles can be added to this highway before capacity is reached, assuming the proportion of vehicle types remains the same but the peak-hour factor increases to 0.95?

**6.19** A four-lane undivided multilane highway (two lanes in each direction) has 3.3-m lanes and 1.5-m shoulders. At one point along the highway there is a 4% upgrade that is 1.0 km long. There are 15 access points along this grade. The peak-hour traffic volume has 2500 passenger cars and 200 trucks and buses (no recreational vehicles), and 720 of these vehicles arrive in the most congested 15-min period. This traffic stream is primarily commuters. The measured free-flow speed is 88.5 km/h. To improve the level of service, the local transportation agency is considering reducing the number of access points by blocking some driveways and rerouting their traffic. How many of the 15 access points must be blocked to achieve LOS C?

**6.20** A Class I two-lane highway is on level terrain, has a measured free-flow speed of 104.6 km/h, and has 50% no-passing zones. During the peak hour, the analysis direction flow rate is 182 veh/h, the opposing direction flow rate is 78 veh/h, and the $PHF = 0.90$. There are 15% large trucks and buses (no RVs). Determine the level of service.

**6.21** A Class I two-lane highway is on rolling terrain and the free-flow speed was measured at 90 km/h, but this was during a two-way flow rate of 275 veh/h. There are 80% no-passing zones. During the peak hour, the analysis direction flow rate is 324 veh/h, the opposing direction flow rate is 216 veh/h, and the $PHF = 0.87$. There are 5% large trucks and buses and 10% recreational vehicles. Determine the level of service.

**6.22** A Class I two-lane highway is on level terrain with passing permitted throughout. The highway has 3.3-m lanes with 1.2-m shoulders. There are 10 access points per kilometer. The base $FFS$ is 96.5 km/h. During the peak hour, 440 vehicles are traveling in the analysis direction and 360 vehicles are traveling in the opposing direction. If the $PHF$ is 0.85 and there are 4% large trucks, 3% buses, and 2% recreational vehicles, what is the level of service?

**6.23** A Class II two-lane highway needs to be redesigned for an area with rolling terrain. During the peak hour, 380 vehicles are traveling in the analysis direction and 300 vehicles are traveling in the opposing direction. The $PHF$ is 0.92. The traffic stream includes 8% large trucks, 2% buses, and no recreational vehicles. What is the maximum percentage of no-passing zones that can be built into the design with LOS C maintained?

**6.24** A Class III two-lane highway is on level terrain, has a measured free-flow speed of 72.4 km/h, and has 100% no-passing zones. During the peak hour, the analysis direction flow rate is 150 veh/h, the opposing direction flow rate is 100 veh/h, and the *PHF* = 0.95. There are 5% large trucks and 10% recreational vehicles. Determine the level of service.

**6.25** A four-lane freeway (two lanes in each direction) segment consists of passenger cars only, a driving population of regular users, a peak-hour directional distribution of 0.70, a peak-hour factor of 0.80, and a measured free-flow speed of 112.6 km/h. Assuming Fig. 6.8 applies, if the AADT is 30,000 veh/day, determine the level of service for the 10th, 50th, and 100th highest annual hourly volumes.

**6.26** A four-lane freeway (two lanes in each direction) operates at capacity during the peak hour. It has 3.3-m lanes, 1.2-m shoulders, and there are three ramps within five kilometers upstream of the segment midpoint and four ramps within five kilometers downstream of the segment midpoint. The freeway has only regular users, there are 8% large trucks and buses (no recreational vehicles), and it is on rolling terrain with a peak-hour factor of 0.85. It is known that 12% of the AADT occurs in the peak hour and that the directional factor is 0.6. What is the freeway's AADT?

**6.27** A six-lane multilane highway (three lanes in each direction) has regular weekday users and currently operates at maximum LOS C conditions. The measured free-flow speed is 88.5 km/h. The highway is on rolling terrain with 12% large trucks and buses and 6% recreational vehicles, and the peak-hour factor is 0.92. If 17% of all directional traffic occurs during the peak hour, determine the total directional traffic volume.

# Chapter 7

---

# Traffic Control and Analysis at Signalized Intersections

## 7.1  INTRODUCTION

Due to conflicting traffic movements, roadway intersections are a source of great concern to traffic engineers. Intersections can be a major source of crashes and vehicle delays (as vehicles yield to avoid conflicts with other vehicles). Most roadway intersections are not signalized due to low traffic volumes and adequate sight distances. However, at some point, traffic volumes and accident frequency/severity (and other factors) reach a level that warrants the installation of a traffic signal.

The installation and operation of a traffic signal to control conflicting traffic and pedestrian movements at an intersection has advantages and disadvantages. The advantages include a potential reduction in some types of crashes (particularly angle crashes), provision for pedestrians to cross the street, provision for side-street vehicles to enter the traffic stream, provision for the progressive flow of traffic in a signal-system corridor, possible improvements in capacity, and possible reductions in delays. However, signals are by no means the perfect solution for delay or accident problems at an intersection. A poorly timed signal or one that is not justified can have a negative impact on the operation of the intersection by increasing vehicle delay, increasing the rate of vehicle accidents (particularly rear-end accidents), causing a disruption in traffic progression (adversely impacting the through movement of traffic), and encouraging the use of routes not intended for through traffic (such as routes through residential neighborhoods). Traffic signals are also costly to install, with some basic signal installations costing in excess of $100,000. Therefore, the decision to install a signal must be weighed and studied carefully. To assist transportation engineers in this process, the Federal Highway Administration of the U.S. Department of Transportation publishes the *Manual on Uniform Traffic Control Devices* (MUTCD) [U.S. Federal Highway Administration 2009], which contains a section on warrants for the installation of a traffic signal. There are a total of nine warrants, which include consideration of vehicle volumes, pedestrian volumes, school crossings, signal coordination, and crash experience. The reader is referred to the MUTCD for details on these warrants.

Unlike uninterrupted flow, in which vehicle movement is affected only by other vehicles and the roadway environment, the introduction of a traffic control device such as a signal exerts a significant influence on the flow of vehicles. Thus, the analysis of traffic flow at signalized intersections can become very complex. This chapter will make several simplifying assumptions to keep the material at an accessible level.

The chapter begins by providing an overview of the physical elements of intersection configuration and traffic signal control. A basic understanding of these principles provides the foundation for designing intersection geometry and traffic movement sequence plans. This is followed by a presentation of concepts, definitions, and analytical techniques that are used in the design and analysis of signal timing plans at signalized intersections.

## 7.2    INTERSECTION AND SIGNAL CONTROL CHARACTERISTICS

An intersection is defined as an at-grade crossing of two or more roadways. For analysis, the roadways entering the intersection are segmented into approaches, which are defined by lane groups (groups of one or more lanes). These lane groups are usually based on the allowed movements (left, through, right) within each lane and the sequencing of allowed movements by the traffic signal. The establishment of lane groups will be discussed in more detail in Section 7.4.2.

To illustrate these concepts, note that approach 1 of the intersection depicted in Fig. 7.1 consists of a lane for the exclusive use of left turns, a lane for the exclusive use of right turns, and two lanes for the exclusive use of through movements. Approach 3 is similar to approach 1 but does not include an exclusive right-turn lane; instead, the right turns share the outside lane with the through movements. Because the lanes for the exclusive use of left and right turns are short, they are usually referred to as bays and are intended to hold a limited number of queued vehicles. Queuing analysis can be used to determine the length of bay necessary to prevent queued turning vehicles from overflowing the bay and blocking the through lanes (known as spillover) and/or the length necessary to prevent queued through vehicles from blocking the entrance of the turn bay (known as spillback). Approach 2 consists of a shared through/right-turn lane and an exclusive left-turn lane (not a bay in this case because it has the same length as the adjacent lane). Approach 4 is similar to approach 2, but the inside lane is a shared through/left-turn lane.

From a driver's perspective, a traffic signal is just a collection of light-emitting devices [usually incandescent bulbs or light-emitting diodes (LEDs)] and lenses that are housed in casings of various configurations (referred to as signal heads) whose purpose is to display red, yellow, and green full circles and/or arrows. Figure 7.2 shows typical configurations of signal heads in the United States. These signal heads are usually mounted to mast arms or wire spanned across the intersection.

The following terminology is commonly used in the design of traffic signal controls.

**Indication.** The illumination of one or more signal lenses (greens, yellows, reds) indicating an allowed or prohibited traffic movement.

**Interval.** A period of time during which all signal indications (greens, yellows, reds) remain the same for all approaches.

**Figure 7.1** Typical signalized intersection elements.

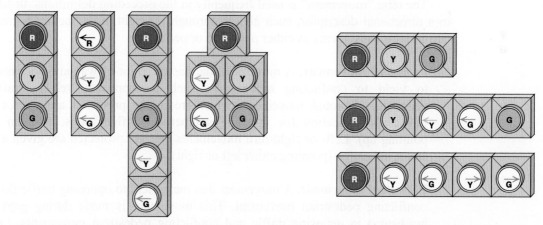

**Figure 7.2** Typical signal head configurations in the United States (G = green; Y = yellow; R = red).

**Cycle.** One complete sequence (for all approaches) of signal indications (greens, yellows, reds).

**Cycle length.** The total time for the signal to complete one cycle (given the symbol C and usually expressed in seconds).

**Green time.** The amount of time within a cycle for which a movement or combination of movements receives a green indication (the illumination of a signal lens). This is expressed in seconds and given the symbol $G$.

**Yellow time**. The amount of time within a cycle for which a movement or combination of movements receives a yellow indication. This is expressed in seconds and given the symbol $Y$. This time is referred to as the change interval, as it alerts drivers that the signal indication is about to change from green to red.

**Red time.** The amount of time within a cycle for which a movement or combination of movements receives a red indication. This is expressed in seconds and given the symbol $R$.

**All-red time.** The time within a cycle in which all approaches have a red indication (expressed in seconds and given the symbol $AR$). This time is referred to as the clearance interval, because it allows vehicles that might have entered at the end of the yellow interval to clear the intersection before the green phase starts for the next conflicting movement(s). This type of interval is becoming increasingly common for safety reasons because the rate of vehicles entering at the end of the yellow and beginning of the red indication has steadily increased in recent years.

**Phase.** The sum of the displayed green, yellow, and red times for a movement or combination of movements that receive the right of way simultaneously during the cycle. The sum of the phase lengths (in seconds) is the cycle length.

The term "movement" is used frequently in the preceding definitions. In addition to a directional descriptor, such as left, through, or right, a distinction is made by categorizing movements as either protected or permitted.

**Protected movement.** A movement that has the right-of-way and does not need to yield to conflicting movements, such as opposing vehicle traffic or pedestrians. Through movements, which are always protected, are given a green full circle indication (or in some geometric configurations, a green arrow pointing up). Left- or right-turn movements that are protected are given a green arrow indication (pointing either left or right).

**Permitted movement.** A movement that must yield to opposing traffic flow or a conflicting pedestrian movement. This movement is made during gaps (time headways) in opposing traffic and conflicting pedestrian movements. Left- or right-turn movements with a green full circle indication are permitted movements. Left-turning vehicles in this situation must wait for gaps in the opposing through and right-turning traffic before making their turns. Right-turning vehicles must yield to pedestrians in the adjacent crosswalk before making their turns.

To understand how these control characteristics are implemented, it is useful to analyze the physical implementation of these concepts. The display of the various signal indications (green, yellow, red, protected, permitted) at an intersection is handled by a signal controller (which is typically located in a cabinet next to the intersection). Modern signal controllers are sophisticated pieces of electronic equipment. These controllers, when combined with a method of vehicle detection, offer great flexibility in controlling phase duration and sequence. Traffic signal controllers are designed to operate in one or more of the following modes: pretimed, semi-actuated, or fully actuated.

**Pretimed.** A signal whose timing (cycle length, green time, etc.) is fixed over specified time periods and does not change in response to changes in traffic flow at the intersection. No vehicle detection is necessary with this mode of operation.

**Semi-actuated.** A signal whose timing (cycle length, green time, etc.) is affected when vehicles are detected (by video, pavement-embedded inductance loop detectors, etc.) on some, but not all, approaches. This mode of operation is usually found where a low-volume road intersects a high-volume road, often referred to as the minor and major streets, respectively. In such cases, green time is allocated to the major street until vehicles are detected on the minor street; then the green indication is briefly allocated to the minor street and then returned to the major street.

**Fully actuated.** A signal whose timing (cycle length, green time, etc.) is completely influenced by the traffic volumes, when detected, on all of the approaches. Fully actuated signals are most commonly used at intersections of two major streets and where substantial variations exist in all approach traffic volumes over the course of a day.

## 7.2.1   Actuated Control

Although pretimed signal control does not require the expense of vehicle detection hardware, it results in signal timing that it is not responsive to real-time traffic demands. The fixed-time values of the pretimed signal are based on expected traffic demands during the time period of interest. However, traffic arrivals can vary significantly from one cycle to the next, as described in Chapter 5. Thus, with fixed timing, a phase may provide excessive green time one cycle (which results in extra delay for the vehicles that move in other phases) and not enough time in another cycle.

Improvements in signal-controller technology over the last 20 years have made possible significant advances in traffic control at signalized intersections. Modern signal controllers are able to accept inputs on traffic demands and utilize this information to adjust the green interval duration and phasing sequence from one cycle to the next.

Actuated control's ability to respond to changes in traffic demands allows the green time to be reduced for a movement when the arrival rate is lower than normal, the green time to be extended when the arrival rate is higher than normal, or a phase to be skipped altogether if no demand is present for that movement. Thus, an intersection operating under actuated control will almost always result in lower overall delays than one operating under pretimed control.

### Vehicle Detection

Actuated control requires vehicle detection technology. The most common form of vehicle detection technology is the inductance loop detector (ILD). The ILD, a simple technology that has been in use for several decades now, consists of a loop (or coil) of wire embedded in the pavement through which an electrical current is circulated. This current is monitored by a device that interfaces with the signal controller, and when a vehicle passes over the ILD, the inductance level of the current is altered. When this change in the inductance level is detected by the monitoring device, it sends an input to the signal controller to indicate the presence of a vehicle.

ILDs can take on a variety of shapes, with different shapes having different advantages/disadvantages for detection ability. Furthermore, the sensitivity of the detectors can be tuned. The challenge in tuning the sensitivity is to find the level that allows detection of smaller vehicles (such as motorcycles or bicycles), yet is not so sensitive as to detect objects that are not vehicles.

The limitation of ILD technology is that vehicles can be detected only where the ILDs are placed. This limits the control options because it is prohibitively expensive to implement a large number of ILDs on an intersection approach.

A newer vehicle detection technology that is increasing in popularity at signalized intersections is video imaging processing (VIP). This technology has three main components: video camera, video digitization and processing unit, and computer. With VIP technology, virtual (software-based) detectors can be placed anywhere within a video camera's field of view. The video processing unit converts the video from the camera into a digital format, and a software algorithm processes the combination of the digitized field of view and virtual detectors to determine vehicle presence, as well as additional measures that cannot be obtained with ILDs (such as queue length). With the additional measures that can be obtained from VIP systems, a greater number of control strategies are possible. Thus, a greater level of traffic responsiveness can be obtained.

### Typical Phase Operation

With the most basic configuration for vehicle detection, signal phases operating under actuated control consist of the following phase periods.

**Initial Green**. This period provides a practical minimum amount of green time to the traffic movement.

**Extended Green**. After the initial green time expires, the phase enters an extension mode. This extension mode provides a continuation of green time as long as a vehicle arrives (at the detector) within a specified amount of time after the arrival of the previous vehicle.

A maximum green time is also specified for the phase. Thus, even if the arrival rate of vehicles is high enough to continue the extension period indefinitely, the phase will eventually be terminated so that other traffic movements can be served. If the phase terminates in this manner, it is referred to as a max-out condition. If the phase terminates prior to reaching the maximum green time, due to the time gap in successive vehicle arrivals exceeding the extension interval (the time allowed between successive vehicle arrivals before it is assumed no more vehicles are arriving), this is referred to as a gap-out condition.

Initial green periods are intended to serve all, or most, of the vehicle queue that develops during that traffic movement's red period. Vehicles that join the queue at the start of green, or arrive just after queue clearance, are usually handled during the extension period. Therefore, a large enough green time should be specified so that the regularly expected initial queue length can be served, but not too large that excessive green time is wasted during cycles with much lower than average vehicle arrivals. Additionally, the green time should be at least several seconds greater than the lost time for a phase. Typical minimum green times are on the order of 10 seconds.

The time allowed between successive vehicle arrivals at vehicle detectors before it is assumed that no more vehicles are arriving (the extension interval) ranges from 2 to 4 seconds. Overall phase operations are quite sensitive to the selected value of this extension interval (also commonly referred to as the unit extension). A smaller value, such as 2 seconds, will generally result in a "snappy" operation; that is, the phase will tend to gap out quickly after the initial queue is served. In practice, a small extension interval can often result in a phase gapping out prematurely due to a driver in the queue hesitating (from being distracted or inattentive) in starting up from green, thus increasing delays and driver frustration. A larger extension interval, such as 4 seconds, will generally result in "sluggish" operation and a greater likelihood of a max-out. Although the longer extension interval decreases the likelihood of premature gap-out, it can also lead to frustration for the drivers of other movements as they may perceive that the green time for the active phase is lasting an unnecessarily long time.

The basic sequence of events for an actuated phase is as follows: The initial green period is provided; the phase enters the extension period; the extension period continues until the phase either gaps out or maxes out; the yellow and all-red intervals commence and then the phase terminates. This sequence is illustrated in Fig. 7.3. This is the most basic operation of an actuated phase. Many more elements can be incorporated, depending on the level of traffic responsiveness desired, the level of coordination with other signalized intersections, and the level of vehicle detection.

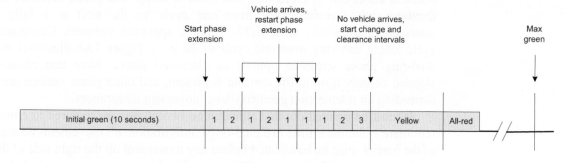

**Figure 7.3** Basic operation of an actuated signal phase.

For a more complete description of actuated control, the reader is referred to the *Manual of Traffic Signal Design* [Kell and Fullerton 1998] and the *Highway Capacity Manual* [Transportation Research Board 2010].

## 7.2.2    Signal Controller Operation

Most modern signal controllers are designed to operate in what is termed a dual-ring configuration. This configuration allows maximum flexibility for controlling phase duration and sequencing, which is necessary for operating in a fully actuated mode. The dual-ring concept can be best explained with a graphical illustration (see Fig. 7.4a).

In this figure, the three-letter notation in each numbered box refers to direction and movement type. For example, WBL means westbound left and NBT means northbound through. At an intersection with four approaches and separate/exclusive left-turn phases for each approach, a total of eight separate movements are possible. The dual-ring terminology comes from four movements (1–4) being represented on the top ring, and four movements (5–8) being represented on the bottom ring. The logic behind this configuration is straightforward. Any movement in ring 1 can occur simultaneously with any movement in ring 2 as long as both are on the same side of the barrier (a term used figuratively to separate conflicting traffic movements). For example, the first phase at an intersection can consist of opposing WB and EB left-turn movements (1 and 5). However, if no vehicles are detected in the EB left-turn lane, this movement can be skipped and the first phase will consist of movement 1 (WBL) and movement 6 (WBT). If no left turns are detected for both the WB and EB directions, then the first phase will consist of movements 2 (EBT) and 6 (WBT). The same applies to movements on the right side of the barrier (3, 4, 7, and 8).

Furthermore, if the volume of one of the movements during a phase subsides sooner than the volumes for the other movement(s), the green time for this movement can be terminated and another movement can be initiated, according to the previously described logic. For example, suppose phase 1 consists of movements 1 and 5, but the volume for movement 1 is considerably larger than the volume for movement 5. If both movements 1 and 5 received enough green time to satisfy the vehicle demand for movement 1, this would result in wasted green time for movement 5. With the dual-ring configuration, movement 5 can be terminated before movement 1, and movement 6 can be initiated while movement 1 continues. This results in a more efficient allocation of green time and reduced delay. The phase sequence and phase durations can therefore vary from one cycle to the next at a fully actuated intersection, especially with highly variable approach volumes. Consequently, the cycle length can vary from one cycle to the next. Figure 7.4b illustrates the typical dual-ring phase sequence options as discussed above. Note that phases can be skipped entirely if no traffic demand is present, and other phase options are possible depending on intersection geometry/lane movement assignment.

It must be pointed out that although no two phases are required to either start or terminate at the same time in a dual-ring configuration, all movements on the left side of the barrier must be terminated before any movement on the right side of the barrier can be initiated, and vice versa. This is a safety feature: no movement on the left side of the barrier can be allowed to move simultaneously with any movement on the right side of the barrier, or else conflicting traffic streams will intersect.

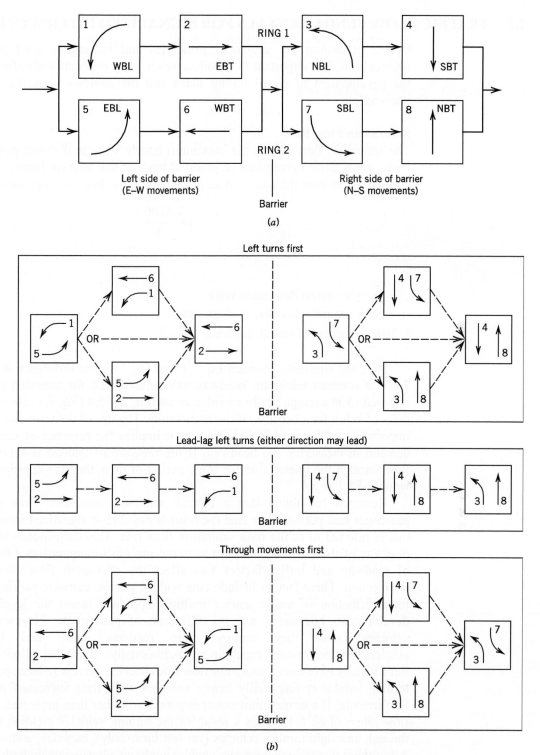

**Figure 7.4** Dual-ring signal control. (*a*) Movement-based representation of dual-ring logic. (*b*) Phase-based representation of dual-ring logic.

Part (*a*) Reproduced with permission of the Transportation Research Board, *Highway Capacity Manual* 2010, Copyright, National Academy of Sciences, Washington, D.C. Adapted from Exhibit 18-3, p. 18-5.

## 7.3    TRAFFIC FLOW FUNDAMENTALS FOR SIGNALIZED INTERSECTIONS

Before presenting the analytical principles and techniques used for signalized intersections, it is important to introduce several key concepts and definitions used in the development of signal timing plans and the analysis of traffic at signalized intersections.

### Saturation Flow Rate

The saturation flow rate is the maximum hourly volume that can pass through an intersection, from a given lane or group of lanes, if that lane (or lanes) were allocated constant green over the course of an hour. Saturation flow rate is given by

$$s = \frac{3600}{h} \tag{7.1}$$

*where*

$$s = \text{saturation flow rate in veh/h,}$$
$$h = \text{saturation headway in s/veh, and}$$
$$3600 = \text{number of seconds per hour.}$$

Note the similarity between Eq. 7.1 and Eq. 5.4. The difference is that $h$ in Eq. 7.1 is a constant minimum headway value maintained for saturated conditions, as opposed to an average headway value as used in Eq. 5.4 (Eq. 7.1 also directly yields units of veh/h for $s$ because of the numerator). The use of the term "saturation" is an important qualifier in this definition, as it implies the presence of constant vehicle demand in measuring the headway. If the measure of interest is simply the traffic flow through the intersection for some period of time, then the appropriate equation would be 5.1 or 5.4.

Research has found that a typical maximum saturation flow rate of 1900 passenger cars per hour per lane (pc/h/ln) is possible at signalized intersections, and this is referred to as the base saturation flow rate. This corresponds to a saturation headway of about 1.9 seconds. Just as in the analysis of uninterrupted flow, a number of roadway and traffic factors can affect the maximum flow rate through an intersection. These factors include lane widths; grades; curbside parking maneuvers; the distribution of traffic among multiple approach lanes; the level of roadside development; bus stops; and the influence of pedestrians, bicycles, and heavy vehicles (since they occupy more roadway space and have poorer acceleration/deceleration capabilities). Additionally, lanes that allow left or right turns usually have lower saturation flow rates because drivers reduce speed to make a turning maneuver (especially heavy vehicles, with their increased turning radii). Furthermore, if a turning movement is permitted rather than protected, its saturation flow rate will be reduced as a result of the turning vehicles yielding to conflicting through and right-turning vehicles (for left turns only), bicycles, and/or pedestrians. All of these factors are accommodated by applying adjustments to the base saturation flow rate. The end result is usually a value less than 1900 pc/h/ln for each approach lane at an intersection, and is referred to as the adjusted saturation flow rate. Additionally, the units are converted to vehicles per hour per lane (veh/h/ln) due to adjustment of the heavy-vehicle volume with passenger car equivalents (in a manner similar to the procedures of Chapter 6). The process for arriving at an adjusted saturation flow rate by making adjustments to the base saturation flow rate, for the preceding factors, is quite involved and beyond the scope of this book (see the *Highway Capacity Manual* [Transportation Research Board 2010]). Of course, saturation flow rates can be measured directly in the field, in which case no further adjustments are necessary. For the rest of this chapter, it should be assumed that the provided saturation flow rates have been adjusted for the given conditions; the term "adjusted" has been dropped just for notational convenience.

**Lost Time**

Due to the traffic signal's function of continuously alternating the right-of-way between conflicting movements, traffic streams are continuously started and stopped. Every time this happens, a portion of the cycle length is not being completely utilized, which translates to lost time (time that is not effectively serving any movement of traffic). Total lost time is a combination of start-up and clearance lost times. Start-up lost times occur because when a signal indication turns from red to green, drivers in the queue do not instantly start moving at the saturation flow rate; there is an initial lag due to drivers reacting to the change of signal indication. This start-up delay results in a portion of the green time for that movement not being completely utilized. This start-up lost time has a typical value of around 2 seconds. This concept is illustrated in Fig. 7.5.

In this figure, note that the headway for the first several vehicles is larger than the saturation headway. The saturation headway is typically reached after the fourth vehicle in the queue. The summation of the amount of headway time greater than the saturation headway for each of the first four vehicles yields the total start-up lost time for the movement.

The stopping of a traffic movement also results in lost time. When the signal indication turns from green to yellow, the latter portion of time during the yellow interval is generally not utilized by traffic. Additionally, if there is an all-red interval, this time period is generally not utilized by traffic. These periods of time during the change and clearance intervals that are not effectively used by traffic are referred to as clearance lost time. Typically, the last second of the yellow interval and the entire all-red interval are included in the estimate of clearance lost time. However, for intersections with significant red-light running, the clearance lost time may be negligible.

Start-up and clearance lost times are summed to arrive at a total lost time for the phase, given as

$$t_L = t_{sl} + t_{cl} \qquad (7.2)$$

*where*

$t_L$ = total lost time for a movement during a cycle in seconds,

$t_{sl}$ = start-up lost time in seconds, and

$t_{cl}$ = clearance lost time in seconds.

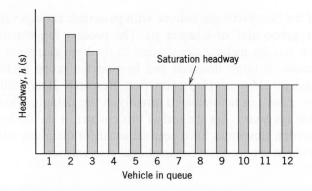

**Figure 7.5** Concept of saturation headway and lost time.

This amount of time remains fixed, regardless of phase or cycle length. Thus, for shorter cycle lengths, the lost time will comprise a larger percentage of the cycle length and will result in a larger amount of lost time over the course of a day compared with longer cycle lengths. However, longer cycle lengths usually have more phases than shorter cycle lengths, which may result in similar proportions of lost time.

### Effective Green and Red Times

For analysis purposes, the time during a cycle that is effectively (or not effectively) utilized by traffic must be used rather than the time for which green, yellow, and red signal indications are actually displayed, because they are most likely different. This results in two measures of interest: the effective green time and the effective red time. The effective green time is the time during which a traffic movement is effectively utilizing the intersection. The effective green time for a given movement or phase is calculated as

$$g = G + Y + AR - t_L \tag{7.3}$$

*where*

$g$ = effective green time for a traffic movement in seconds,
$G$ = displayed green time for a traffic movement in seconds,
$Y$ = displayed yellow time for a traffic movement in seconds,
$AR$ = displayed all-red time in seconds, and
$t_L$ = total lost time for a movement during a cycle in seconds.

The effective red time is the time during which a traffic movement is not effectively utilizing the intersection. The effective red time for a given movement or phase is calculated as

$$r = R + t_L \tag{7.4}$$

*where*

$r$ = effective red time for a traffic movement in seconds,

$R$ = displayed red time for a traffic movement in seconds, and

$t_L$ = total lost time for a movement during a cycle in seconds.

Alternatively, the effective red time can be calculated as follows, assuming the cycle length and effective green time have already been determined:

$$r = C - g \qquad (7.5)$$

*where*

$C$ = cycle length in seconds, and

Other terms are as defined previously.

Likewise, the effective green time can be calculated by subtracting the effective red time from the cycle length.

**Capacity**

Because movements on an intersection approach do not receive a constant green indication (as assumed in the definition for saturation flow rate), another measure must be defined that accounts for the hourly volume that can be accommodated on an intersection approach given that the approach will receive less than 100% green time. This measure is capacity and is given by

$$c = s \times g/C \qquad (7.6)$$

*where*

$c$ = capacity (the maximum hourly volume that can pass through an intersection from a lane or group of lanes under prevailing roadway, traffic, and control conditions) in veh/h,

$s$ = saturation flow rate in veh/h, and

$g/C$ = ratio of effective green time to cycle length.

## 7.4    DEVELOPMENT OF A TRAFFIC SIGNAL PHASING AND TIMING PLAN

Assuming the decision to install a traffic signal at an intersection has been made, an appropriate phasing and timing plan must be developed. The development of a traffic signal phasing and timing plan can be complex, particularly if the intersection has multiple-lane approaches and requires protected turning movements (a turn arrow). However, the timing plan analysis can be simplified by dealing with each approach separately. This section provides the basic process and fundamentals needed to develop a phasing and timing plan for an isolated, fixed-time (pretimed) traffic signal. As timing plans become more complex, they simply build on these fundamental principles. The reader is encouraged to review the material in other references to see how actuated and progressive timing plans are developed [Kell and Fullerton 1991; Transportation Research Board 2010]. This section describes the basic process that results in the development of a signal phasing and timing plan.

### 7.4.1    Select Signal Phasing

Recall that a cycle is the sum of individual phases. The most basic traffic signal cycle is made up of two phases, as shown in Fig. 7.6. In this case, phase 1 accommodates the movement of the northbound and southbound vehicles, and phase 2 accommodates the movement of the eastbound and westbound vehicles. These phases will alternate during the continuous operation of the signal. This phasing scheme, however, could prove to be very inefficient if one or more of the approaches includes a high left-turn volume.

Given that each approach consists of one lane, vehicles will be delayed behind a left-turning vehicle waiting for a gap in the opposing traffic stream. If the high volume of left turns is present on both the northbound and southbound approaches, for example, each of these approaches could be given a separate phase. This would be more efficient because left-turning vehicles on these two approaches would not have to wait for gaps in the opposing traffic stream, thus greatly reducing delays for all vehicles. This would result in a three-phase operation, as shown in Fig. 7.6. When movements on opposing approaches are given separate phases, as in this case, it is referred to as split phasing.

Some common signal phasing configurations and sequences are shown in Fig. 7.7. In this figure, note that the dashed lines represent permitted movements and the solid lines represent protected movements. When the left turns precede the through and right-turn movements in the phasing sequence for an approach, they are referred to as leading left turns. When the left turns follow the through and right-turn movements, they are referred to as lagging left turns. Although not shown in Fig. 7.7, it is also possible for a movement to be protected for a period of time and then permitted for a period of time, or vice versa. This is most commonly seen with left-turn movements, and is referred to as protected plus permitted or permitted plus protected, depending on the sequence.

It is important to remember that there is lost time (start-up and clearance) associated with each phase. Thus, with each phase added to a cycle, the lost time increases. Although the lost time may be only 3 to 5 seconds per phase, the accumulated lost time throughout the day can be significant.

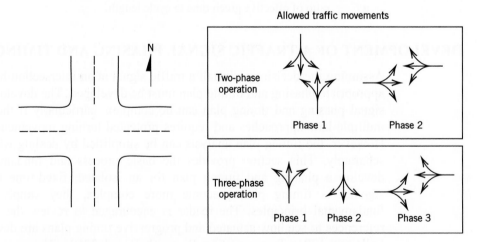

**Figure 7.6** Illustration of two-phase and three-phase signal operation.

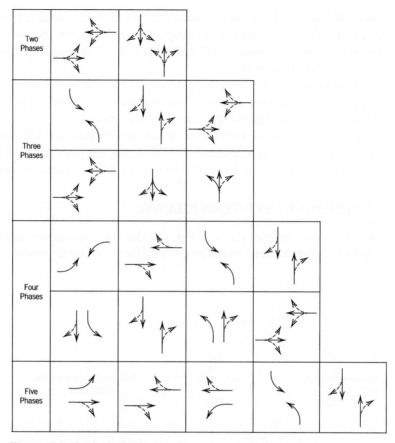

**Figure 7.7** Typical phasing configurations and sequencing.

Given this, a point of diminishing returns is reached with the addition of phases as the efficiencies gained by separating traffic movements eventually become outweighed by the inefficiencies of increased lost time. Thus, a primary concern in signal timing is to keep the number of phases to a minimum. Because protected-turn phases add to lost time, they should be used only when warranted. Because of opposing motor vehicle traffic, left-turn movements typically require a protected-turn phase much more often than right turns. There are no nationally established guidelines on when protected left-turn phasing should be used, so local policies and practices should be consulted before a decision is made about whether to provide a protected left-turn phase. In general, decisions on whether to provide a protected left-turn phase are based on one or more of the following factors:

- Volume (just left turn or combination of left turn and opposing volume)
- Delay
- Queuing (spillover)
- Traffic progression
- Opposing traffic speeds
- Geometry (number of left-turn lanes, crossing distance, sight distance)
- Crash experience (which may also be related to any of the above factors)

More specific guidance on this issue can be found in several references, including the *Highway Capacity Manual* [Transportation Research Board 2010], the *Traffic Control Devices Handbook* [ITE 2001], and the *Manual of Traffic Signal Design* [Kell and Fullerton 1998].

One of the more common guidelines is the use of the cross product of left-turn volume and opposing through and right-turn volumes. The *Highway Capacity Manual* offers the following criteria for this guideline: The use of a protected left-turn phase should be considered when the product of left-turning vehicles and opposing traffic volume exceeds 50,000 during the peak hour for one opposing lane, 90,000 for two opposing lanes, or 110,000 for three or more opposing lanes.

## EXAMPLE 7.1 DETERMINE LEFT-TURN PHASING

Refer to the intersection shown in Fig. 7.8. Use the cross-product guideline to determine if protected left-turn phases should be provided for any of the approaches.

**Figure 7.8** Intersection geometry and peak traffic volumes for example problems.

## SOLUTION

There are 250 westbound vehicles that turn left during the peak hour. The product of the westbound left-turning vehicles and the opposing eastbound traffic (right-turn and straight-through vehicles) is 275,000 [250 × (900 + 200)]. There are 300 eastbound vehicles that turn left during the peak hour. The product of the eastbound left-turning vehicles and the opposing westbound traffic (right-turn and straight-through vehicles) is 345,000 [300 × (1000 + 150)].

Because the cross product for each of these approaches is greater than 90,000 (the requirement for two opposing lanes), a protected left-turn phase is suggested for the WB and EB left-turn movements. The NB and SB approaches do not require a protected left-turn phase using this criterion because the cross products for these approaches are less than 50,000 (for one opposing lane). Therefore, a three-phase traffic-signal plan is recommended, as shown in Fig. 7.9.

**Figure 7.9** Recommended signal phasing plan for the intersection in Example 7.1.

### 7.4.2 Establish Analysis Lane Groups

Each intersection approach is initially treated separately, and the results are later aggregated. Thus, each approach must be subdivided into logical groupings of traffic movements for analysis purposes.

The *Highway Capacity Manual* [Transportation Research Board 2010] provides detailed guidelines on this process. Generally, the process consists of placing like movements into *movement groups* (left-turn, right-turn, and through movements would be identified) and then translating *movement groups* to *lane groups* based on the allowable movements from each lane. The only time these two group designations differ is when a shared lane (left-turn and through movements allowed from the same lane) is present on an approach with two or more lanes. When shared lanes are present on an approach with two or more lanes, the *Highway Capacity Manual* [Transportation Research Board 2010] employs an iterative procedure to identify the expected distribution of each movement type in each lane, based on the principle that drivers will choose the lane that they perceive will minimize their travel time through the intersection (delay). This procedure is beyond the scope of this book and, as a result, the subsequent example problems and end-of-chapter problems will provide specific lane distributions of traffic as necessary. Consequently, it is only necessary to make reference to lane groups for the remainder of the chapter.

Based on the lane and traffic movement distribution on an approach, lane groups can be readily determined. The following general guidelines are offered for establishing lane groups [Transportation Research Board 2010]:

- If an exclusive turn lane (or lanes) is present, it should be treated as a separate lane group.
- Each shared lane on an approach should be treated as a separate lane group.
- Any remaining lanes, which would be exclusive through lanes, should be treated as a separate lane group.

Figure 7.10 shows some typical lane groupings for analysis purposes. Note that when multiple lanes are combined into a lane group, the subsequent analysis calculations for this lane group should treat these lanes as a single unit. Because of the aforementioned complexity of dealing with shared lanes, the subsequent example and end-of-chapter problems in this book assume that adjacent exclusive through lanes and shared through/right-turn lanes will be combined into one lane group (such as is done in Example 7.2). This assumption will often introduce only very minimal error when compared to treating the exclusive through lane and shared through/right-turn lane as separate lane groups and does not affect learning the basic concept of aggregating lane group results into overall approach results (which is covered later in this chapter).

| Number of Approach Lanes | Movements by Lane and Corresponding Lane Groups | Number of Approach Lanes | Movements by Lane and Corresponding Lane Groups |
|---|---|---|---|
| 1 | LT + TH + RT | 3 | EXC LT<br>EXC TH<br>TH + RT |
| 2 | EXC LT<br>TH + RT | 3 | EXC LT<br>EXC TH<br>EXC RT |
| 2 | LT + TH<br>TH + RT | | |

**Figure 7.10** Example lane groupings for analysis (LT = left turn; TH = through; RT = right turn; EXC = exclusive).

Adapted from Exhibit 18-12, p. 18-34, *Highway Capacity Manual*, Transportation Research Board 2010.

**EXAMPLE 7.2    DETERMINE LANE GROUPS**

Determine the lane groups to use for analysis of the Maple and Vine Streets intersection (Fig. 7.8).

*SOLUTION*

The EB and WB left-turn movements will each be a lane group because they have a separate lane and move in a separate phase from the through/right-turn movements. Likewise, the EB and WB through/right-turn movements proceed together in a separate phase and will therefore be separate lane groups. Although the right turns use only the outside lane, this movement's impact on the saturation flow rate for the two lanes combined will be determined. The NB and SB left turns will also each be a separate lane group. Even though they move during the same phase as the adjacent through and right-turn movements, these left turns are permitted and will have very different operating characteristics from the through and right-turn movements. Because the through and right-turn movements use the same lane, they will be an individual lane group for both the NB and SB approaches. The recommended lane groups for analysis for each of the approaches are shown in Fig. 7.11.

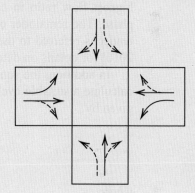

**Figure 7.11** Analysis lane groups for the three-phase design at the intersection of Maple Street and Vine Street.

## 7.4.3    Calculate Analysis Flow Rates and Adjusted Saturation Flow Rates

Just as for the analysis of uninterrupted flow, the hourly traffic volume arriving on each intersection approach must be converted to an analysis flow rate that accounts for the peak 15-minute flow within that hour (typically the peak hour). This is accomplished by calculating the peak-hour factor (*PHF*) and dividing this into the hourly volume (as shown in Chapter 6), which yields the analysis flow rate.

One note about adjusting for the *PHF*. With the multiple traffic streams entering an intersection, a separate *PHF* can be calculated for each approach's traffic stream. However, adjusting each approach volume by its specific *PHF* can yield unrealistically high combined analysis volumes, because the different approach volumes usually do not peak during the same 15-minute period. Applying a single *PHF* determined for the intersection as a whole will result in more reasonable analysis volumes.

The adjustment of the saturation flow rate was discussed in Section 7.3. Again, it is assumed that the approach volumes and saturation flow rates provided in this chapter have already been adjusted.

## 7.4.4    Determine Critical Lane Groups and Total Cycle Lost Time

For any combination of lane group movements during a particular phase, one of these lane groups will control the necessary green time for that phase. This lane group is referred to as the critical lane group. When the traffic movements of each lane group occur during only one phase of the signal cycle, the determination of the critical lane group for each phase is straightforward. In this case, the critical lane group for each phase is simply the lane group with the highest ratio of vehicle arrival rate to vehicle departure rate ($\lambda/\mu$). This quantity is referred to as the flow ratio (which was called the traffic intensity, $\rho$, in Chapter 5) and is designated $v/s$ (arrival flow rate divided by saturation flow rate). If the allocation of green time for each phase is based on the flow ratio of the critical lane group, then the noncritical lane group movements will be accommodated as well.

As previously discussed, with dual-ring controllers, a wide variety of phasing sequences is possible. The situation where a movement starts in one signal phase and continues in the next signal phase is referred to as an overlapping phase (see the last two rows of Fig. 7.7). For the case of overlapping phases in a signal cycle, the identification of the critical lane groups is more complex, as the lane group with the highest flow ratio in each phase is not necessarily the critical lane group for that phase. The remainder of this chapter will focus only on nonoverlapping phases. The reader is referred to the *Highway Capacity Manual* [Transportation Research Board 2010] for details on determining critical lane groups for overlapping phases.

In addition, the sum of the flow ratios for the critical lane groups can be used to calculate a suitable cycle length, which will be discussed in the next section. This is given by

$$Y_c = \sum_{i=1}^{n} \left(\frac{v}{s}\right)_{ci} \tag{7.7}$$

*where*

$$
\begin{aligned}
Y_c &= \text{sum of flow ratios for critical lane groups,} \\
(v/s)_{ci} &= \text{flow ratio for critical lane group } i, \text{ and} \\
n &= \text{number of critical lane groups.}
\end{aligned}
$$

The total lost time for the cycle will also be used in the calculation of cycle length. In determining the total lost time for the cycle, the general rule is to apply the lost time for a critical lane group when its movements are initiated (the start of its green interval). The total cycle lost time is given as

$$L = \sum_{i=1}^{n} \left(t_L\right)_{ci} \tag{7.8}$$

*where*

$$
\begin{aligned}
L &= \text{total lost time for cycle in seconds,} \\
(t_L)_{ci} &= \text{total lost time for critical lane group } i \text{ in seconds, and} \\
n &= \text{number of critical lane groups.}
\end{aligned}
$$

**EXAMPLE 7.3     CALCULATE SUM OF FLOW RATIOS**

Calculate the sum of the flow ratios for the critical lane groups for the three-phase timing plan determined in Example 7.1 given the saturation flow rates in Table 7.1.

**Table 7.1** Saturation Flow Rates for Three-Phase Design at Intersection of Maple Street and Vine Street

| Phase 1 | Phase 2 | Phase 3 |
|---------|---------|---------|
| EB L: 1750 veh/h | EB T/R: 3400 veh/h | SB L: 450 veh/h |
| | | NB L: 475 veh/h |
| WB L: 1750 veh/h | WB T/R: 3400 veh/h | SB T/R: 1800 veh/h |
| | | NB T/R: 1800 veh/h |

*SOLUTION*

Note that the saturation flow rates are relatively low for the SB and NB left (L) turns because they are permitted only, and the opposing through and right-turn (T/R) vehicles limit the number of usable gaps for these vehicles. The saturation flow rates for the WB and EB through and right-turn (T/R) movements account for both through lanes.

The flow ratios will now be calculated, with the critical lane group for each phase indicated with a check mark in Table 7.2. As indicated in the table, the critical lane group for phases 1, 2, and 3, respectively, are the EB left turn, the WB through and right turn, and the NB through and right turn.

**Table 7.2** Flow Ratios and Critical Lane Groups for Three-Phase Design at Intersection of Maple Street and Vine Street

| Phase 1 | Phase 2 | Phase 3 |
|---------|---------|---------|
| EB L: $\dfrac{300}{1750} = 0.171$ ✓ | EB T/R: $\dfrac{1100}{3400} = 0.324$ | SB L: $\dfrac{70}{450} = 0.156$ |
| | | NB L: $\dfrac{90}{475} = 0.189$ |
| WB L: $\dfrac{250}{1750} = 0.143$ | WB T/R: $\dfrac{1150}{3400} = 0.338$ ✓ | SB T/R: $\dfrac{370}{1800} = 0.206$ |
| | | NB T/R: $\dfrac{390}{1800} = 0.217$ ✓ |

The sum of the flow ratios for the critical lane groups for this phasing plan will be needed for the next section. Since this phasing plan does not include any overlapping phases, this value is simply the sum of the highest lane group $v/s$ ratios for the three phases, as follows:

$$Y_c = \sum_{i=1}^{n} \left(\frac{v}{s}\right)_{ci}$$
$$= 0.171 + 0.338 + 0.217 = \underline{\underline{0.726}}$$

Assuming 2 seconds of start-up lost time and 2 seconds of clearance lost time (1 second of yellow time plus 1 second of all-red time), for each critical lane group, gives a lost time of 4 s/phase. The total lost time for the cycle is then 12 seconds (3 phases × 4 s/phase).

## EXAMPLE 7.4 CALCULATE SUM OF FLOW RATIOS AND TOTAL LOST TIME

Suppose it is necessary to run the NB and SB movements in a split-phase configuration (with phase 3 for SB movements and a new phase 4 for NB movements). Calculate the sum of the flow ratios for the critical lane groups and total cycle lost time for this situation, assuming the EB and WB movement phasing remains the same. Table 7.3 summarizes the calculation of the flow ratios and the identification of the critical lane groups.

**SOLUTION**

The sum of the flow ratios for the critical lane groups for this phasing plan is

$$\sum_{i=1}^{n}\left(\frac{v}{s}\right)_{ci} = 0.171+0.338+0.206+0.217 = \underline{\underline{0.932}}$$

The total lost time for the cycle is <u>16 seconds</u> (4 phases × 4 s/phase).

**Table 7.3** Flow Ratios and Critical Lane Groups for Four-Phase Design (Split Phase for N-S Movements) at Intersection of Maple Street and Vine Street

| Phase 1 | Phase 2 | Phase 3 | Phase 4 |
|---|---|---|---|
| EB L: $\frac{300}{1750} = 0.171$ ✓ | EB T/R: $\frac{1100}{3400} = 0.324$ | SB L: $\frac{70}{1750} = 0.040$ | NB L: $\frac{90}{1750} = 0.051$ |
| WB L: $\frac{250}{1750} = 0.143$ | WB T/R: $\frac{1150}{3400} = 0.338$ ✓ | SB T/R: $\frac{370}{1800} = 0.206$ ✓ | NB T/R: $\frac{390}{1800} = 0.217$ ✓ |

### 7.4.5 Calculate Cycle Length

The cycle length is simply the summation of the individual phase lengths. In practice, cycle lengths are generally kept as short as possible, typically between 60 and 75 seconds. However, complex intersections with five or more phases can have cycle lengths of 120 seconds or more. The minimum cycle length necessary for the lane group volumes and phasing plan of an intersection is given by

$$C_{min} = \frac{L \times X_c}{X_c - \sum_{i=1}^{n}\left(\frac{v}{s}\right)_{ci}} \tag{7.9}$$

*where*

$C_{min}$ = minimum necessary cycle length in seconds (typically rounded up to the nearest 5-second increment in practice),

$L$ = total lost time for cycle in seconds,

$X_c$ = critical v/c ratio for the intersection,

$(v/s)_{ci}$ = flow ratio for critical lane group $i$, and

$n$ = number of critical lane groups.

In this equation, the total lost time for the cycle and the sum of the flow ratios for the critical lane groups are predetermined. However, a critical intersection volume/capacity ratio, $X_c$, must be chosen for the desired degree of utilization. In other words, if it is desired that the intersection operate at its full capacity, a value of 1.0 is used for $X_c$. A value of 1.0 is not generally recommended, however, due to the randomness of vehicle arrivals, which can result in occasional cycle failures. Note that this equation gives the minimum cycle length necessary for the intersection to operate at a specified degree of capacity utilization. This cycle length does not necessarily minimize the average vehicle delay experienced by motorists at the intersection.

A practical equation for the calculation of the cycle length that seeks to minimize vehicle delay was developed by Webster [1958]. Webster's optimum cycle length formula is

$$C_{opt} = \frac{1.5 \times L + 5}{1.0 - \sum_{i=1}^{n} \left(\frac{v}{s}\right)_{ci}} \qquad (7.10)$$

*where*

$C_{opt}$ = cycle length to minimize delay in seconds, and

Other terms are as defined previously.

The cycle length determined from this calculation is only approximate. Webster noted that values between $0.75C_{opt}$ and $1.5C_{opt}$ will likely give similar values of delay. Calculating an accurate optimal cycle length (and phase length) can be a very computationally intensive exercise for all but the simplest signalized intersections, especially if coordination among multiple signals is involved.

It should be noted that regardless of the minimum or optimal cycle length calculated, practical maximum cycle lengths must generally be observed. Public acceptance or tolerance of large cycle lengths will vary by location (urban vs. rural), but as a rule, cycle lengths in excess of 3 minutes (180 seconds) should be used only in exceptional circumstances.

**EXAMPLE 7.5    CALCULATE MINIMUM AND OPTIMAL CYCLE LENGTHS**

Calculate the minimum and optimal cycle lengths for the intersection of Maple and Vine Streets, using the information provided in the preceding examples, for both the three-phase and four-phase design.

## SOLUTION

For the three-phase design (Example 7.3), the sum of the flow ratios for the critical lane groups and the total cycle lost time were determined to be 0.726 and 12 seconds, respectively. For the minimum cycle length, a somewhat conservative value of 0.9 will be used for the critical intersection $v/c$ ratio to minimize the potential of cycle failures due to occasionally high arrival volumes. Using these values in Eq. 7.9 gives

$$C_{min} = \frac{12 \times 0.9}{0.9 - 0.726} = 62.1 \rightarrow \underline{\underline{65}} \text{ s (rounding up to nearest 5 seconds)}$$

Using Eq. 7.10 for the optimal cycle length gives

$$C_{opt} = \frac{1.5 \times 12 + 5}{1.0 - 0.726} = 83.9 \rightarrow \underline{\underline{85}} \text{ s (rounding up to nearest 5 seconds)}$$

For the four-phase design (Example 7.4), the sum of the critical flow ratios and the total cycle lost time were determined to be 0.932 and 16 seconds, respectively. The first issue with this design is that a higher $X_c$ will need to be used because the sum of flow ratios for critical lane groups is higher than the 0.90 used for the three-phase design (otherwise the denominator of Eq. 7.9 will be negative). To minimize the cycle length, the maximum value of 1.0 will be used for $X_c$ in Eq. 7.9, as follows:

$$C_{min} = \frac{16 \times 1.0}{1.0 - 0.932} = \underline{\underline{235.3}} \text{ s}$$

The second issue is that despite the use of an $X_c$ value of 1.0 (the intersection operating at capacity) to minimize the cycle length, an unreasonably high cycle length is still required for this design. Thus, this design is not nearly as desirable as the three-phase design.

Generally a split-phase design is recommended only under one or more of the following conditions:

- The left turns are the dominant movement.
- The left turns share a lane with the through movement.
- There is a large difference in the total approach volumes.
- There are unusual opposing approach geometrics.

It should also be noted that serving pedestrians in an efficient manner on split-phase approaches can be difficult.

### 7.4.6   Allocate Green Time

After a cycle length has been calculated, the next step in the traffic signal timing process is to determine how much green time should be allocated to each phase. The cycle length is the sum of all effective green times plus the total lost time. Thus, after subtracting the total lost time from the cycle length, the remaining time can be distributed as green time among the phases of the cycle.

There are several strategies for allocating the green time to the various phases. One of the most popular and simplest is to distribute the green time so that the $v/c$ ratios are equalized for the critical lane groups, as by the following equation:

$$g_i = \left(\frac{v}{s}\right)_{ci}\left(\frac{C}{X_i}\right)$$   (7.11)

where

$g_i$ = effective green time for phase $i$,

$(v/s)_{ci}$ = flow ratio for critical lane group $i$,

$C$ = cycle length in seconds, and

$X_i$ = $v/c$ ratio for lane group $i$.

## EXAMPLE 7.6   DETERMINE GREEN TIMES

Determine the green-time allocations for the 65-second cycle length found in Example 7.5, using the method of $v/c$ ratio equalization.

### SOLUTION

Because the calculated cycle length was rounded up a few seconds, the critical intersection $v/c$ ratio for this rounded cycle length will be calculated for use in the green-time allocation calculations. Equation 7.9 can be rearranged to solve for $X_c$ as follows:

$$X_c = \frac{\sum_{i=1}^{n}\left(\frac{v}{s}\right)_i \times C}{C - L}$$

Using this equation with

$\sum(v/s)_{ci}$ = 0.726 (Example 7.3)

$C$ = 65 s (Example 7.5)

$L$ = 12 s (Example 7.4)

gives

$$X_c = \frac{0.726 \times 65}{65 - 12} = 0.890$$

Therefore, the cycle length of 65 seconds and $X_c$ of 0.890 are used to calculate the effective green times for the three phases, as follows:

$$g_1 = \left(\frac{v}{s}\right)_{c1}\left(\frac{C}{X_1}\right)$$   (EB and WB left-turn movements)

$$= 0.171 \times \frac{65}{0.890} = 12.5\,\text{s}$$

$$g_2 = \left(\frac{v}{s}\right)_{c2}\left(\frac{C}{X_2}\right)$$   (EB and WB through and right-turn movements)

$$= 0.338 \times \frac{65}{0.890} = 24.7\,\text{s}$$

$$g_3 = \left(\frac{v}{s}\right)_{c3}\left(\frac{C}{X_3}\right) \quad \text{(NB and SB left-, through, and right-turn movements)}$$

$$= 0.217 \times \frac{65}{0.890} = \underline{\underline{15.8\,\text{s}}}$$

The cycle length is checked by summing these effective green times and the lost time, giving

$$C = g_1 + g_2 + g_3 + L$$
$$= 12.5 + 24.7 + 15.8 + 12 = 65.0$$

Therefore, all calculations are correct.

### 7.4.7 Calculate Change and Clearance Intervals

Recall that the change interval corresponds to the yellow time and the clearance interval corresponds to the all-red time. If an all-red interval does not exist, then the yellow time is considered as both the change and clearance intervals. The change interval alerts drivers that the green interval is about to end and that they should come to stop before entering the intersection, or continue through the intersection if they are too close to come to a safe stop. The clearance interval allows those vehicles that might have entered the intersection at the end of the yellow to clear the intersection before conflicting traffic movements are given a green signal indication. In the past, the yellow indication was intended to also allow for clearance time. Today, however, there is routine red-indication abuse and frequent running of red indications after the yellow time. As a result, the all-red indication is often implemented.

Typically, the yellow time is in the range of 3 to 5 seconds. Warning times that are shorter than 3 seconds and longer than 5 seconds are not practical because long warning times encourage motorists to continue to enter the intersection whereas short times can place the driver in a dilemma zone. A dilemma zone is created for the driver if a safe stop before the intersection cannot be accomplished, and continuing through the intersection at a constant speed (without accelerating) will result in the vehicle entering the intersection during a red indication. If a dilemma zone exists, drivers always make the wrong decision, whether they decide to stop or to continue through the intersection. Figure 7.12 illustrates the dilemma zone. Referring to this figure, suppose a vehicle traveling at a constant speed requires distance $x_s$ to stop. If the vehicle is closer to the intersection than distance $d_d$, then it can enter before the all-red indication. If the vehicle is in the shaded area ($x_s - d_d$ from the intersection) when the yellow light is displayed, the driver is in the dilemma zone and can neither stop in time nor continue through the intersection at a constant speed without passing through a red indication.

Formulas and policies for calculating yellow ($Y$) and all-red ($AR$) times vary by agency, but one set of commonly accepted formulas is provided in the *Traffic Engineering Handbook* [ITE 1999] and are as follows:

$$Y = t_r + \frac{V}{2a + 2gG} \tag{7.12}$$

*where*

$Y$ = yellow time (usually rounded up to the nearest 0.5 second),

$t_r$ = driver perception/reaction time, usually taken as 1.0 second,

$V$ = speed of approaching traffic in m/s,

$a$ = deceleration rate for the vehicle, usually taken as 3.05 m/s$^2$,

$g$ = acceleration due to gravity (9.807 m/s$^2$), and

$G$ = percent grade divided by 100.

**Figure 7.12** The dilemma zone for traffic approaching a signalized intersection.

and

$$AR = \frac{w+l}{V} \qquad (7.13)$$

*where*

$AR$ = all-red time (usually rounded up to the nearest 0.5 second),

$w$ = width of the cross street in m,

$l$ = length of the vehicle, usually taken as 6 m, and

$V$ = speed of approaching traffic in m/s.

To avoid a dilemma zone and the possibility of a vehicle being in the intersection when a conflicting movement receives a green-signal indication, the total of the change and clearance intervals (yellow plus all-red times) should always be equal to or greater than the sum of Eqs. 7.12 and 7.13.

**EXAMPLE 7.7    DETERMINE YELLOW AND ALL-RED TIMES**

Determine the yellow and all-red times for vehicles traveling on Vine and Maple Streets as shown in Fig 7.8.

*SOLUTION*

For the Vine Street phasing (applying Eqs. 7.12 and 7.13),

$$Y = 1.0 + \frac{(55 \times 1000/3600)}{2(3.05)}$$

$$= 3.51 \quad \rightarrow \quad \underline{\underline{4.0\,\text{s}}}\,(\text{rounding to the nearest }0.5\,\text{s})$$

$$AR = \frac{18.3 + 6}{55 \times 1000/3600}$$

$$= 1.59 \quad \rightarrow \quad \underline{\underline{2.0\,\text{s}}}\,(\text{rounding to the nearest }0.5\,\text{s})$$

For the Maple Street phasing (applying Eqs. 7.12 and 7.13),

$$Y = 1.0 + \frac{(65 \times 1000/3600)}{2(3.05)}$$

$$= 3.96 \quad \rightarrow \quad \underline{\underline{4.0\,\text{s}}}\,(\text{rounding to the nearest }0.5\,\text{s})$$

$$AR = \frac{11 + 6}{65 \times 1000/3600}$$

$$= 0.94 \quad \rightarrow \quad \underline{\underline{1.0\,\text{s}}}\,(\text{rounding to the nearest }0.5\,\text{s})$$

Note that separate calculations are usually required for exclusive left-turn phases, as vehicle approach speeds are often lower than for through vehicles and intersection crossing distances may be longer (due to the width of the opposing direction and the circular travel path).

## 7.4.8 Check Pedestrian Crossing Time

In urban areas and other locations where pedestrians are present, the signal-timing plan should be checked for its ability to provide adequate pedestrian crossing time. At locations where streets are wide and green times are short, it is possible that pedestrians can be caught in the middle of the intersection when the phase changes. To avoid this problem, the minimum green time required for pedestrian crossing should be checked against the apportioned green time for the phase. If there is not enough green time for a pedestrian to safely cross the street, the apportioned green time should be increased to meet the pedestrian needs. If pedestrian pushbuttons are provided at an intersection (for actuated control), the green time can be increased to meet pedestrian crossing needs only when the pushbuttons are activated.

The minimum pedestrian green time is given by

$$G_p = 3.2 + \frac{L}{S_p} + \left(0.27 N_{ped}\right) \quad \text{for } W_E \leq 3.05 \text{ m} \tag{7.14}$$

$$G_p = 3.2 + \frac{L}{S_p} + \left(2.7 \frac{N_{ped}}{W_E}\right) \quad \text{for } W_E > 3.05 \text{ m} \tag{7.15}$$

*where*

$G_p$ = minimum pedestrian green time in seconds,

3.2 = pedestrian start-up time in seconds,

$L$ = crosswalk length in m,

$S_p$ = walking speed of pedestrians, usually taken as 1 m/s,

$N_{ped}$ = number of pedestrians crossing during an interval, and

$W_E$ = effective crosswalk width in m.

The generally recommended walking speed of 1 m/s [U.S. Federal Highway Administration 2009] represents a slower-than-average speed. However, at intersections where a significant number of slower pedestrians (elderly, vision impaired, etc.) are served, the use of a slower walking speed may be warranted.

**EXAMPLE 7.8    DETERMINE PEDESTRIAN GREEN TIME**

Determine the minimum amount of pedestrian green time required for the intersection of Vine and Maple Streets. Assume a maximum of 15 pedestrians crossing either street during any one phase and a crosswalk width of 2.4 m.

*SOLUTION*

A pedestrian who crosses Maple Street will cross while Vine Street has a green interval. The minimum pedestrian green time needed on Vine Street is (using Eq. 7.14, as the effective crosswalk width is less than or equal to 3.05 m)

$$G_p = 3.2 + \frac{18.3}{1} + (0.27 \times 15) = \underline{\underline{25.6 \text{ s}}}$$

In Example 7.6, Vine Street was assigned 15.8 seconds of effective green time [13.8 seconds of displayed green time (from Eq. 7.3)]. This amount of time is insufficient for pedestrians crossing Maple Street. Therefore, the green time for this phase will have to be increased to accommodate crossing pedestrians, and the overall signal timing plan adjusted accordingly (although we will continue to use the previously computed green time in subsequent examples). The minimum pedestrian green time needed on Maple Street (for the through/right-turn phase, when pedestrian movement would be permitted) is

$$G_p = 3.2 + \frac{11}{1} + (0.27 \times 15) = \underline{\underline{18.3 \text{ s}}}$$

In Example 7.6, Maple Street was assigned 24.7 seconds of effective green time (23.7 seconds of displayed green time) for this phase, so this green time is adequate for pedestrians crossing Vine Street.

## 7.5    ANALYSIS OF TRAFFIC AT SIGNALIZED INTERSECTIONS

This section will utilize and build upon the elements of traffic flow theory introduced in Chapter 5 and Section 7.3 to make possible the basic analysis of traffic flow at signalized intersections.

### 7.5.1  Signalized Intersection Analysis with *D*/*D*/1 Queuing

The assumption of *D*/*D*/1 queuing (as discussed in Chapter 5) provides a strong intuitive appeal that helps in understanding the analytical fundamentals underlying traffic analysis at signalized intersections. To begin applying *D*/*D*/1 queuing to signalized intersections, we consider the case where the approach capacity exceeds the approach arrivals. Under these conditions, and the assumption of uniform arrivals throughout the cycle and uniform departures during green, a *D*/*D*/1 queuing system as shown in Fig. 7.13 will result.

Note that this chapter will use the variables $v$ (for arrival rate) and $s$ (for departure/saturation flow rate), rather than the variables $\lambda$ and $\mu$ used in Chapter 5, as these variables are more commonly used in signalized intersection analyses.

The "Arrivals $v \times t$" line gives the total number of vehicle arrivals at time $t$, and the "Departures $s \times t$" line gives the slope of vehicle departures (number of vehicles that depart) during the effective greens. Note that the per-cycle approach arrivals will be $vC$ and the corresponding approach capacity (maximum departures) per cycle will be $sg$. Figure 7.13 is predicated on the assumption that $sg$ exceeds $vC$ for all cycles (no queues exist at the beginning or end of a cycle).

**Figure 7.13** *D*/*D*/1 signalized intersection queuing with approach capacity ($sg$) exceeding arrivals ($vC$) for all cycles.

$v$ = arrival rate, typically in veh/s,

$s$ = departure rate, typically in veh/s,

$t$ = elapsed time since a reference time, typically the start of green or red, in seconds,

$t_c$ = time from the start of the effective green until queue clearance in seconds,

$r$ = effective red time in seconds,

$g$ = effective green time in seconds, and

$C$ = cycle length in seconds.

Given the properties of *D*/*D*/1 queues presented in Chapter 5, a number of general equations can be derived by simple inspection of Fig. 7.13:

1. The time to queue clearance after the start of the effective green, $t_c$ (note that $v(r + t_c) = st_c$),

$$t_c = \frac{vr}{s - v} \tag{7.16}$$

2. The proportion of the cycle with a queue, $P_q$,

$$P_q = \frac{r + t_c}{C} \tag{7.17}$$

3. The proportion of vehicles stopped, $P_s$,

$$P_s = \frac{v(r + t_c)}{v(r + g)} = \frac{r + t_c}{C} = P_q$$

$$\text{also, } P_s = \frac{v(r + t_c)}{v(r + g)} = \frac{st_c}{vC} = \frac{t_c}{\dfrac{v}{s}C} \tag{7.18}$$

4. The maximum number of vehicles in the queue, $Q_{max}$,

$$Q_{max} = vr \tag{7.19}$$

5. The total vehicle delay per cycle, $D_t$,

$$D_t = \frac{vr^2}{2(1 - v/s)} \tag{7.20}$$

6. The average delay per vehicle, $d_{avg}$,

$$d_{avg} = \frac{vr^2}{2(1 - v/s)} \times \frac{1}{vC} = \frac{r^2}{2C(1 - v/s)}$$

$$\text{also, } d_{avg} = \frac{0.5C\left(1 - \dfrac{g}{C}\right)^2}{1 - \left(\dfrac{v}{c} \times \dfrac{g}{C}\right)} \tag{7.21}$$

7. The maximum delay of any vehicle, assuming a FIFO queuing discipline, $d_{max}$,

$$d_{max} = r \tag{7.22}$$

**EXAMPLE 7.9    SIGNALIZED INTERSECTION ANALYSIS USING EQS. 7.16–7.22**

An approach at a pretimed signalized intersection has a constant saturation flow rate of 2400 veh/h and is allocated 24 seconds of effective green time in an 80-second signal cycle. If the total approach flow rate is 500 veh/h and arrivals are uniform throughout the cycle, provide an analysis of the intersection assuming $D/D/1$ queuing.

## SOLUTION

Putting arrival and departure rates into common units of vehicles per second,

$$v = \frac{500 \text{ veh/h}}{3600 \text{ s/h}} = 0.139 \text{ veh/s}$$

$$s = \frac{2400 \text{ veh/h}}{3600 \text{ s/h}} = 0.667 \text{ veh/s}$$

Checking to make certain that capacity exceeds arrivals, note that the capacity ($sg$) is 16 veh/cycle ($0.667 \times 24$), which is greater than (permitting fractions of vehicles for the sake of clarity) the 11.12 arrivals ($vC = 0.139 \times 80$). Therefore, Eqs. 7.16 to 7.22 are valid. By definition,

$$r = C - g$$
$$= 80 - 24 = 56 \text{ s}$$

This leads to the following values:

1.  Time to queue clearance after the start of the effective green (Eq. 7.16),

$$t_c = \frac{0.139(56)}{(0.667 - 0.139)}$$
$$= 14.74 \text{ s}$$

2.  Proportion of the cycle with a queue (Eq. 7.17),

$$P_q = \frac{56 + 14.74}{80}$$
$$= 0.884$$

3.  Proportion of vehicles stopped (Eq. 7.18),

$$P_s = \frac{14.74}{0.139/0.667(80)}$$
$$= 0.884$$

4.  Maximum number of vehicles in the queue (Eq. 7.19),

$$Q_{max} = 0.139(56)$$
$$= 7.78 \text{ veh}$$

5.  Total vehicle delay per cycle (Eq. 7.20),

$$D_t = \frac{0.139(56)^2}{2(1 - 0.139/0.667)}$$
$$= 275.33 \text{ veh-s}$$

6.  Average delay per vehicle (Eq. 7.21),

$$d_{avg} = \frac{56^2}{2(80)(1-0.139/0.667)}$$
$$= 24.76 \text{ s/veh}$$

7.  Maximum delay of any vehicle (Eq. 7.22),

$$d_{max} = r$$
$$= 56 \text{ s}$$

## EXAMPLE 7.10   SIGNALIZED INTERSECTION ANALYSIS WITH $D/D/1$ QUEUING

Confirm the average delay result from Example 7.9 using a $D/D/1$ queuing diagram.

### SOLUTION

Again, with arrival and departure rates in units of vehicles per second,

$$v = 0.139 \text{ veh/s}$$
$$s = 0.667 \text{ veh/s}$$

and 24 seconds of effective green time in an 80-second signal cycle yields the following graph (Fig. 7.14) of cumulative arrivals and cumulative departures for one cycle.

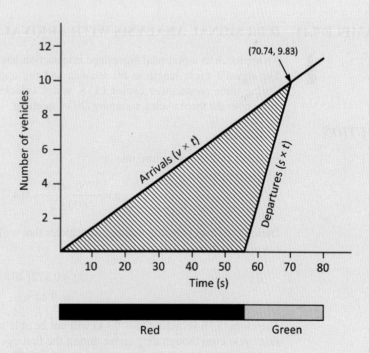

**Figure 7.14** $D/D/1$ queuing diagram for Example 7.10.

This leads to the following calculations for average delay:

1. Time for queue clearance after the start of the effective green remains unchanged at 14.74 s,

2. Total vehicle delay per cycle,

$$D_t = 0.5 \times r \times \left[ v \times (r + t_c) \right]$$
$$= 0.5 \times 56 \times \left[ 0.139 \times (56 + 14.74) \right]$$
$$= \underline{\underline{275.32 \text{ veh-s}}}$$

3. Average delay per vehicle,

$$d_{avg} = \frac{D_t}{vC} = \frac{275.32}{0.139 \times 80}$$
$$= \underline{\underline{24.76 \text{ s/veh}}}$$

This average delay value matches with that from Example 7.9.

Recall that Eqs. 7.16 through 7.22 are valid only when the arrivals are uniform throughout the cycle, the saturation flow rate is constant during the effective green period, and the approach capacity exceeds approach arrivals. For the case when approach arrivals exceed capacity for some signal cycles, $D/D/1$ queuing can again be used, as illustrated in the following example.

**EXAMPLE 7.11**   $D/D/1$ **SIGNAL ANALYSIS WITH ARRIVALS EXCEEDING CAPACITY**

An approach to a pretimed signalized intersection has a saturation flow rate of 1700 veh/h. The signal's cycle length is 60 seconds and the approach's effective red is 40 seconds. During three consecutive cycles 15, 8, and 4 vehicles arrive. Determine the total vehicle delay over the three cycles assuming $D/D/1$ queuing.

*SOLUTION*

For all cycles, the departure rate is

$$s = \frac{1700 \text{ veh/h}}{3600 \text{ s/h}} = 0.472 \text{ veh/s}$$

During the first cycle, the number of vehicles that will depart from the signal is (permitting fractions for the sake of clarity)

$$sg = 0.472(20)$$
$$= 9.44 \text{ veh}$$

Therefore, 5.56 vehicles (15 − 9.44) will not be able to pass through the intersection on the first cycle even though they arrive during the first cycle. At the end of the second cycle, 23 vehicles (15 + 8) will have arrived, but only 18.88 (2$sg$) will have departed, leaving 4.12 vehicles waiting at the beginning of the third cycle. At the end of the third cycle, a total of

27 vehicles will have arrived and as many as 28.32 ($3sg$) could have departed, so the queue that began to form during the first cycle will dissipate at some time during the third cycle. This process is shown graphically in Fig. 7.15.

From this figure, the total vehicle delay of the first cycle is (the area between arrival and departure curves) is

$$D_1 = 0.5(60)(15) - 0.5(20)(9.44)$$
$$= 355.6 \text{ veh-s}$$

Similarly, the delay in the second cycle is

$$D_2 = 0.5(60)(15+23) - (40)(9.44) - 0.5(20)(9.44+18.88)$$
$$= 479.2 \text{ veh-s}$$

To determine the delay in the third cycle, it is necessary to first know exactly when, in this cycle, the queue dissipates. The time to queue clearance after the start of the effective green, $t_c$, is (with $\lambda_3$ being the arrival rate during the third cycle and $n_3$ being the number of vehicles in the queue at the start of the third cycle)

$$n_3 + v_3(r+t_c) = st_c$$

where

$$v_3 = \frac{4 \text{ veh}}{60 \text{ sec}} = 0.067 \text{ veh/s}$$

Therefore,

$$(23 - 18.88) + 0.067(40 + t_c) = 0.472t_c$$

which gives $t_c = 16.8$ seconds. Thus the queue will clear 56.8 seconds (40 + 16.8) after the start of the third cycle, at which time a total of 26.8 vehicles (0.067 × 56.8 + 15 + 8) will have arrived at, and departed from, the intersection. The vehicle delay for the third cycle is

$$D_3 = 0.5(56.8)(23+26.8) - (40)(18.88) - 0.5(16.8)(18.88+26.8)$$
$$= 275.4 \text{ veh-s}$$

giving the total delay over all three cycles as

$$D_t = D_1 + D_2 + D_3$$
$$= 355.6 + 479.2 + 275.4$$
$$= 1110.2 \text{ veh-s}$$

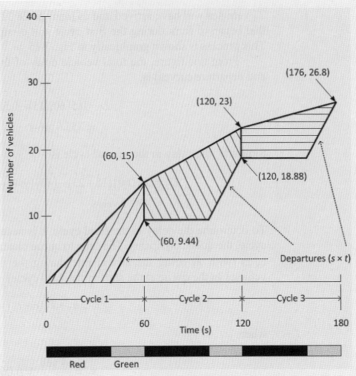

**Figure 7.15** $D/D/1$ queuing diagram for Example 7.11.

It is also possible to handle the case where intersection arrivals and/or departures are deterministic but time-varying in a fashion similar to that shown in Chapter 5. An example of time-varying arrivals is presented next.

**EXAMPLE 7.12**   **$D/D/1$ SIGNAL ANALYSIS WITH TIME-VARYING ARRIVALS**

The saturation flow rate of an approach to a pretimed signal is 6000 veh/h. The signal has a 60-second cycle with 20 seconds of effective red allocated to the approach. At the beginning of an effective red (with no vehicles remaining in the queue from a previous cycle), vehicles start arriving at a rate $v(t) = 0.4 + 0.01t + 0.00057t^2$ [where $v(t)$ is in vehicles per second and $t$ is the number of seconds from the beginning of the cycle]. Thirty seconds into the cycle the arrival rate remains constant at its 30-second level and stays at that rate until the end of the cycle. What is the total vehicle delay over the cycle (in vehicle-seconds), assuming $D/D/1$ queuing?

**SOLUTION**

Vehicle arrivals for the first 30 seconds (again allowing fractions of vehicles for the sake of clarity) are

$$\int_0^{30} 0.4 + 0.01t + 0.00057t^2 = 0.4t + 0.005t^2 + 0.00019t^3 \Big|_0^{30}$$

$$= 21.63 \text{ veh}$$

The arrival rate after 30 seconds is 1.213 veh/s [$0.4 + 0.01(30) + 0.00057(30)^2$] and is no longer time-varying. During the effective green, the departure rate is 1.667 veh/s

(6000/3600). To determine when the queue will clear, let $t'$ be the time after 30 seconds (the time after which the arrival rate is no longer time-varying), so

$$21.63 + 1.213t' = 1.667(t' + 10)$$

which gives $t' = 10.93$ seconds. Thus the queue clears at 40.93 seconds after the beginning of the cycle. For delay, the area under the arrival curve is

$$\int_0^{30} 0.4t + 0.005t^2 + 0.00019t^3 + \frac{1}{2}\Big[21.63 + \big(1.213(10.93) + 21.63\big)\Big](10.93)$$

$$= 0.2t^2 + 0.00167t^3 + 0.0000475t^4 \Big|_0^{30} + 308.87$$

$$= 263.48 + 308.87$$

$$= 572.35 \text{ veh-s}$$

The area under the departure curve is

$$\frac{1}{2}\big[34.90 \times 20.94\big] = 365.42 \text{ veh-s}$$

So total vehicle delay over the cycle is <u>206.93 veh-s</u> (572.35 − 365.42).

## 7.5.2   Signal Coordination

The main limitation with equations 7.16-7.22 is that their derivations are based on the assumption of uniform traffic arrivals throughout the cycle and a constant saturation flow rate during the effective green period, as well as no queue at the start of the effective red period or at the end of the effective green period. As discussed in Chapter 5, non-uniform arrivals are very likely in most traffic flow situations, and are frequently the case for signalized arterials. Thus, the assumption of uniform arrivals throughout the cycle is generally unrealistic and is likely to yield considerable error compared with values measured in the field.

One of the biggest influences on the arrival pattern of vehicles at each approach to the intersection is the level of signal coordination along the roadway. The term signal coordination generally refers to the level of timing coordination, or synchronization, between adjacent signals on the roadway.

Ideally, traffic signals should be timed so that as many vehicles as possible arrive at the signalized intersection when the signal indication is green, which would indicate good signal coordination. If a large proportion of vehicles arrive at an intersection during the red interval, the signal coordination would be considered poor. From a delay perspective, the larger the proportion of vehicles arriving during green, the lower the signal delay, and the larger the proportion of vehicles arriving during red, the higher the signal delay.

The effect of signal coordination on traffic arrival patterns is referred to as progression quality. Quantitatively, progression quality is expressed as the number of vehicles that arrive at an intersection approach while the signal indication is green for that approach, relative to all vehicles that arrive at that intersection approach during the entire signal cycle. This value is denoted as *PVG*, for Proportion of Vehicles arriving on Green. A general overview of signal coordination is provided in the rest of this section, as well as an example delay calculation for different arrivals rates during the green and red periods.

### Fundamental Relationships

The three most significant factors affecting progression quality are signal spacing, vehicle speed, and cycle length. The relationship between signal spacing and vehicle speed is most easily illustrated by considering a one-way arterial. Consider two intersections on a street running east-west separated by some distance $d_o$. With traffic traveling westbound, the time at which the signal phase of the westernmost signal (downstream signal) turns green after the easternmost (upstream) signal phase turns green should be equal to the travel time between the two intersections. The time difference between the start of the green between corresponding phases at adjacent signalized intersections is referred to as the offset, and is calculated as

$$offset = \frac{d_o}{V} \qquad (7.23)$$

*where*

$offset$ = start of green phase for downstream intersection relative to upstream intersection, for the same traffic movement, in seconds,

$d_o$ = distance between upstream and downstream intersection for offset calculation, in meters, and

$V$ = travel speed between upstream and downstream intersection, in m/s.

In practice, to take driver perception-reaction times into account, the downstream signal should turn green a few seconds before the lead vehicles reach the stop bar. For simplification, this time is not considered in the forthcoming example.

---

**EXAMPLE 7.13    CALCULATE OFFSET FOR IDEAL ONE-WAY PROGRESSION**

Two intersections on a one-way arterial are separated by 800 m. The average running speed of vehicles along this segment is 64 km/h. Calculate the offset necessary for ideal progression between the two intersections.

**SOLUTION**

Applying Eq. 7.23,

$$offset = \frac{800 \text{ m}}{64 \text{ km/h} \times \dfrac{1000 \text{ m/km}}{3600 \text{ s/h}}}$$

$$= 45 \text{ s}$$

---

It is worth noting that coordinating signal timing between adjacent intersections for good progression in only one direction is very straightforward. However, for an arterial with traffic in both directions, the setting of the offset for ideal progression in one direction may lead to poor progression in the other direction.

**EXAMPLE 7.14    DRAW A TIME-SPACE DIAGRAM**

For Example 7.13, draw the time-space diagram, assuming that the cycle length is 60 seconds and the $g/C$ ratio is 0.5, for both intersections.

*SOLUTION*

For a 60-second cycle length and a $g/C$ of 0.5, the major street effective green and effective red times will each be 30 seconds. The time-space diagram for this situation is shown in Fig. 7.16. In this figure, the solid diagonal lines represent vehicles traveling from intersection 1 to intersection 2, and the dashed diagonal lines represent vehicles traveling from intersection 2 to intersection 1. The slope of these lines represent the vehicle speed.

In Fig. 7.16, the traffic flowing from intersection 1 to intersection 2 has ideal progression. However, also notice that the traffic flowing from intersection 2 to intersection 1 has the worst-case progression; that is, all traffic arriving on green at intersection 2 arrives on the red at intersection 1.

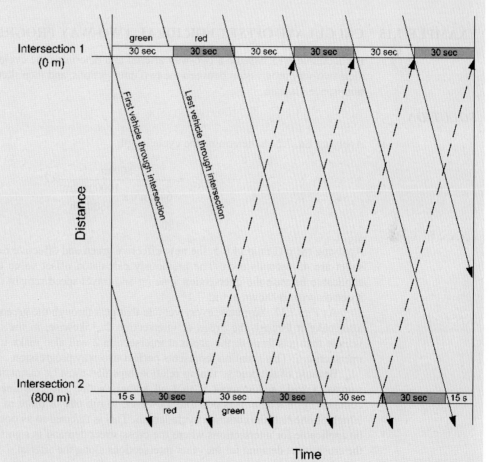

**Figure 7.16**.  Time-space diagram illustrating good coordination for only one direction, for Example 7.14.

To obtain good progression for both directions of travel in the previous example problem, the cycle length must be considered. For good progression in both directions, the cycle length (for both intersections) needs to be twice the travel time from Intersection 1 to Intersection 2. That is, Eq. 7.23 is multiplied by 2, as shown in Eq. 7.24 (assuming that the travel speed is the same in both directions).

$$C_{prog} = \frac{d_o}{V} \times 2 \tag{7.24}$$

where

$C_{prog}$ = cycle length necessary for ideal two-way progression, in seconds, and
Other terms as previously defined.

When planning new construction, this relationship can help guide the location of new signalized intersections. For an analysis of an existing roadway, however, the objective is to usually find a common cycle length and corresponding offset values that produce the best progression through the existing intersections.

### EXAMPLE 7.15    CALCULATE OFFSET FOR IDEAL TWO-WAY PROGRESSION

For Example 7.13, assume a two-way arterial and determine the cycle length necessary for ideal two-way progression between the two intersections, and then sketch the corresponding time-space diagram.

**SOLUTION**

Applying Eq. 7.24 to determine the cycle length,

$$C_{prog} = \frac{800 \text{ m}}{64 \text{ km/h} \times \dfrac{1000 \text{ m/km}}{3600 \text{ s/h}}} \times 2$$

$$= 90 \text{ s}$$

Applying the $g/C$ ratio of 0.5, the new effective green and effective red times for the major street are 45 seconds each. The previously calculated offset value of 45 seconds is still applicable because the intersection spacing and travel speed remain unchanged. The time-space diagram is shown in Fig. 7.17.

As Fig. 7.17 illustrates, every vehicle that gets through the green at intersection 1 will also make it through the green at intersection 2. Likewise, in the other direction, every vehicle that gets through the green at intersection 2 will also make it through the green at intersection 1. This situation represents perfect two-way progression.

Because of the precise timing relationships that must be maintained between adjacent signals, a basic requirement is that all signals within the coordinated system run on a common cycle length. It is possible, however, to run one or more of the signals at a cycle length of one-half the common cycle length. This is referred to as double-cycling, and may be applicable for intersections where the cross-street demand is significantly different from the cross-street demand for the other intersections along the arterial.

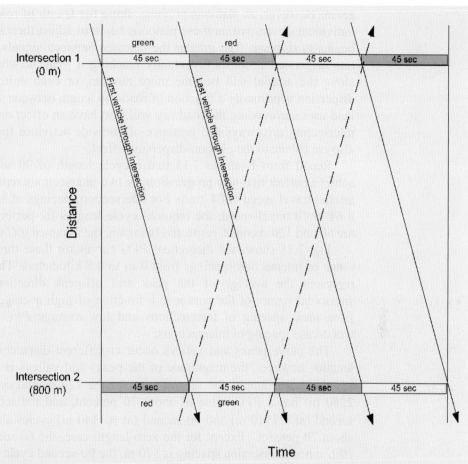

**Figure 7.17.** Time-space diagram illustrating good coordination for both directions, for Example 7.15.

Two other factors that can significantly affect progression quality are $g/C$ ratio and platoon dispersion, as described in the following sections.

### Effective Green to Cycle Length Ratio (g/C)

To illustrate the influence of the $g/C$ ratio, first consider an intersection approach that has 100% $g/C$ ratio (constant green). For this situation, the *PVG* would be 100% because there would never be a red indication. Next, consider the opposite case where an approach has a $g/C$ ratio of 0% (constant red). In this case the *PVG* would be 0%, because there would never be a green indication for vehicle arrivals. Thus, for the unlikely case of uniform vehicle arrivals, the *PVG* will be equal to the g/C ratio. In more realistic vehicle distributions, the *PVG* will not equal the $g/C$ ratio, but the $g/C$ ratio still serves as a limiting condition, as shown later in Fig. 7.18.

### Platoon Dispersion

When queued vehicles depart an intersection after the start of a green phase, they are usually closely spaced. These closely spaced groupings of vehicles are referred to as platoons. One of the goals of signal coordination is to maintain these platoons of vehicles and allow them to arrive at successive downstream intersections on the green. However, as platoons progress along the length of roadway between signals, individual drivers within these platoons begin to adjust their speeds, and the platoon begins to disperse. The greater the distance between signals, the more pronounced this dispersion becomes, eventually reaching a point at which the flow of traffic along the arterial will become more random, or even uniform. Although platoon dispersion is primarily a function of roadway length between signals, the character of land uses surrounding the roadway will also have an effect on platoon dispersion, as intersecting driveways and presence of curbside activities (parking, bus stops, etc.) also contribute to the platoon-dispersing effect.

Recall from Example 7.15 that a cycle length of 90 seconds was required to achieve perfect two-way progression for two intersections separated by 800 m and an arterial travel speed of 64 km/h. For intersection spacings of 570 m and 1140 m, with a 64 km/h travel speed, the required cycle lengths for perfect two-way progression are 60 and 120 seconds, respectively (again, the assumed $g/C$ value is 0.5).

Fig. 7.18 shows the theoretical $PVG$ curves for these three cycle lengths over a range of intersection spacings from 0 m to 4.8 kilometers. The graphed $PVG$ values represent the average of the peak and off-peak directional $PVG$ values. The sinusoidal nature of the curves is a function of high average $PVG$ values resulting from ideal spacing of intersections and low average $PVG$ values resulting from worst-case spacing of intersections.

The curve peaks and valleys occur at different distances for each of the cycle lengths; however, the magnitude of the peaks and valleys is consistent across cycle lengths. For example, the second peak for the 120-second cycle (which occurs at $\approx$ 2280 m) has a $PVG$ value of about 70 percent, and the second peaks for the 90-second (at $\approx$ 1710 m) and 60-second (at $\approx$ 1140 m) cycles also have $PVG$ values of about 70 percent. Except for the zero-length case, the 60-second cycle has the best $PVG$ when intersection spacing is 570 m, the 90-second cycle has the best $PVG$ when spacing is 840 m, and the 120-second cycle has the best $PVG$ when spacing is 1140 m.

Note that $PVG$ values eventually "dampen out" as the length between signals gets large, reflecting the effect of platoon dispersion. Also note that the $PVG$ values converge on the $g/C$ ratio of 0.5.

**Figure 7.18** Proportion of vehicles arriving on green ($PVG$) for different cycle lengths over a range of intersection spacings.

The following example demonstrates how $D/D/1$ queuing can be used to estimate signal delay for non-uniform arrivals.

**EXAMPLE 7.16   CALCULATE SIGNAL DELAY CONSIDERING PROGRESSION**

Consider Example 7.10, but instead of uniform arrivals throughout the cycle, 60% of the traffic volume arrives during the green indication. Determine the average delay for this condition.

*SOLUTION*

In this case, the arrivals are not uniform throughout the cycle; therefore, the arrival rate during the effective green and effective red periods will be different. These rates are readily calculated by using the overall arrival rate, the effective green time, the cycle length, and the proportion of vehicles arriving on green, as follows:

$$v_{green} = \frac{v \times PVG}{g/C} = \frac{0.139 \times 0.6}{24/80}$$
$$= 0.278 \text{ veh/s}$$

$$v_{red} = \frac{v \times (1 - PVG)}{1 - g/C} = \frac{0.139 \times (1 - 0.6)}{1 - 24/80}$$
$$= 0.0794 \text{ veh/s}$$

With these arrival rates and the other values for Example 7.10, the graph in Fig. 7.19 of cumulative arrivals and cumulative departures for one cycle results.

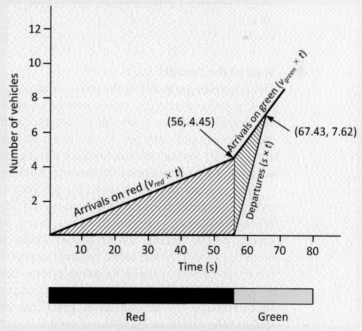

**Figure 7.19** $D/D/1$ queuing diagram for Example 7.16.

This leads to the following calculations for average delay:

1.  Time to queue clearance after the start of the effective green,

$$t_c = \frac{v_{red} \times r}{\left(s - v_{green}\right)} = \frac{0.0794(56)}{(0.667 - 0.278)}$$

$$= \underline{\underline{11.43\,\text{s}}}$$

2.  Total vehicle delay per cycle,

$$D_t = \left[0.5 \times \left(v_{red} \times r\right) \times r\right] + \left[0.5 \times \left(v_{red} \times r\right) \times t_c\right]$$

$$= \left[0.5 \times \left(0.0794 \times 56\right) \times 56\right] + \left[0.5 \times \left(0.0794 \times 56\right) \times 11.43\right]$$

$$= 124.50 + 25.41$$

$$= \underline{\underline{149.91\,\text{veh-s}}}$$

3.  Average delay per vehicle,

$$d_{avg} = \frac{D_t}{vC} = \frac{149.91}{0.139 \times 80}$$

$$= \underline{\underline{13.48\,\text{s/veh}}}$$

By timing the signal system so that 60% of the vehicles arrive on green at this approach, the average delay would be reduced by 11.28 s (24.76 – 13.48). Note that the *PVG* value for the uniform arrival case would be 0.3 (i.e., the same as the *g/C* ratio).

### State of the Practice

The examples presented in the previous section were highly simplified and idealized. Real signalized arterials rarely consist of only two intersections, and the spacing between intersections is seldom consistent. Determining the appropriate offset values, green times, and common cycle length to minimize signal delay, for a two-way arterial with several intersections, is a complex problem. A closed-form analytical solution is usually not possible; thus, an iterative solution technique must be used.

The number of combinations for the above variable values for two directions of traffic flow through several intersections quickly reaches an extremely large number. When considering a network of streets, where several major arterials may intersect at several points, finding optimal signal timing settings (to support coordination and minimize delay) for the entire network cannot be done from a practical perspective. For these network-wide optimization efforts, the number of iterations required to find an optimal solution becomes so extreme that the solution algorithm must be run on a supercomputer for several days. Practicing engineers performing these kinds of studies rarely have access to these resources, or the available time.

Accordingly, modern software packages that perform signal timing "optimization" for arterial or network coordination employ a search algorithm that attempts to find the best solution amongst a greatly reduced number of variable-value combinations. This allows a solution to be found within a reasonable amount of time, using commonly available computing hardware. However, because these algorithms do not perform an exhaustive search of all of the variable-value combinations (as is generally required to find a truly optimal solution for this type of problem), the identified solution is considered to be only a "local" optimum rather than a "global" optimum. So while the local optimum solution may provide reasonable results, it is unlikely that this particular set of variable settings will provide the best results. Nonetheless, with the tremendous increase in desktop computing power over recent decades, the differences between a local optimum and global optimum solution have decreased, as the search space (the set of variable values considered by the solution algorithm) has accordingly been expanded.

While computer software can identify optimal, or nearly optimal, signal timing settings, there are practical situations that can still prevent a signal system from running at its highest possible efficiency, such as accommodating emergency and transit vehicle preemption (priority). Additionally, recall that signals operating under actuated control do not have a fixed cycle length. To include such signals in a coordinated system, they must be reprogrammed to run on a fixed cycle length, which could possibly result in a net loss of efficiency for those signals. One alternative that is commonly employed in this situation is to convert the signal operation from fully actuated to semi-actuated, where any green time not used for the minor street phases is allocated to the major street phases to keep the cycle length constant.

It should also be pointed out that sometimes signal timing may be optimized for measures other than signal delay, such as number of stopped vehicles, emissions output, and so on. Because different optimization criteria almost always provide different results, many signal timing strategies seek to balance the values of more than one factor (such as vehicle stops and vehicle delays). Needless to say, the optimization-measurement problem makes this aspect of traffic analysis a fruitful area for future research.

### 7.5.3   Control Delay Calculation for Level of Service Analysis

The level-of-service concept was introduced in Chapter 6 for uninterrupted-flow facilities. This same concept also applies to interrupted-flow facilities. Although a variety of performance measures can be calculated for signalized intersections, as demonstrated previously, only one measure has been chosen as the service measure in the *Highway Capacity Manual* [Transportation Research Board 2010]. This measure is control delay, and it applies to both signalized and unsignalized intersections. Control delay (i.e., signal delay for the case of a signalized intersection) represents the total delay experienced by the driver as a result of the control, which includes delay due to deceleration time, queue move-up time, stop time, and acceleration time, as illustrated in Fig. 7.20.

Analytic methods for estimating delay, such as the $D/D/1$ queuing approach described previously, are generally not able to capture the delay due to deceleration and acceleration and thus usually underestimate the actual delay. Furthermore, while the assumption of uniform arrivals leads to the intuitive and straightforward $D/D/1$ queuing analysis approach, it has been found to underestimate delay when the $v/c$ ratio for an approach exceeds 0.5. This is because as the traffic intensity increases from moderate to a level nearing the capacity of the intersection, the probability of having cycle failures, where not all queued vehicles get through during a particular cycle, increases substantially. These cycle failures are random occurrences for the most part, but must be accounted for in the estimation of overall delay to achieve reasonably accurate results under higher flow conditions. Whereas Eq. 7.21 is based on a purely theoretical derivation, the need to account for stochastic vehicle arrivals which lead to occasional oversaturation adds considerable complexity to the analysis of delay at signalized intersections. Numerous researchers over the last several decades have proposed delay formulas and refinements to meet this need, based on combinations of analytical, empirical, and simulation-based methods. The following formulation, denoted as $d_2$, is one that has been adopted by the *Highway Capacity Manual* [Transportation Research Board 2010], given as

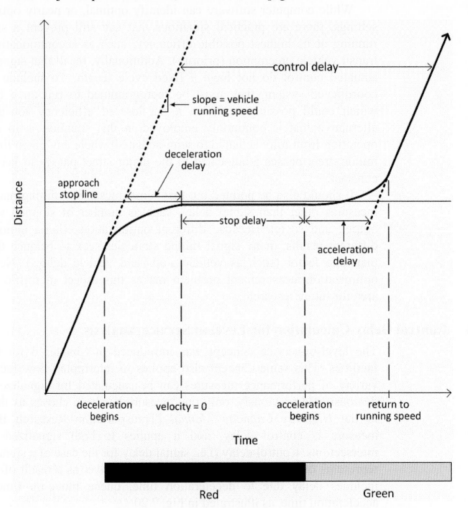

**Figure 7.20** Illustration of control delay for a single vehicle traveling through a signalized intersection.
Reproduced with Permission of the Transport Research Board, *Highway Capacity Manual* 2010, Copyright, National Academy of Science, Washington, D.C. Exhibit 31-5, p. 31-7.

$$d_2 = 900T\left[(X-1) + \sqrt{(X-1)^2 + \frac{8kIX}{cT}}\right] \qquad (7.25)$$

*where*

$d_2$ = average incremental delay per vehicle due to random arrivals and occasional oversaturation in seconds,

$T$ = duration of analysis period in $h$,

$X$ = $v/c$ ratio for lane group,

$k$ = delay adjustment factor that is dependent on signal controller mode,

$I$ = upstream filtering/metering adjustment factor, and

$c$ = lane group capacity in veh/h.

Assuming the analysis flow rate is based on the peak 15-minute traffic flow within the analysis hour, $T$ is equal to 0.25.

The signal-controller–mode delay adjustment factor, $k$, accounts for whether the intersection is operating in an actuated or pretimed mode. A value of 0.5 is used for intersections under pretimed control. Given that actuated intersection control usually results in more efficient handling of traffic volumes, $k$ can take on values less than 0.5 to account for this efficiency and resultant reduced delay. For actuated control, $k$ depends upon the $v/c$ ratio and the passage time (an actuated controller setting related to the maximum amount of time the green display can be extended due to a vehicle detector actuation). For the purposes of the example and end-of-chapter problems, pretimed signal control is assumed.

The upstream filtering/metering factor is used to adjust for the effect that an upstream signal has on the randomness of the arrival pattern at a downstream intersection. An upstream signal will typically have the effect of reducing the variance of the number of arrivals at the downstream intersection. $I$ is defined as

$$I = \frac{\text{variance of the number of arrivals per cycle}}{\text{mean number of arrivals per cycle}} \qquad (7.26)$$

Recall from Chapter 5 that for a Poisson distribution (which is used to represent random arrivals), the variance is equal to the mean. Thus, if the downstream arrival pattern of vehicles conforms to the Poisson distribution, this factor will be equal to 1.0. This is considered to be the case for signalized intersections operating in an isolated mode (intersections sufficiently distant from adjacent signalized intersections). For nonisolated signalized intersections, the $I$ value will be less than 1.0 due to the reduced variance of vehicle arrivals. The $I$ value is dependent on the $v/c$ ratio of the upstream movements that contribute to the downstream intersection volume. Again, for the purposes of this chapter, it will be assumed that a signal is operating in isolated mode, and thus the $I$ value will be equal to 1.0.

The *Highway Capacity Manual* [Transportation Research Board 2010] also includes a formula for estimating the delay caused by an initial queue of vehicles at the beginning of the analysis time period, denoted as $d_3$. This equation and its inputs are beyond the scope if this book; however, a simple illustration of this issue was included in Example 7.11. If no initial queue is present at the start of the analysis period, the $d_3$ term is simply set to zero.

Thus, estimation of signal delay as prescribed in the *Highway Capacity Manual* [Transportation Research Board 2010] includes two terms in addition to a term for calculating delay due to uniform arrivals, as follows:

$$d = d_1 + d_2 + d_3 \qquad (7.27)$$

*where*

> $d$ = average signal delay per vehicle in seconds,
> $d_1$ = average delay per vehicle due to uniform arrivals in seconds,
> $d_2$ = average delay per vehicle due to random arrivals in seconds, and
> $d_3$ = average delay per vehicle due to initial queue at start of analysis time period, in seconds.

For signalized intersections where vehicle arrivals are not influenced by coordination, isolated signalized intersections for one, the *PVG* is considered to be equal to the $g/C$ ratio. In this case, Eq. 7.21 can be applied to determine $d_1$. If signal coordination results in different arrival rates for the green and red intervals, a calculation process similar to that used in Example 7.16 must be applied to determine $d_1$.

Note that Eq. 7.27 is applied to each established lane group in the intersection. Individual lane group delays must be aggregated to arrive at an overall intersection delay, which will be discussed in the following section.

## EXAMPLE 7.17   CALCULATE SIGNAL DELAY USING EQ. 7.27

Compute the average approach delay per cycle using Eq. 7.27, given the conditions described in Example 7.9. Assume the intersection is isolated, the traffic flow accounts for the peak 15-min period, and that there is no initial queue at the start of the analysis period.

**SOLUTION**

In this example, the uniform delay is computed as (using the alternative version of Eq. 7.21)

$$d_1 = \frac{0.5C\left(1 - \dfrac{g}{C}\right)^2}{1 - \left(\dfrac{v}{c} \times \dfrac{g}{C}\right)}$$

*with*

$$C = 80 \text{ s}$$
$$g = 24 \text{ s}$$

The *v/c* ratio is calculated as

$$\frac{v}{c} = \frac{v}{s \times g/C}$$

*with*

$v$ = 500 veh/h

$s$ = 2400 veh/h

giving

$$\frac{v}{c} = \frac{500}{2400 \times 24/80} = \frac{500}{720} = 0.694$$

Uniform delay is calculated as

$$d_1 = \frac{0.5(80)\left(1 - \dfrac{24}{80}\right)^2}{1 - \left[0.694 \times \dfrac{24}{80}\right]} = 24.75 \text{ s}$$

This value matches the delay value computed in Example 7.9 (the small difference is due to rounding of arrival and departure rates in Example 7.9). Random delay will be computed as (using Eq. 7.25)

$$d_2 = 900T\left[(X-1) + \sqrt{(X-1)^2 + \frac{8kIX}{cT}}\right]$$

*with*

$T$ = 0.25 (15 min, from problem statement)

$X$ = 0.694 (from above)

$k$ = 0.5 (pretimed control)

$I$ = 1.0 (isolated mode)

$c$ = 720 veh/h (from above)

$$d_2 = 900(0.25)\left[(0.694 - 1) + \sqrt{(0.694 - 1)^2 + \frac{8(0.5)(1.0)0.694}{(720)0.25}}\right]$$

$$= 5.45 \text{ s}$$

Now the total signal delay is computed using Eq. 7.27:

$$d = d_1 + d_2 + d_3$$

*with*

$d_1$ = 24.75 s

$d_2$ = 5.45 s

$d_3$ = 0 s (given)

$$d = 24.75 + 5.45 + 0 = \underline{\underline{30.20 \text{ s/veh}}}$$

As this example illustrates, ignoring the $d_2$ component of delay when the $v/c$ ratio is above 0.5 results in a significant underestimation of average delay.

### 7.5.4 Level-of-Service Determination

Before the implementation of any developed signal phasing and timing plan, the level of service should be determined to assess whether the intersection will operate at an acceptable level under this plan. As previously mentioned, the service measure (the performance measure by which level of service is assessed) for signalized intersections is delay. In previous examples we calculated the delay for a specific lane group, but now we must do this for all lane groups and then aggregate the delay values to arrive at an overall intersection delay measure and corresponding level of service.

The first step is to aggregate the delays of all lane groups for an approach, and then repeat the procedure for each approach of the intersection. This will result in approach-specific delays and levels of service. The aggregated lane group delay for each approach is given by

$$d_A = \frac{\sum_i d_i v_i}{\sum_i v_i} \tag{7.28}$$

*where*

    $d_A$ = average delay per vehicle for approach $A$ in seconds,
    $d_i$ = average delay per vehicle for lane group $i$ (on approach $A$) in seconds, and
    $v_i$ = analysis flow rate for lane group $i$ in veh/h.

Once all the approach delays have been calculated, they can be aggregated to arrive at the overall intersection delay. The aggregated approach delay for the intersection is given by

$$d_I = \frac{\sum_A d_A v_A}{\sum_A v_A} \tag{7.29}$$

*where*

    $d_I$ = average delay per vehicle for the intersection in seconds,
    $d_A$ = average delay per vehicle for approach $A$ in seconds, and
    $v_A$ = analysis flow rate for approach $A$ in veh/h.

The delay level-of-service criteria for signalized intersections are specified in the *Highway Capacity Manual* [Transportation Research Board 2010] and are given in

Table 7.4. These delay criteria can be used to determine the level of service for a lane group, an approach, and the intersection.

**Table 7.4** Level-of-Service Criteria for Signalized Intersections

Reproduced with permission of the Transportation Research Board, *Highway Capacity Manual* 2010, Copyright, National Academy of Sciences, Washington, D.C. Exhibit 18-4, p. 18-6.

| LOS | Control delay per vehicle (s/veh) |
|-----|-----------------------------------|
| A | $\leq 10$ |
| B | $> 10$–20 |
| C | $> 20$–35 |
| D | $> 35$–55 |
| E | $> 55$–80 |
| F | $> 80$ |

## EXAMPLE 7.18   DETERMINE LEVEL OF SERVICE FOR APPROACH

Determine the level of service for the eastbound approach of Maple Street assuming no initial queue at the start of the analysis period.

*SOLUTION*

Lane groups for this intersection were established in Example 7.2. Two lane groups were established for the eastbound (EB) approach, one for the left-turn movement and the other for the combined through and right-turn movements. The delay will be calculated for the left-turn lane group first, followed by the through/right-turn lane group.

For the left-turn lane group, the uniform delay is computed using Eq. 7.21 with

$$C = 65 \text{ s}$$

$$g = 12.5 \text{ s}$$

$$\frac{v}{c} = \frac{v}{s \times g/C} = \frac{300}{1750 \times 12.5/65} = \frac{300}{337} = 0.891$$

giving

$$d_1 = \frac{0.5(65)\left(1 - \dfrac{12.5}{65}\right)^2}{1 - \left[0.891 \times \dfrac{12.5}{65}\right]} = 25.6 \text{ s}$$

Random delay is computed using Eq. 7.25 with

$T = 0.25$ (15 min)

$X = 0.891$ (from above)

$k = 0.5$ (pretimed control)

$I = 1.0$ (isolated mode)

$c = 337$ veh/h (from above)

giving

$$d_2 = 900(0.25)\left[(0.891-1)+\sqrt{(0.891-1)^2+\frac{8(0.5)(1.0)0.891}{(337)0.25}}\right]$$

$$= 27.8 \text{ s}$$

Now, the total signal delay is computed using Eq. 7.27 with

$d_1$ = 25.6 s
$d_2$ = 27.8 s
$d_3$ = 0 s (given)

$$d = 25.6+27.8+0 = 53.4 \text{ s}$$

From Table 7.4, this lane group delay corresponds to a level of service of D.

For the through/right-turn lane group, the uniform delay is computed using Eq. 7.21 with

$$C = 65 \text{ s}$$
$$g = 24.7 \text{ s}$$
$$\frac{v}{c} = \frac{v}{s \times g/C} = \frac{1100}{3400 \times 24.7/65} = \frac{1100}{1292} = 0.851$$

giving

$$d_1 = \frac{0.5(65)\left(1-\frac{24.7}{65}\right)^2}{1-\left[0.851\times\frac{24.7}{65}\right]} = 18.5 \text{ s}$$

Random delay is computed using Eq. 7.25 with

$T$ = 0.25 (15 min)
$X$ = 0.851 (from above)
$k$ = 0.5 (pretimed control)
$I$ = 1.0 (isolated mode)
$c$ = 1292 veh/h (from above)

giving

$$d_2 = 900(0.25)\left[(0.851-1)+\sqrt{(0.851-1)^2+\frac{8(0.5)(1.0)0.851}{(1292)0.25}}\right]$$

$$= 7.2 \text{ s}$$

Now, the average signal delay for this lane group is computed using Eq. 7.27 with

$$d_1 = 18.5 \text{ s}$$
$$d_2 = 7.2 \text{ s}$$
$$d_3 = 0 \text{ s (given)}$$

$$d = 18.5 + 7.2 + 0 = 25.7 \text{ s}$$

From Table 7.4, this lane group delay corresponds to a level of service of C.

Now, to compute the volume-weighted aggregate delay for the approach, we use Eq. 7.28:

$$d_A = \frac{\sum d_i v_i}{\sum v_i}$$

*with*

$$d_{LT} = \text{delay for left-turn lane group}$$
$$d_{T/R} = \text{delay for through/right-turn lane group}$$
$$v_{LT} = \text{analysis flow rate for left-turn lane group}$$
$$v_{T/R} = \text{analysis flow rate for through/right-turn lane group}$$

giving

$$d_{EB} = \frac{v_{LT} \times d_{LT} + v_{T/R} \times d_{T/R}}{v_{LT} + v_{T/R}}$$
$$= \frac{300 \times 53.4 + 1100 \times 25.7}{300 + 1100}$$
$$= \frac{44,290}{1400}$$
$$= \underline{\underline{31.6 \text{ s}}}$$

From Table 7.4, this approach delay corresponds to a <u>level of service of C</u>.

## EXAMPLE 7.19   DETERMINE LEVEL OF SERVICE FOR INTERSECTION

Determine the level of service for the intersection of Maple and Vine Streets.

### SOLUTION

The delay for each of the other three approaches (WB, NB, SB) can be determined by the exact same process used in Example 7.18 for the EB approach. Due to the length of the calculations involved for the remaining three approaches, the results are summarized in Table 7.5.

In this table, note that the $v/c$ ratios for the critical lane groups match (rounding differences aside) the calculated critical intersection $v/c$ ratio, $X_c$, as they should, because green time was allocated based on the strategy of equalizing $v/c$ ratios for the critical lane groups in each phase (using Eq. 7.11).

The overall intersection delay calculation will be shown for the sake of clarity. Using Eq. 7.29, the intersection delay is given by

$$d_I = \frac{31.6 \times 1400 + 30.2 \times 1400 + 49.7 \times 480 + 42.3 \times 440}{1400 + 1400 + 480 + 440}$$

$$= \frac{128,988}{3,720}$$

$$= \underline{34.7 \text{ s}}$$

It is worth pointing out that, although all but two lane groups (EB T/R and WB T/R) have a level of service of D or worse, the much higher volumes for the EB T/R and WB T/R lane groups relative to the other lane groups keeps the <u>level of service at C</u> (albeit barely) for the intersection due to the volume weighting in the delay aggregation.

**Table 7.5** Summary of Delay and Level-of-Service Calculations for the Intersection of Maple and Vine Streets

| Approach | EB | | WB | | NB | | SB | |
|---|---|---|---|---|---|---|---|---|
| Lane group | LT | T/R | LT | T/R | LT | T/R | LT | T/R |
| Analysis flow rate ($v$) | 300 | 1100 | 250 | 1150 | 90 | 390 | 70 | 370 |
| Saturation flow rate ($s$) | 1750 | 3400 | 1750 | 3400 | 475 | 1800 | 450 | 1800 |
| Flow ratio ($v/s$) | 0.171 | 0.324 | 0.143 | 0.338 | 0.189 | 0.217 | 0.156 | 0.206 |
| Critical lane group (✓) | ✓ | | | ✓ | | ✓ | | |
| $Y_c$ | | | | 0.726 | | | | |
| Lost time/phase | | | | 4 | | | | |
| Total lost time | | | | 12 | | | | |
| Cycle length ($C_{min}$) | | | | 65 | | | | |
| $X_c$ | | | | 0.891 | | | | |
| Eff. green time ($g$) | 12.5 | 24.7 | 12.5 | 24.7 | 15.8 | 15.8 | 15.8 | 15.8 |
| $g/C$ | 0.192 | 0.380 | 0.192 | 0.380 | 0.243 | 0.243 | 0.243 | 0.243 |
| Lane group capacity ($c$) | 337 | 1292 | 337 | 1292 | 115 | 438 | 109 | 438 |
| $v/c$ ($X$) | 0.891 | 0.851 | 0.743 | 0.890 | 0.779 | 0.891 | 0.640 | 0.846 |
| $d_1$ | 25.6 | 18.5 | 24.7 | 18.9 | 23.0 | 23.8 | 22.1 | 23.4 |
| $k$ | | | | 0.5 | | | | |
| $I$ | | | | 1.0 | | | | |
| $T$ | | | | 0.25 | | | | |
| $d_2$ | 27.8 | 7.2 | 13.8 | 9.5 | 39.4 | 23.0 | 25.3 | 17.9 |
| Lane group delay ($d$) | 53.4 | 25.7 | 38.5 | 28.3 | 62.4 | 46.7 | 47.3 | 41.4 |
| Lane group LOS | D | C | D | C | E | D | D | D |
| Approach delay | 31.6 | | 30.2 | | 49.7 | | 42.3 | |
| Approach LOS | C | | C | | D | | D | |
| Intersection delay | | | | 34.7 | | | | |
| Intersection LOS | | | | C | | | | |

## NOMENCLATURE FOR CHAPTER 7

$a$     deceleration rate for vehicle at an intersection

$AR$     all-red time

$C$     cycle length

$C_{min}$     minimum cycle length

$C_{opt}$     optimum cycle length

$c$     capacity

$d_{avg}$     average vehicle delay per cycle ($D/D/1$ queuing)

$d_1$     average vehicle delay per cycle assuming uniform arrivals

$d_2$     average vehicle delay per cycle due to random arrivals and occasional oversaturation

$d_3$     average vehicle delay per cycle due to initial queue at start of analysis time period

$d_d$     distance from the intersection for which the dilemma zone is avoided

$d_{max}$     maximum delay of any vehicle ($D/D/1$ queuing)

$D$     deterministic arrivals or departures

$D_t$     total vehicle delay ($D/D/1$ queuing)

$g$     effective green time or acceleration due to gravity

$G$     displayed green time or grade of roadway

$G_p$     pedestrian green time

$I$     upstream filtering/metering adjustment factor

$k$     delay adjustment factor dependent on signal controller mode

$l$     vehicle length

$L$     total cycle lost time or crosswalk length

$n$     number of phases or number of vehicles or number of critical lane groups

$N_{ped}$     number of crossing pedestrians per phase

$P_q$     proportion of the signal cycle with a queue ($D/D/1$ queuing)

$P_s$     proportion of stopped vehicles ($D/D/1$ queuing)

$Q$     number of vehicles in the queue

$Q_{max}$     maximum number of vehicles in queue ($D/D/1$ queuing)

$r$     effective red time

$R$     displayed red time

$s$     saturation flow rate

$S_p$     pedestrian walking speed

$t$     time

$t_c$     time after the start of effective green until queue clearance ($D/D/1$ queuing)

$t_L$     total lost time for a movement during a cycle

$t_r$     driver perception/reaction time

$v$     analysis flow rate (also referred to as volume, arrival rate)

$V$     travel speed of vehicle

$w$     width of street

$W_E$     effective crosswalk width

$X_i$     volume-to-capacity ratio for lane group $i$

$X_c$     critical volume-to-capacity ratio for the intersection

$x_s$     distance required to stop

$Y$     displayed yellow time

$Y_c$     sum of flow ratios for critical lane groups

$\lambda$     arrival rate

$\mu$     departure rate

$\rho$     traffic intensity

## REFERENCES

Institute of Transportation Engineers (ITE). *Traffic Engineering Handbook*, 5th ed. Washington, DC, 1999.

Institute of Transportation Engineers (ITE). *Traffic Control Devices*, 2nd ed. Washington, DC, 2001.

Kell, James H., and Iris J. Fullerton. *Manual of Traffic Signal Design*, 2nd ed. Washington, DC: Institute of Transportation Engineers, 1998.

Transportation Research Board. *Traffic Flow Theory: A Monograph*. Special Report 165. Washington, DC: National Research Council, 1975.

Transportation Research Board. *Highway Capacity Manual*, Washington, DC: National Research Council, 2010.

U.S. Federal Highway Administration. *Manual on Uniform Traffic Control Devices for Streets and Highways*. Washington, DC: U.S. Government Printing Office, 2009.

Webster, F. V. *Traffic Signal Settings*, Road Research Technical Paper No. 39. London: Great Britain Road Research Laboratory, 1958.

## PROBLEMS

**7.1** An intersection has a three-phase signal with the movements allowed in each phase and corresponding analysis and saturation flow rates shown in Table 7.6. Calculate the sum of the flow ratios for the critical lane groups.

**7.2** An intersection has a four-phase signal with the movements allowed in each phase and corresponding analysis and saturation flow rates shown in Table 7.7. Calculate the sum of the flow ratios for the critical lane groups.

**7.3** The minimum cycle length for an intersection is determined to be 95 seconds. The critical lane group flow ratios were calculated as 0.235, 0.250, 0.170, and 0.125 for phases 1–4, respectively. What $X_c$ was used in the determination of this cycle length, assuming a lost time of 5 seconds per phase?

**7.4** A pretimed four-phase signal has critical lane group flow rates for the first three phases of 200, 187, and 210 veh/h (saturation flow rates are 1800 veh/h/ln for all phases). The lost time is known to be 4 seconds for each phase. If the cycle length is 60 seconds, what is the estimated effective green time of the fourth phase?

**7.5** A four-phase traffic signal has critical lane group flow ratios of 0.225, 0.175, 0.200, and 0.150. If the lost time per phase is 5 seconds and a critical intersection $v/c$ of 0.85 is desired, calculate the minimum cycle length and the phase effective green times such that the lane group $v/c$ ratios are equalized.

**7.6** For Problem 7.1, calculate the minimum cycle length and the effective green time for each phase (balancing $v/c$ for the critical lane groups). Assume the

lost time is 4 seconds per phase and a critical intersection $v/c$ of 0.90 is desired.

**7.7** For Problem 7.1, calculate the optimal cycle length (Webster's formulation) and the corresponding effective green times (based on lane group $v/c$ equalization). Assume lost time is 4 seconds per phase.

**7.8** For Problem 7.2, calculate the minimum cycle length and the effective green time for each phase (balancing $v/c$ for the critical movements). Assume the lost time is 4 seconds per phase and a critical intersection $v/c$ of 0.95 is desired.

**7.9** For Problem 7.2, calculate the optimal cycle length (Webster's formulation) and the corresponding effective green times (based on lane group $v/c$ equalization). Assume lost time is 4 seconds per phase.

**7.10** Consider Example 7.3. Two additional 3.6-m through lanes are added to Vine Street (the street in the intersection shown in Fig. 7.8), one lane in each direction. If the peak-hour traffic volumes are unchanged but the Vine Street left-turn saturation flow rates increase by 100 veh/h because of the added through lanes, what would the revised effective green time, yellow time, and all-red time be for each phase? Assume minimum cycle length and a critical intersection $v/c$ of 0.90 is desired.

**7.11** Consider the intersection of Vine and Maple Streets as shown in Fig. 7.8. Suppose Vine Street's northbound and southbound approaches are both on an 8% upgrade, and the assumed vehicle approach speed is 50 km/h. What should the yellow and all-red times be?

**Table 7.6** Data for Problem 7.1

| Phase | 1 | 2 | 3 | |
|---|---|---|---|---|
| Allowed movements | NB L, SB L | NB T/R, SB T/R | EB L, WB L | EB T/R, WB T/R |
| Analysis flow rate | 330, 365 veh/h | 1125, 1075 veh/h | 110, 80 veh/h | 250, 285 veh/h |
| Saturation flow rate | 1700, 1750 veh/h | 3400, 3300 veh/h | 650, 600 veh/h | 1750, 1800 veh/h |

**Table 7.7** Data for Problem 7.2

| Phase | 1 | 2 | 3 | 4 |
|---|---|---|---|---|
| Allowed movements | EB L, WB L | EB T/R, WB T/R | SB L, SB T/R | NB L, NB T/R |
| Analysis flow rate | 245, 230 veh/h | 975, 1030 veh/h | 255, 235 veh/h | 225, 215 veh/h |
| Saturation flow rate | 1750, 1725 veh/h | 3350, 3400 veh/h | 1725, 1750 veh/h | 1700, 1750 veh/h |

**7.12** Consider Problem 7.10. Calculate the new minimum pedestrian green time, assuming the effective crosswalk width is 2 m and the maximum number of crossing pedestrians in any phase is 20.

**7.13** An intersection approach has a saturation flow rate of 1500 veh/h, and vehicles arrive at the approach at the rate of 800 veh/h. The approach is controlled by a pretimed signal with a cycle length of 60 seconds and $D/D/1$ queuing holds. Local standards dictate that signals should be set such that all approach queues dissipate 10 seconds before the end of the effective green portion of the cycle. Assuming that approach capacity exceeds arrivals, determine the maximum length of effective red that will satisfy the local standards.

**7.14** An approach to a pretimed signal has 30 seconds of effective red, and $D/D/1$ queuing holds. The total delay at the approach is 83.33 veh-s/cycle and the saturation flow rate is 1000 veh/h. If the capacity of the approach equals the number of arrivals per cycle, determine the approach flow rate and cycle length.

**7.15** An approach to a pretimed signal has 25 seconds of effective green in a 60-second cycle. The approach volume is 500 veh/h and the saturation flow rate is 1400 veh/h. Calculate the average vehicle delay assuming $D/D/1$ queuing.

**7.16** An observer notes that an approach to a pretimed signal has a maximum of eight vehicles in a queue in a given cycle. If the saturation flow rate is 1440 veh/h and the effective red time is 40 seconds, how much time will it take this queue to clear after the start of the effective green (assuming that approach capacity exceeds arrivals and $D/D/1$ queuing applies)?

**7.17** An approach to a pretimed signal with a 60-second cycle has nine vehicles in the queue at the beginning of the effective green. Four of the nine vehicles in the queue are left over from the previous cycle (at the end of the previous cycle's effective green). The saturation flow rate of the approach is 1500 veh/h, total delay for the cycle is 5.78 vehicle-minutes, and at the end of the effective green there are 2 vehicles left in the queue. Determine the arrival rate assuming that it is unchanged over the duration of the observation period (from the beginning to the end of this 5.78–vehicle-minute delay cycle). (Assume $D/D/1$ queuing.)

**7.18** At the beginning of an effective red, vehicles are arriving at an approach at the rate of 500 veh/h and 16 vehicles are left in the queue from the previous cycle (at the end of the previous cycle's effective green). However, due to the end of a major sporting event, the arrival rate is continuously increasing at a constant rate of 200 veh/h/min (after 1 minute the arrival rate will be 700 veh/h, after 2 minutes 900 veh/h, etc.). The saturation flow rate of the approach is 1800 veh/h, the cycle length is 60 seconds, and the effective green time is 40 seconds. Determine the total vehicle delay until complete queue clearance. (Assume $D/D/1$ queuing.)

**7.19** The saturation flow rate for an intersection approach is 3600 veh/h. At the beginning of a cycle (effective red) no vehicles are queued. The signal is timed so that when the queue (from the continuously arriving vehicles) is 13 vehicles long, the effective green begins. If the queue dissipates 8 seconds before the end of the cycle and the cycle length is 60 seconds, what is the arrival rate, assuming $D/D/1$ queuing?

**7.20** The saturation flow rate for a pretimed signalized intersection approach is 1800 veh/h. The cycle length is 80 seconds. It is known that the arrival rate during the effective green is twice the arrival rate during the effective red. During one cycle, there are two vehicles in the queue at the beginning of the cycle (the beginning of the effective red) and there are eight vehicles in the queue at the end of the effective red (the beginning of the effective green). If the queue clears exactly at the end of the effective green and $D/D/1$ queuing applies, determine the total vehicle delay in the cycle (in veh-s).

**7.21** An approach to a signalized intersection has a saturation flow rate of 1800 veh/h. At the beginning of an effective red, there are six vehicles in the queue and vehicles arrive at 900 veh/h. The signal has a 60-second cycle with 25 seconds of effective red. What is the total vehicle delay after one cycle (assume $D/D/1$ queuing)?

**7.22** An approach to a signalized intersection has a saturation flow rate of 2640 veh/h. For one cycle, the approach has three vehicles in queue at the beginning of an effective red, and vehicles arrive at 1064 veh/h. The signal for the approach is timed such that the effective green starts eight seconds after the approach's vehicle queue reaches 10 vehicles, and lasts 15 seconds. What is the total vehicle delay for this signal cycle?

**7.23** An approach to a signal has a saturation flow rate of 1800 veh/h. During one 80-second cycle, there are four vehicles queued at the beginning of the cycle (the start of the effective red) and two vehicles queued at the end of the cycle (the end of the effective green). At the beginning of the effective green there are 10 vehicles in the queue. The arrival rate is constant and the process is $D/D/1$. If the effective red is known to be less than 40 seconds, what is the total vehicle delay for this signal cycle?

**7.24** Vehicles arrive at an approach to a pretimed signalized intersection. The arrival rate over the cycle is given by the function $v(t) = 0.22 + 0.012t$ [$v(t)$ is in veh/s and $t$ is in seconds]. There are no vehicles in the queue when the cycle (effective red) begins. The cycle length is 60 seconds and the saturation flow rate is 3600 veh/h. Determine the effective green and red times that will allow the queue to clear exactly at the end of the cycle (the end of the effective green), and determine the total vehicle delay over the cycle (assuming $D/D/1$ queuing).

**7.25** At the start of the effective red at an intersection approach to a pretimed signal, vehicles begin to arrive at a rate of 800 veh/h for the first 40 seconds and 500 veh/h from then on. The approach has a saturation flow rate of 1200 veh/h and an effective green of 20 seconds, and the cycle length is 40 seconds. What is the total vehicle delay two full cycles after the 800-veh/h arrival rate begins? (Assume $D/D/1$ queuing.)

**7.26** A left-turn movement has a maximum arrival rate of 200 veh/h. The saturation flow of this movement is 1400 veh/h. For this approach, the yellow time is 4 seconds, all red time is 2 seconds, and total lost time is 3 seconds. The cycle length is 120 seconds. What minimum displayed green time must be provided to ensure that the queue in each cycle clears, and what is the total delay per cycle and delay per vehicle for this green time? (Assume $D/D/1$ queuing.)

**7.27** Vehicles begin to arrive at a signal approach at a rate of $v(t) = 0.2286 + 0.0008t$ [with $v(t)$ in veh/s and $t$ in seconds] at the beginning of the cycle (the beginning of an effective red) and there are two vehicles already in the queue that are left over from the effective green of the previous cycle. The signal is designed so that the effective green starts when there are 10 vehicles in the queue. The saturation flow rate is 1800 veh/h. What is the total vehicle delay after one cycle (cycles are 60 seconds long) and when will the effective green start in cycle #2 (or, equivalently, how long will the effective red be in cycle #2)? (Assume $D/D/1$ queuing.)

**7.28** At the beginning of a signal approach's effective red there are 8 vehicles in queue. The arrival rate over the cycle is $v(t) = 0.05 + 0.001t$ [with $v(t)$ in veh/s and $t$ in seconds]. If the saturation flow rate is 1800 veh/h and the cycle length is 80 seconds, what is the minimum effective green time needed for this cycle to have zero vehicles in the queue when the effective red of the next cycle starts and what is the total delay in

this cycle with this green time. (Assume $D/D/1$ queuing.)

**7.29** A signal approach has 20 seconds of displayed green, 4 seconds of yellow, and 3 seconds of all red (start-up lost time and clearance times are typical). The cycle length is 60 seconds. At the beginning of an effective red there are no vehicles in queue and vehicles arrive at three-quarters of the saturation flow rate for 30 seconds. Then there is zero flow for 10 seconds, and then one-half of the saturation flow rate from 40 seconds until the end of the cycle. What is the total delay for this cycle if the saturation flow rate is 1400 veh/h? (Assume $D/D/1$ queuing.)

**7.30** An approach with a saturation flow rate of 1800 veh/h has 3 vehicles in queue at the start of an effective red. For the first cycle, the approach arrival rate is given by the function $v(t) = 0.5 - 0.005t$ [with $v(t)$ in veh/s and $t$ in seconds measured from the beginning of the effective red]. From the second cycle onward (starting at the beginning of the second effective red) vehicles arrive at a fixed rate of 720 veh/h. The approach has 26 seconds of effective red and a 60 second cycle for all cycles. How many cycles will it take to have no vehicles in the queue at the start of an effective red and what would be the total delay until this happens? (Assume $D/D/1$ queuing.)

**7.31** Vehicles arrive at a signal approach at a rate of $v(t) = 0.3 - 0.001t$ [with $v(t)$ in veh/s and $t$ in seconds measured from the beginning of the effective red of the first cycle]. The signal has a 70-second cycle length with 40 seconds of effective red. The saturation flow rate of the approach is 1800 veh/h. What is the total vehicle delay after two cycles (when $t = 140$ seconds) and when will the queue clear (measured from the beginning of the first cycle) during an effective green (that is, the $t$ at which there will no longer be a queue)? (Assume $D/D/1$ queuing.)

**7.32** An approach to a signalized intersection has a green time of 35 seconds, and all-red time of 2 seconds, a yellow time of 3 seconds, and a total lost time of 3 seconds. The arrival rate is $v(t) = 0.5 + 0.002t$ [with $v(t)$ in veh/s and $t$ in seconds measured from the beginning of the effective red of the cycle]. The saturation flow rate of the approach is 3400 veh/h. How long must the cycle length be so that the queue that forms at the beginning of the cycle (effective red) dissipates exactly at the end of the cycle (end of the effective green) and what would be the average delay per vehicle over the cycle? (Assume $D/D/1$ queuing.)

**7.33** Recent computations at an approach to a pretimed-signalized intersection indicate that the volume-to-capacity ratio is 0.8, the saturation flow rate is 1600 veh/h, and the effective green time is 50 seconds. If the uniform delay (assuming $D/D/1$ queuing) is 11.25 seconds per vehicle, determine the arrival flow rate (in veh/h) and the cycle length.

**7.34** At one signalized intersection approach, 20 vehicles, on average, arrive during the 30-second effective green time. During the rest of the cycle, 45 vehicles, on average, arrive at the intersection from this approach. The cycle length is 90 seconds. What is the proportion of vehicles arriving on green for this approach?

**7.35** Consider Problem 7.34. Assume the saturation flow rate is 8000 veh/h. Determine the average uniform delay for this approach.

**7.36** Consider Problems 7.34 and 7.35. Assume that improvements were made to the signal timing such that the PVG for this approach is now 0.45. Also assume that the average number of vehicle arrivals per cycle is still 65 (but the arrivals on green and red will be different from Prob. 7.34). Determine the new average uniform delay for this approach.

**7.37** An approach to a pretimed signal has 25 seconds of effective green, a saturation flow rate of 1300 veh/h, and a volume-to-capacity ratio less than 1. If the cycle length is 60 seconds and the overall delay formula (Eq. 7.27) estimates an average delay that is 34 s greater than that estimated by using just the uniform delay formula, determine the vehicle arrival rate. (Assume the signal is isolated and $d_3 = 0$.)

**7.38** For Problem 7.6, calculate the northbound average approach delay (using Eq. 7.27) and level of service.

**7.39** For Problem 7.6, calculate the southbound average approach delay (using Eq. 7.27) and level of service.

**7.40** For Problem 7.6, calculate the westbound average approach delay (using Eq. 7.27) and level of service.

**7.41** For Problem 7.6, calculate the eastbound average approach delay (using Eq. 7.27) and level of service.

**7.42** For Problem 7.6, calculate the overall intersection average delay (using Eqs. 7.27-7.29) and level of service.

**7.43** For Problem 7.8, calculate the northbound average approach delay (using Eq. 7.27) and level of service.

**7.44** For Problem 7.8, calculate the southbound average approach delay (using Eq. 7.27) and level of service.

**7.45** For Problem 7.8, calculate the westbound average approach delay (using Eq. 7.27) and level of service.

**7.46** For Problem 7.8, calculate the eastbound average approach delay (using Eq. 7.27) and level of service.

**7.47** For Problem 7.8, calculate the overall intersection average delay (using Eqs. 7.27-7.29) and level of service.

**7.48** A new shopping center opens near the intersection of Vine and Maple Streets (the intersection shown in Fig. 7.8). The net effect is to increase the approaching traffic volumes by 10%. Calculate the new level of service for the westbound approach, assuming all else remains the same. Assume other required input values are as in Table 7.5. Use Eq. 7.27 for the delay calculation.

**7.49** For Problem 7.48, calculate the new level of service for the northbound approach.

**7.50** Calculate the overall intersection level of service for Problem 7.10. Assume other required input values are as in Table 7.5. Use Eqs. 7.27-7.29 for the delay calculations.

**7.51** Consider Problem 7.10. How much traffic volume can be added to the southbound approach (assuming the same turning movement percentage) before LOS D is reached for the approach? Use Eq. 7.27 for the delay calculation.

**7.52** Consider Problem 7.10. How much traffic volume must be diverted from the eastbound approach (assuming the same turning movement percentage) to achieve LOS B for the approach? Use Eq. 7.27 for the delay calculation.

## Multiple Choice Problems (Multiple Sections)

**7.53** A signalized intersection has a cycle length of 70 seconds. For one traffic movement, the displayed all-red time is set to two seconds while the displayed yellow time is five seconds. The effective red time is 37 seconds and the total lost time per cycle for the movement is four seconds. What is the displayed green time for the traffic movement?

    a)   30 s
    b)   31 s
    c)   33 s
    d)   65 s

**7.54** An isolated pretimed signalized intersection has an approach with a saturation flow rate of 1900 veh/h. For this approach, the displayed red time is 58 seconds, the displayed yellow time is three seconds, the all-red time is two seconds, the effective green time is 28 seconds, and the total lost time is four sec/phase. What is the average uniform delay per vehicle when the approach flow rate is 550 veh/h?

a) 26.3 s
b) 29.7 s
c) 30.1 s
d) 413.3 s

**7.55** An isolated pretimed signalized intersection has an approach with a traffic flow rate of 750 veh/h and a saturation flow rate of 3200 veh/h. This approach is allocated 32 seconds of effective green time. The cycle length is 100 seconds. Determine the average approach delay (using Eq. 7.27).

a) 4.6 s
b) 30.2 s
c) 34.8 s
d) 35.0 s

**7.56** Calculate the sum of flow ratios for the critical lane groups for the three-phase timing plan, with traffic and saturation flow rates shown in the following tables.

Traffic Flow Rates (veh/h) for Problem 7.49

| Phase 1 | Phase 2 | Phase 3 |
|---------|---------|---------|
| EB L: 250 | EB T/R: 1200 | SB L: 75 |
|  |  | NB L: 100 |
| WB L: 300 | WB T/R: 1350 | SB T/R: 420 |
|  |  | NB T/R: 425 |

Saturation Flow Rates (veh/h) for Problem 7.49

| Phase 1 | Phase 2 | Phase 3 |
|---------|---------|---------|
| EB L: 1800 | EB T/R: 3600 | SB L: 500 |
|  |  | NB L: 525 |
| WB L: 1800 | WB T/R: 3600 | SB T/R: 1950 |
|  |  | NB T/R: 1950 |

a) 0.950
b) 0.760
c) 0.690
d) 0.622

**7.57** A signalized intersection has a sum of critical flow ratios of 0.72 and a total cycle lost time of 12 seconds. Assuming a critical intersection $v/c$ ratio of 0.9, calculate the minimum necessary cycle length.

a) 48.0 s
b) 60.0 s
c) 82.1 s
d) 42.9 s

**7.58** A signalized intersection approach has an upgrade of 4%. The total width of the cross street at this intersection is 18 m. The average vehicle length of approaching traffic is 5 m. The speed of approaching traffic is 64 km/h. Determine the sum of the minimum necessary change and clearance intervals.

a) 3.58 s
b) 4.93 s
c) 4.88 s
d) 2.50 s

# Chapter 8

# Travel Demand and Traffic Forecasting

## 8.1 INTRODUCTION

Traffic volumes will change over time in response to changes in economic activity, individual travel patterns and preferences, and travelers' social/recreational activities. In addition, traffic volumes will be affected by any significant modification of a highway network, which would include items such as new road construction or operational changes on existing roads (use or retiming of traffic signals). Analysts therefore must develop methodological approaches for forecasting changes in traffic volumes. For new road construction, traffic forecasts are needed to determine an appropriate pavement design (number of equivalent axle loads, as discussed in Chapter 4) and geometric design (number of lanes, shoulder widths, and so on) that will provide an acceptable level of service. For operational improvements, traffic forecasts are needed to estimate the effectiveness of alternate improvement options.

In forecasting vehicle traffic, two interrelated elements must be considered: the overall regional traffic growth/decline and possible traffic diversion. Overall traffic growth/decline is clearly an important concern, because projects such as highway construction and operational improvements will be undertaken in a dynamic environment with continual change in economic activity and individual traveler activities and preferences. In the design of highway projects, engineers must seek to provide a sufficient highway level of service and an acceptable pavement ride quality for future traffic volumes. One would expect that factors affecting long-term regional traffic growth/decline trends are primarily economic and, to a historically lesser extent, social in nature. The economics of the region in which a highway project is being undertaken determine the amount of traffic-generating activities (work, social/recreational, and shopping) and the spatial distribution of residential, industrial, and commercial areas. The social aspects of the population determine attitudes and behavioral tendencies with regard to possible traffic-generating decisions. For example, some regional populations may have social characteristics that make them more likely than other regional populations to make fewer trips, to carpool, to vanpool, or to take public transportation (buses or subways), all of which significantly impact the amount of highway traffic.

In addition to overall regional traffic growth/decline, there is the more microscopic, short-term phenomenon of traffic diversion. As new roads are constructed, as operational improvements are made, and/or as roads gradually become more congested, traffic will divert as drivers change routes or trip-departure times in an effort to avoid congestion and improve the level of service that they experience. Thus the highway network must be viewed as a system with the realization that a capacity or level of service change on any one segment of the highway network will impact traffic flows on many of the surrounding highway segments.

Travel demand and traffic forecasting is a formidable problem because it requires accurate regional economic forecasts as well as accurate forecasts of highway users' social and behavioral attitudes regarding trip-oriented decisions, in order to predict growth/decline trends and traffic diversion. Virtually everyone is aware how inaccurate economic forecasts can be, which is testament to the complexity and uncertainty associated with such forecasts. Similarly, one can readily imagine the difficulty associated with forecasting individuals' travel decisions.

Despite the difficulties involved in accurately forecasting traffic, over the years analysts have persisted in the development and refinement of a wide variety of travel demand and traffic forecasting techniques. An overriding consideration has always been the ease with which such techniques can be implemented in terms of data requirements and the ability of users to comprehend the underlying methodological approach. The field has evolved such that many traffic analysts can legitimately argue that the more recent developments in travel demand and traffic forecasting are largely beyond the reach of practice-oriented implementation. In many respects, this is an expected evolution because, due to the complexity of the problem, there will always be a tendency for theoretical work to exceed the limits of practical implementability. Unfortunately, the outgrowth of the methodological gap between theory and practice has resulted in the use of a wide variety of travel demand and traffic forecasting techniques, the selection of which is a function of the technical expertise of forecasting agencies' personnel as well as time and financial concerns.

In the past, textbooks have attempted to cover the full range of travel demand and traffic forecasting techniques from the readily implementable, simplistic approaches to the more theoretically refined methods. In so doing, such textbooks have often sacrificed depth of coverage, and as a consequence, travel demand and traffic forecasting frequently had the appearance of being confusing and disjointed. This chapter attempts to convey the basic principles underlying travel demand and traffic forecasting as opposed to reviewing the many techniques available to forecast traffic. This is achieved by focusing on an approach that is fairly advanced technically and effectively and efficiently conveys the fundamental concepts of travel demand and traffic forecasting. For more information on travel demand and traffic forecasting techniques that are more implementable or more theoretically advanced than the concepts provided in this chapter, the reader is referred to other sources [Meyer and Miller 2001; Sheffi 1985; Washington et al. 2011].

## 8.2   TRAVELER DECISIONS

Forecasts of highway traffic should, at least in theory, be predicated on some understanding of traveler decisions, because the various decisions that travelers make regarding trips will ultimately determine the quantity, spatial distribution (by route), and temporal distribution of vehicles on a highway network. Within this context, travelers can be viewed as making four distinct but interrelated decisions regarding trips: temporal decisions, destination decisions, modal decisions, and spatial or route decisions. The temporal decision includes the decision to travel and, more importantly, when to travel. The destination decision is concerned with the selection of a specific destination (shopping center, recreational facility, etc.), and the modal decision relates to how the trip is to be made (by automobile, bus, walking, or bicycling). Finally, spatial decisions focus on which route is to be taken from the traveler's origin (the traveler's initial location) to the desired destination. Being able to understand, let alone predict, such decisions is a monumental task. The remaining sections of this chapter seek to define the dimensions of this decision-prediction task and, through examples and illustrations, to demonstrate methods of forecasting traveler decisions and, ultimately, traffic volumes.

## 8.3   SCOPE OF THE TRAVEL DEMAND AND TRAFFIC FORECASTING PROBLEM

Because travel demand and traffic forecasting are predicated on the accurate forecasting of traveler decisions, two factors must be addressed in the development of an effective travel demand and traffic forecasting methodology: the complexity of the traveler decision-making process and system equilibration.

To begin the development of a fuller understanding of the complexity of traveler decisions, consider the schematic presented in Fig. 8.1. This figure indicates that traveler socioeconomics and activity patterns constitute a major driving force in the decision-making process. Socioeconomics, including factors such as household income, number of household members, and traveler age, affect the types of activities that the traveler is likely to be involved in (work, yoga classes, shopping, children's day care, dancing lessons, community meetings, etc.), which in turn are primary factors in many travel decisions. Socioeconomics can also have a direct effect on travel-related decisions by, for example, limiting modal availability (e.g., travelers in low-income households may be forced to take a bus due to the unavailability of a household automobile).

If we look more directly at the decision to travel, mode/destination choice, and highway route choice, Fig. 8.1 indicates that both long-term and short-term factors affect these decisions. For the decision to travel as well as mode/destination choice, the long-term factors of modal availability, residential and commercial distributions, and modal infrastructure play a significant role. These factors are considered long term because they change relatively slowly over time. For example, the development and/or relocation of residential neighborhoods and commercial centers is a process that may take years. Changes in modal infrastructure (construction/relocation of highways, subways, commuter rail systems) and modal availability (changes in automobile ownership, bus routing/scheduling) are also factors that evolve over relatively long periods of time. In contrast, a short-term factor, such as modal traffic, is one that can vary within a short period of time, as discussed in Chapter 6.

Moving down the illustration presented in Fig. 8.1, we see that the choice of a traveler's highway route is also determined by both long-term (highway infrastructure) and short-term (highway traffic) factors.

LT: Long-term factors.
ST: Short-term factors.

**Figure 8.1** Overview of the process by which highway traffic is determined.

The outcome of the combination of these traveler decisions is, of course, highway traffic, the prediction of which is the objective of travel demand and traffic forecasting.

Aside from the complexities involved in the traveler decision-making process, the issue of system equilibration (mentioned at the beginning of this section) must also be considered. Note that Fig. 8.1 indicates not only that long- and short-term factors affect traveler decisions and choices, but also that these decisions and choices in turn affect the long- and short-term factors. Such a reciprocal relationship is most apparent in considering the relationship between traveler choices and short-term factors. For example, consider a traveler's choice of highway route. One would expect that the traveler would be more likely to select a route between origin and destination that provides a shorter travel time. The travel time on various routes will be a function of route distance (highway infrastructure, long-term) and route traffic (higher traffic volumes reduce travel speed and increase travel time, as discussed in Chapter 5). But travelers' decisions to take specific routes ultimately determine the route traffic on which their route decisions are based. This interdependence between traveler decisions and modal traffic is schematically presented in Fig. 8.2. In addition to these short-term effects, persistently high traffic volumes may lead to a change in the highway infrastructure (construction of additional lanes and/or new highways to reduce congestion), again resulting in an interdependence. This interdependence creates the problem of equilibration, which is common to many modeling applications.

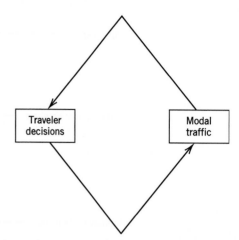

**Figure 8.2** Interdependence of traveler decisions and traffic flow.

Perhaps the most recognizable equilibration problem is determination of price in a classic model of economic supply and demand for a product. From a modeling perspective, as will be shown, equilibration adds yet another dimension of difficulty to an already complex travel demand and traffic forecasting problem. It is safe to say that no existing methodology has come close to accurately capturing the complexities involved in traveler decisions or fully addressing the issue of equilibration. However, within rather obvious limitations, the field of travel demand and traffic forecasting has, over the years, made progress toward more accurately modeling traveler decision complexities and equilibration concerns. This evolution of travel demand and traffic forecasting methodology has led to the popular approach of viewing traveler decisions as a sequence of three distinct decisions, as shown in Fig. 8.3, the result of which is forecasted traffic flow (a direct outgrowth of the highway route choice decision). Clearly, the sequential structure of traveler decisions is a considerable simplification of the actual decision-making process in which all trip-related decisions are considered simultaneously by the traveler. However, this sequential simplification permits the development of a sequence of mathematical models of traveler behavior that can be applied to forecast traffic flow. The following sections of this chapter present and discuss typical functional forms of the mathematical models used to forecast the three sequential traveler decisions shown in Fig. 8.3.

## 8.4  TRIP GENERATION

The first traveler decision to be modeled in the sequential approach to travel demand and traffic forecasting is trip generation. The objective of trip generation modeling is to develop an expression that predicts exactly when a trip is to be made. This is an inherently difficult task due to the wide variety of trip types (working, social/recreational, shopping, etc.) and activities (eating lunch, exercising, visiting friends, etc.) undertaken by a traveler in a sample day, as is schematically shown in Fig. 8.4. To address the complexity of the trip generation decision, the following approach is typically taken:

Figure 8.3 Overview of the sequential approach to traffic estimation.

Figure 8.4 Weekday trip generation for a typical traveler.

1. *Aggregation of decision-making units.* Predicting trip generation behavior is simplified by considering the trip generation behavior of a household (a group of travelers sharing the same domicile) as opposed to the behavior of individual travelers. Such an aggregation of traveler decisions is justified on the basis of the comparatively homogeneous nature of household members (socially and economically) and household members' often intertwined trip-generating activities (joint shopping trips, etc.).

2. *Segmentation of trips by type.* Different types of trips have different characteristics that make them more or less likely to be taken at various times of the day. For example, work trips are more likely to be taken in the morning hours than are shopping trips, which are more likely to be taken during the evening hours. Also, it is more likely that a traveler will take multiple shopping trips during the course of a day as opposed to multiple

work trips. To account for this, three distinct trip types are used: (1) work trips, including trips to and from work; (2) shopping trips; and (3) social/recreational trips, which include vacations, visiting friends, church meetings, sporting events, and so on.

3. *Temporal aggregation.* Although research has been undertaken to develop mathematical expressions that predict when a traveler is likely to make a trip (Hamed and Mannering 1993), trip generation more often focuses on the number of trips made over some period of time. Thus trips are aggregated temporally, and trip generation models seek to predict the number of trips per hour or per day.

## 8.4.1 Typical Trip Generation Models

Trip generation models generally assume a linear form in which the number of vehicle-based (automobile, bus, or subway) trips is a function of various socioeconomic and/or distributional (residential and commercial) characteristics. An example of such a model, for a given trip type, is

$$T_i = b_0 + b_1 z_{1i} + b_2 z_{2i} + \ldots + b_k z_{ki} \tag{8.1}$$

*where*

$T_i$ = number of vehicle-based trips of a given type (shopping or social/recreational) in some specified time period made by household $i$,

$b_k$ = coefficient estimated from traveler survey data and corresponding to characteristic $k$, and

$z_{ki}$ = characteristic $k$ (income, employment in neighborhood, number of household members) of household $i$.

The estimated coefficients ($b$'s) are usually estimated by the method of least squares regression (linear regression) using data collected from traveler surveys. A brief description and example of this method are presented in Appendix 8A.

---

**EXAMPLE 8.1    SHOPPING-TRIP GENERATION**

A simple linear regression model is estimated for shopping-trip generation during a shopping-trip peak hour. The model is

Number of peak-hour vehicle-based shopping trips per household
      = 0.12 + 0.09(household size)
      + 0.011(annual household income in thousands of dollars)
      − 0.15(employment in the household's neighborhood, in hundreds)

A particular household has six members and an annual income of $50,000. They currently live in a neighborhood with 450 retail employees, but are moving to a new home in a neighborhood with 150 retail employees. Calculate the predicted number of vehicle-based peak-hour shopping trips the household makes before and after the move.

## SOLUTION

Note that the signs of the model coefficients ($b$'s, +0.09, and +0.011) indicate that as household size and income increase, the number of shopping trips also increases. This is reasonable because wealthier, larger households can be expected to make more vehicle-based shopping trips. The negative sign of the employment coefficient (−0.15) indicates that as retail employment in a household's neighborhood increases, fewer vehicle-based shopping trips will be generated. This reflects the fact that larger retail employment in a neighborhood implies more shopping opportunities nearer to the household, thereby increasing the possibility that a shopping trip can be conducted without the use of a vehicle (a non–vehicle-based trip, such as walking).

Turning to the problem solution, before the household moves,

$$\text{Number of vehicle trips} = 0.12 + 0.09(6) + 0.011(50) - 0.15(4.5) = \underline{0.535}$$

After the household moves,

$$\text{Number of vehicle trips} = 0.12 + 0.09(6) + 0.011(50) - 0.15(1.5) = \underline{0.985}$$

Thus the model predicts that the move will result in 0.45 additional peak-hour vehicle-based shopping trips due to the decline in neighborhood shopping opportunities as reflected by the decline in neighborhood retail employment.

---

**EXAMPLE 8.2**   **SOCIAL/RECREATIONAL TRIP GENERATION**

A model for social/recreational trip generation is estimated, with data collected during a major holiday, as

Number of peak-hour vehicle-based social/recreational trips per household
= 0.04 + 0.018(household size)
+ 0.009(annual household income in thousands of dollars)
+ 0.16(number of nonworking household members)

If the household described in Example 8.1 has one working member, how many peak-hour social/recreational trips are predicted?

## SOLUTION

The positive signs of the model coefficients indicate that increasing household size, income, and number of nonworking household members result in more social/recreational trips. Again, wealthier and larger households can be expected to be involved in more vehicle-based trip-generating activities, and the larger the number of nonworking household members, the larger the number of people available at home to make peak-hour social/recreational trips.

The solution to this problem is

$$\text{Number of vehicle trips} = 0.04 + 0.018(6) + 0.009(50) + 0.16(5) = \underline{1.398}$$

**EXAMPLE 8.3**     **TOTAL TRIP GENERATION**

A neighborhood has 205 retail employees and 700 households that can be categorized into four types, with each type having characteristics as follows:

| Type | Household size | Annual income | Number of nonworkers in the peak hour | Workers departing |
|------|----------------|---------------|----------------------------------------|-------------------|
| 1 | 2 | $40,000 | 1 | 1 |
| 2 | 3 | $50,000 | 2 | 1 |
| 3 | 3 | $55,000 | 1 | 2 |
| 4 | 4 | $40,000 | 3 | 1 |

There are 100 type 1, 200 type 2, 350 type 3, and 50 type 4 households. Assuming that shopping, social/recreational, and work vehicle-based trips all peak at the same time (for exposition purposes), determine the total number of peak-hour trips (work, shopping, social/recreational) using the generation models described in Examples 8.1 and 8.2.

**SOLUTION**

For vehicle-based shopping trips,

$Type\ 1$: $0.12 + 0.09(2) + 0.011(40) - 0.15(2.05)$     = 0.4325 trips/household
                                                                                          × 100 households
                                                                                          = 43.25 trips

$Type\ 2$: $0.12 + 0.09(3) + 0.011(50) - 0.15(2.05)$     = 0.6325 trips/household
                                                                                          × 200 households
                                                                                          = 126.5 trips

$Type\ 3$: $0.12 + 0.09(3) + 0.011(55) - 0.15(2.05)$     = 0.6875 trips/household
                                                                                          × 350 households
                                                                                          = 240.625 trips

$Type\ 4$: $0.12 + 0.09(4) + 0.011(40) - 0.15(2.05)$     = 0.6125 trips/household
                                                                                          × 50 households
                                                                                          = 30.625 trips

Therefore, there will be a total of <u>441</u> vehicle-based shopping trips.
        For vehicle-based social/recreational trips,

$Type\ 1$: $0.04 + 0.018(2) + 0.009(40) + 0.16(1)$     = 0.596 trips/household
                                                                                      × 100 households
                                                                                      = 59.6 trips

$Type\ 2$: $0.04 + 0.018(3) + 0.009(50) + 0.16(2)$     = 0.864 trips/household
                                                                                      × 200 households
                                                                                      = 172.8 trips

$Type\ 3$: $0.04 + 0.018(3) + 0.009(55) + 0.16(1)$     = 0.749 trips/household
                                                                                      × 350 households
                                                                                      = 262.15 trips

*Type 4*: 0.04 + 0.018(4) + 0.009(40) + 0.16(3)       = 0.952 trips/household
$\qquad\qquad\qquad\qquad\qquad\qquad\qquad\qquad\qquad$ × 50 households
$\qquad\qquad\qquad\qquad\qquad\qquad\qquad\qquad\qquad$ = 47.6 trips

Therefore, there will be a total of 542.15 vehicle-based social/recreational trips.

For vehicle-based work trips, there will be 100 generated from type 1 households (1 × 100), 200 from type 2 (1 × 200), 700 from type 3 (2 × 350), and 50 from type 4 (1 × 50), for a total of 1050 vehicle-based work trips. Summing the totals for the three trip types gives 2033 peak-hour vehicle-based trips.

It should be noted that the trip generation models used in Examples 8.1, 8.2, and 8.3 are simplified representations of the actual trip generation decision-making process. First, there are many more traveler and household characteristics that affect trip-generating behavior (age, lifestyles, etc.), and second, the models have no variables to capture the equilibration concept discussed earlier. The equilibration concern is important, because if the highway system is heavily congested, travelers are likely to make fewer peak-hour trips as a result of either canceling trips or postponing them until a less congested time period. Unfortunately, such obvious model defects must often be accepted due to data and resource limitations.

### 8.4.2   Trip Generation with Count Data Models

Although linear regression has been a popular method for estimating trip generation models, there is a problem in that the estimated linear regression models can produce fractions of trips for a given time period. As an example, the model presented in Example 8.2 predicted that the household presented in Example 8.1 with one working member would produce 1.398 peak-hour social/recreational trips during the major holiday. Because fractions of trips are not realistic, a modeling approach that gives the probability of making a nonnegative-integer number of trips (0, 1, 2, 3, . . .) may be more appropriate [Washington et al. 2011]. One such model is the Poisson regression, which can be formulated for trip generation (for a given trip type) as

$$P(T_i) = \frac{e^{-\lambda_i}\lambda_i^{T_i}}{T_i!} \tag{8.2}$$

*where*

$T_i$ = number of vehicle-based trips of a given type (shopping or social/recreational) made in some specified time period by household $i$,

$P(T_i)$ = probability of household $i$ making exactly $T_i$ trips (where $T_i$ is a nonnegative integer),

$e$ = base of the natural logarithm ($e = 2.718$), and

$\lambda_i$ = Poisson parameter for household $i$, which is equal to household $i$'s expected number of vehicle-based trips in some specified time period, $E[T_i]$.

Poisson regressions are estimated by specifying the Poisson parameter $\lambda_i$ (the expected number of trips of a specific type made by household $i$ over some time period). The most common relationship between explanatory variables (variables that determine the Poisson parameter) and the Poisson parameter is the log-linear relationship

$$\lambda_i = e^{BZ_i} \tag{8.3}$$

*where*

    $B$ = vector of estimable coefficients,

    $Z_i$ = vector of household characteristics determining trip generation, and

    Other terms are as defined previously.

Note that the Poisson parameter $\lambda_i$ (the expected number of trips of a specific type made by household $i$ over some time period) is a real number (with fractions of trips) but when applied in Eq. 8.2 gives the probability of making a specified nonnegative-integer number of trips ($T_i$).

In Poisson regressions, the coefficient vector $B$ is estimated by maximum-likelihood procedures. A brief description and example of this estimation procedure are presented in Appendix 8B.

---

**EXAMPLE 8.4**    **SHOPPING-TRIP GENERATION WITH THE POISSON MODEL**

Following Example 8.1, a Poisson regression is estimated for shopping-trip generation during a shopping-trip peak hour. The estimated coefficients are

$BZ_i =$    −0.35 + 0.03(household size)
          + 0.004(annual household income in thousands of dollars)
          − 0.10(employment in the household's neighborhood in hundreds)

Given that the household has six members, has an annual income of $50,000, and lives in their new neighborhood with a retail employment of 150, what is the expected number of peak-hour shopping trips and what is the probability that the household will not make a peak-hour shopping trip?

**SOLUTION**

For the expected number of peak-hour shopping trips (a real number),

$$E[T_i] = \lambda_i = e^{BZ_i} = e^{-0.35+0.03(6)+0.004(50)-0.1(1.5)} = \underline{\underline{0.887}} \text{ vehicle trips}$$

For the probability of making zero peak-hour shopping trips (a nonnegative integer), Eq. 8.2 is used to give

$$P(0) = \frac{e^{-0.887} 0.887^0}{0!} = \underline{\underline{0.412}}$$

## 8.5   MODE AND DESTINATION CHOICE

Once the number of trips generated per unit time is known, the next step in the sequential approach to travel demand and traffic forecasting is to determine traveler mode and destination. As was the case with trip generation, trips are classified as work, shopping, and social/recreational. For both shopping and social/recreational trips, a traveler will have the option to choose a mode of travel (automobile, vanpool, or bus) as well as a destination (different shopping centers). In contrast, work trips offer only the mode option, because the choice of work location (destination) is usually a long-term decision that is beyond the time range of most traffic forecasts.

### 8.5.1   Methodological Approach

Following recent advances in the travel demand and traffic forecasting field, development of a model for mode/destination choice necessitates the use of some consistent theory of travelers' decision-making processes. Of the decision-making theories available, one that is based on the microeconomic concept of utility maximization has enjoyed widespread acceptance in mode/destination choice modeling. The basic assumption is that a traveler will select the combination of mode and destination that provides the most utility. The problem then becomes one of developing an expression for the utility provided by various mode and destination alternatives. Because it is unlikely that individual travelers' utility functions can ever be specified with certainty, the unspecifiable portion is assumed to be random. To illustrate this approach, consider a utility function of the following form:

$$V_{im} = \sum_k b_{mk} z_{imk} + \varepsilon_{im} \qquad (8.4)$$

*where*

$V_{im}$ = total utility (specifiable and unspecifiable) provided by mode/destination alternative $m$ to a traveler $i$,

$b_{mk}$ = coefficient estimated from traveler survey data for mode/destination alternative $m$ corresponding to mode/destination or traveler characteristic $k$,

$z_{imk}$ = traveler or mode/destination characteristic $k$ (income, travel time of mode, commercial floor space at destination, etc.) for mode/destination alternative $m$ for traveler $i$, and

$\varepsilon_{im}$ = unspecifiable portion of the utility of mode/destination alternative $m$ for traveler $i$, which will be assumed to be random.

For notational convenience, define the specifiable nonrandom portion of utility $V_{im}$ as

$$U_{im} = \sum_k b_{mk} z_{imk} \qquad (8.5)$$

With these definitions of utility, the probability that a traveler will choose some alternative, say $m$, is equal to the probability that the given alternative's utility is greater than the utility of all other possible alternatives. The probabilistic component arises from the fact that the unspecifiable portion of the utility expression is not known and is assumed to be a random variable. The basic probability statement is

$$P_{im} = \text{prob}\left[U_{im} + \varepsilon_{im} > U_{is} + \varepsilon_{is}\right] \text{ for all } s \neq m \tag{8.6}$$

*where*

$P_{im}$ = probability that traveler $i$ will select alternative $m$,

$\text{prob}[\cdot]$ = notation for probability,

$s$ = notation for available alternatives, and

Other terms are as defined previously.

With this basic probability and utility expression and an assumed random distribution of the unspecifiable components of utility ($\varepsilon_{im}$), a probabilistic choice model can be derived and the coefficients in the utility function ($b_{mk}$'s in Eqs. 8.4 and 8.5) can be estimated with data collected from traveler surveys, along the same lines as for the coefficients in the trip generation models. A popular approach to deriving such a probabilistic choice model is to assume that the random, unspecifiable component of utility ($\varepsilon_{im}$ in Eq. 8.4) is generalized-extreme-value distributed. With this assumption, a rather lengthy and involved derivation gives rise to the logit model formulation [McFadden 1981],

$$P_{im} = \frac{e^{U_{im}}}{\sum_s e^{U_{is}}} \tag{8.7}$$

where $e$ is the base of the natural logarithm ($e = 2.718$).

The coefficients that comprise the specifiable portion of utility ($b_{mk}$'s in Eq. 8.5) are estimated by the method of maximum likelihood (see Appendix 8B). For further information on logit model coefficient estimation and maximum-likelihood estimation techniques, refer to more specialized references [Washington et al. 2011; McFadden 1981].

## 8.5.2    Logit Model Applications

With the total number of vehicle-based trips made in specific time periods known (from trip generation models), the allocation of trips to vehicle-based modes and likely destinations can be undertaken by applying appropriate logit models. This process is best demonstrated by example.

**EXAMPLE 8.5    LOGIT MODEL OF WORK-MODE-CHOICE**

A simple work-mode–choice model is estimated from data in a small urban area to determine the probabilities of individual travelers selecting various modes. The mode choices include automobile drive-alone (*DL*), automobile shared-ride (*SR*), and bus (*B*), and the utility functions are estimated as

$$U_{DL} = 2.2 - 0.2(\text{cost}_{DL}) - 0.03(\text{travel time}_{DL})$$
$$U_{SR} = 0.8 - 0.2(\text{cost}_{SR}) - 0.03(\text{travel time}_{SR})$$
$$U_B = -0.2(\text{cost}_B) - 0.01(\text{travel time}_B)$$

where cost is in dollars and time is in minutes. Between a residential area and an industrial complex, 4000 workers (generating vehicle-based trips) depart for work during the peak hour. For all workers, the cost of driving an automobile is $6.00 with a travel time of 20 minutes, and the bus fare is $1.00 with a travel time of 25 minutes. If the shared-ride option always consists of two travelers sharing costs equally, how many workers will take each mode?

**SOLUTION**

Note that the utility function coefficients logically indicate that as modal costs and travel times increase, modal utilities decline and, consequently, so do modal selection probabilities (see Eq. 8.7). Substitution of cost and travel time values into the utility expressions gives

$$U_{DL} = 2.2 - 0.2(6) - 0.03(20) = 0.4$$
$$U_{SR} = 0.8 - 0.2(3) - 0.03(20) = -0.4$$
$$U_B = -0.2(1.0) - 0.01(25) = -0.45$$

Substituting these values into Eq. 8.7 yields

$$P_{DL} = \frac{e^{0.4}}{e^{0.4} + e^{-0.4} + e^{-0.45}} = \frac{1.492}{1.492 + 0.670 + 0.638} = \frac{1.492}{2.80} = 0.533$$

$$P_{SR} = \frac{0.670}{2.80} = 0.239$$

$$P_B = \frac{0.638}{2.80} = 0.228$$

Multiplying these probabilities by 4000 (the total number of workers departing in the peak hour) gives 2132 workers driving alone, 956 sharing a ride, and 912 using a bus.

**EXAMPLE 8.6    FORECASTING MODE CHOICE WITH THE LOGIT MODEL**

A bus company is making costly efforts in an attempt to increase work-trip bus usage for the travel conditions described in Example 8.5. An exclusive bus lane is constructed that reduces bus travel time to 10 minutes.
 a. Determine the modal distribution of trips after the lane is constructed.
 b. If shared-ride vehicles are also permitted to use the facility, and travel time for bus and shared-ride modes is 10 min, determine the modal distribution.
 c. Given the conditions described in part (b), determine the modal distribution if the bus company offers free bus service.

**SOLUTION**

**a.** After the bus lane construction, the modal utilities of drive-alone and shared-ride are unchanged from those in Example 8.5. However, the bus modal utility becomes

$$U_B = -0.2(1.0) - 0.01(10) = -0.3$$

From Eq. 8.7 with 4000 work trips,

$$P_{DL} = \frac{e^{0.4}}{e^{0.4} + e^{-0.4} + e^{-0.2}} = \frac{1.492}{2.981} = 0.500 \quad \text{and} \quad 0.500(4000) = 2000 \text{ trips}$$

$$P_{SR} = \frac{0.670}{2.981} = 0.225 \quad \text{and} \quad 0.225(4000) = 900 \text{ trips}$$

$$P_{B} = \frac{0.819}{2.981} = 0.275 \quad \text{and} \quad 0.275(4000) = 1100 \text{ trips}$$

or an increase of 188 bus patrons from the prediction of Example 8.5.

**b.** With the bus lane opened to shared-ride vehicles, only the modal utility of shared rides will change from those in part (a) to:

$$U_{SR} = 0.8 - 0.2(3) - 0.03(10) = -0.1$$

From Eq. 8.7 with 4000 work trips,

$$P_{DL} = \frac{e^{0.4}}{e^{0.4} + e^{-0.1} + e^{-0.2}} = \frac{1.492}{3.216} = 0.464 \quad \text{and} \quad 0.464(4000) = 1856 \text{ trips}$$

$$P_{SR} = \frac{0.905}{3.216} = 0.281 \quad \text{and} \quad 0.281(4000) = 1124 \text{ trips}$$

$$P_{B} = \frac{0.819}{3.216} = 0.255 \quad \text{and} \quad 0.255(4000) = 1020 \text{ trips}$$

or a loss of 80 bus patrons and a gain of 224 shared-ride users relative to part (a).

**c.** With free bus fare, the bus modal utility becomes [with other utilities unchanged from part (b)],

$$U_{B} = -0.2(0) - 0.01(10) = -0.1$$

From Eq. 8.7 with 4000 work trips,

$$P_{DL} = \frac{e^{0.4}}{e^{0.4} + e^{-0.1} + e^{-0.1}} = \frac{1.492}{3.301} = 0.452 \quad \text{and} \quad 0.452(4000) = 1808 \text{ trips}$$

$$P_{SR} = \frac{0.905}{3.301} = 0.274 \quad \text{and} \quad 0.274(4000) = 1096 \text{ trips}$$

$$P_{B} = \frac{0.905}{3.301} = 0.274 \quad \text{and} \quad 0.274(4000) = 1096 \text{ trips}$$

or 76 more bus patrons compared with part (b).

**EXAMPLE 8.7    LOGIT MODEL OF SHOPPING MODE/DESTINATION CHOICE**

Consider a residential area and two shopping centers that are possible destinations. From 7:00 to 8:00 P.M. on Friday night, 900 vehicle-based shopping trips leave the residential area for the two shopping centers. A joint shopping-trip mode-destination choice logit model (choice of either auto or bus) is estimated, giving the following coefficients:

| Variable | Auto coefficient | Bus coefficient |
|---|---|---|
| Auto constant | 0.6 | 0.0 |
| Travel time in minutes | −0.3 | −0.3 |
| Commercial floor space (in thousands of m$^2$) | 0.12 | 0.12 |

Initial travel times to shopping centers 1 and 2 are as follows:

| | By auto | By bus |
|---|---|---|
| Travel time to shopping center 1 (in minutes) | 8 | 14 |
| Travel time to shopping center 2 (in minutes) | 15 | 22 |

If shopping center 2 has 40,000 m$^2$ of commercial floor space and shopping center 1 has 25,000 m$^2$, determine the distribution of Friday night shopping trips by destination and mode.

**SOLUTION**

The utility function coefficients indicate that as modal travel times increase, the likelihood of selecting the mode-destination combination declines. Also, as the destination's floor space increases, the probability of selecting that destination will increase, as suggested by the positive coefficient (+0.12). This reflects the fact that bigger shopping centers tend to have a greater variety of merchandise and hence are more attractive shopping destinations. Note that because this is a joint mode-destination choice model, there are four mode-destination combinations and four corresponding utility functions. Let $U_{A1}$ be the utility of the auto mode to shopping center 1, $U_{A2}$ the utility of the auto mode to shopping center 2, and $U_{B1}$ and $U_{B2}$ the utility of the bus mode to shopping centers 1 and 2, respectively. The utilities are

$$U_{A1} = 0.6 - 0.3(8) + 0.12(25) = 1.2$$
$$U_{B1} = 20.3(14) + 0.12(25) = -1.2$$
$$U_{A2} = 0.6\ 2\ 0.3(15) + 0.12(40) = 0.9$$
$$U_{B2} = 20.3(22) + 0.12(40) = -1.8$$

Substituting these values into Eq. 8.7 gives

$$P_{A1} = \frac{3.32}{6.246} = 0.532$$

$$P_{B1} = \frac{0.301}{6.246} = 0.048$$

$$P_{A2} = \frac{2.46}{6.246} = 0.394$$

$$P_{B2} = \frac{0.165}{6.246} = 0.026$$

Multiplying these probabilities by the 900 trips gives <u>479 trips</u> by auto to shopping center 1, <u>43 trips</u> by bus to shopping center 1, <u>355 trips</u> by auto to shopping center 2, and <u>23 trips</u> by bus to shopping center 2.

**EXAMPLE 8.8**    **LOGIT MODEL OF SOCIAL/RECREATIONAL MODE/DESTINATION CHOICE**

A joint mode-destination vehicle-based social/recreational trip logit model is estimated with the following coefficients:

| Variable | Auto coefficient | Bus coefficient |
|---|---|---|
| Auto constant | 0.9 | 0.0 |
| Travel time in minutes | −0.22 | −0.22 |
| Population in thousands | 0.16 | 0.16 |
| Amusement floor space (in thousands of m$^2$) | 1.1 | 1.1 |

It is known that 500 social/recreational trips will depart from a residential area during the peak hour. There are three possible trip destinations with the following characteristics:

| | Travel time (in minutes) | | Population (in thousands) | Amusement floor space (in thousands of m$^2$) |
|---|---|---|---|---|
| | Auto | Bus | | |
| Destination 1 | 14 | 17 | 12.4 | 1.3 |
| Destination 2 | 5 | 8 | 8.2 | 0.92 |
| Destination 3 | 18 | 24 | 5.8 | 2.1 |

Determine the distribution of trips by mode and destination.

*SOLUTION*

As was the case for the shopping mode-destination model presented in Example 8.7, the signs of the coefficient estimates indicate that increasing travel time decreases an alternative's selection probability. Also, increasing population (reflecting an increase in social opportunities) and increasing amusement floor space (reflecting more recreational opportunities) both increase the probability of an alternative being selected. With two modes and three destinations, there are six alternatives, providing the following utilities (using the same subscripting notation as in Example 8.7):

$$U_{A1} = 0.9 - 0.22(14) + 0.16(12.4) + 1.1(1.3) = 1.234$$
$$U_{B1} = -0.22(17) + 0.16(12.4) + 1.1(1.3) = -0.326$$
$$U_{A2} = 0.9 - 0.22(5) + 0.16(8.2) + 1.1(0.92) = 2.124$$

$$U_{B2} = -0.22(8) + 0.16(8.2) + 1.1(0.92) = 0.564$$
$$U_{A3} = 0.9 - 0.22(18) + 0.16(5.8) + 1.1(2.1) = 0.178$$
$$U_{B3} = -0.22(24) + 0.16(5.8) + 1.1(2.1) = -2.042$$

Using Eq. 8.7 with 500 trips, the total number of trips to the six mode-destination alternatives are

$$P_{A1} = \frac{3.435}{15.607} = 0.220 \quad \text{and} \quad 0.220 \times 500 = \underline{\underline{110 \text{ trips}}}$$

$$P_{B1} = \frac{0.722}{15.607} = 0.046 \quad \text{and} \quad 0.046 \times 500 = \underline{\underline{23 \text{ trips}}}$$

$$P_{A2} = \frac{8.365}{15.607} = 0.536 \quad \text{and} \quad 0.536 \times 500 = \underline{\underline{268 \text{ trips}}}$$

$$P_{B2} = \frac{1.76}{15.607} = 0.113 \quad \text{and} \quad 0.113 \times 500 = \underline{\underline{57 \text{ trips}}}$$

$$P_{A3} = \frac{1.195}{15.607} = 0.077 \quad \text{and} \quad 0.077 \times 500 = \underline{\underline{38 \text{ trips}}}$$

$$P_{B3} = \frac{0.13}{15.607} = 0.008 \quad \text{and} \quad 0.008 \times 500 = \underline{\underline{4 \text{ trips}}}$$

## EXAMPLE 8.9    FORECASTING SOCIAL/RECREATIONAL MODE/DESTINATION CHOICE

Consider the situation described in Example 8.8. A labor dispute results in a bus union slowdown that increases travel times from the origin by 4, 2, and 8 minutes to destinations 1, 2, and 3, respectively. If the total number of trips remains constant, determine the resulting distribution of trips by mode and destination.

**SOLUTION**

The mode-destination utilities are computed as

$$U_{A1} = 1.234 \text{ (as in Example 8.8)}$$
$$U_{B1} = -0.22(21) + 0.16(12.4) + 1.1(1.3) = -1.206$$
$$U_{A2} = 2.124 \text{ (as in Example 8.8)}$$
$$U_{B2} = -0.22(10) + 0.16(8.2) + 1.1(0.92) = 0.124$$
$$U_{A3} = 0.178 \text{ (as in Example 8.8)}$$
$$U_{B3} = -0.22(32) + 0.16(5.8) + 1.1(2.1) = -3.802$$

Applying Eq. 8.7 with 500 trips gives the following distribution of trips among mode-destination alternatives:

$$P_{A1} = \frac{3.435}{14.45} = 0.238 \quad \text{and} \quad 0.238 \times 500 = \underline{\underline{119 \text{ trips}}}$$

$$P_{B1} = \frac{0.299}{14.45} = 0.021 \quad \text{and} \quad 0.021 \times 500 = \underline{\underline{10 \text{ trips}}}$$

$$P_{A2} = \frac{8.365}{14.45} = 0.579 \quad \text{and} \quad 0.579 \times 500 = \underline{\underline{290 \text{ trips}}}$$

$$P_{B2} = \frac{1.132}{14.45} = 0.078 \quad \text{and} \quad 0.078 \times 500 = \underline{\underline{39 \text{ trips}}}$$

$$P_{A3} = \frac{1.195}{14.45} = 0.083 \quad \text{and} \quad 0.083 \times 500 = \underline{\underline{41 \text{ trips}}}$$

$$P_{B3} = \frac{0.022}{14.45} = 0.002 \quad \text{and} \quad 0.002 \times 500 = \underline{\underline{1 \text{ trip}}}$$

## 8.6    HIGHWAY ROUTE CHOICE

To summarize, the trip generation and mode-destination choice models give total highway traffic demand between a specified origin (the neighborhood from which trips originate) and a destination (the geographic area to which trips are destined), in terms of vehicles per some time period (usually vehicles per hour). With this information in hand, the final step in the sequential approach to travel demand and traffic forecasting—route choice—can be addressed. The result of the route choice decision will be traffic flow (generally in units of vehicles per hour) on specific highway routes, which is the desired output from the traffic forecasting process.

### 8.6.1    Highway Performance Functions

Route choice presents a classic equilibrium problem, because travelers' route choice decisions are primarily a function of route travel times, which are determined by traffic flow—itself a product of route choice decisions. This interrelationship between route choice decisions and traffic flow forms the basis of route choice theory and model development.

To begin modeling traveler route choice, a mathematical relationship between route travel time and route traffic flow is needed. Such a relationship is commonly referred to as a highway performance function. The most simplistic approach to formalizing this relationship is to assume a linear highway performance function in which travel time increases linearly with flow. An example of such a function is illustrated in Fig. 8.5. In this figure, the free-flow travel time refers to the travel time that a traveler would experience if no other vehicles were present to impede travel speed (as discussed in Chapter 5). This free-flow travel time is generally computed with the assumption that a vehicle travels at the posted speed limit of the route.

Although the linear highway performance function has the appeal of simplicity, it is not a particularly realistic representation of the travel time–traffic flow relationship. Recall that Chapter 5 presented a relationship between traffic speed and flow that is parabolic in nature, with significant reductions in travel speed occurring as the traffic flow approaches the roadway's capacity. This parabolic speed-flow

**Figure 8.5** Linear travel time–traffic flow relationship.

**Figure 8.6** Nonlinear travel time–traffic flow relationship.

relationship suggests a nonlinear highway performance function, such as that illustrated in Fig. 8.6. This figure shows route travel time increasing more quickly as traffic flow approaches capacity, which is consistent with the parabolic relationship presented in Chapter 5.

Both linear and nonlinear highway performance functions will be demonstrated through example, using two theories of travel route choice: user equilibrium and system optimization. For other theories of route choice, refer to Sheffi [1985].

### 8.6.2   User Equilibrium

In developing theories of traveler route choice, two important assumptions are usually made. First, it is assumed that travelers will select routes between origins and destinations on the basis of route travel times only (they will tend to select the route with the shortest travel time). This assumption is not terribly restrictive, because travel time obviously plays the dominant role in route choice; however, other, more subtle factors that may influence route choice (scenery, pavement conditions, etc.) are not accounted for. The second assumption is that travelers know the travel times that would be encountered on all available routes between their origin and destination. This is potentially a strong assumption, because a traveler may not have actually traveled on all available routes between an origin and destination and may repeatedly (day after day) choose one route based only on the perception that travel times on alternative routes are higher. However, in support of this assumption, studies have shown that travelers' perceptions of alternative route travel times are reasonably close to actual observed travel times [Mannering 1989].

With these assumptions, the theory of user-equilibrium route choice can be made operational. The rule of choice underlying user equilibrium is that travelers will select a route so as to minimize their personal travel time between the origin and destination. User equilibrium is said to exist when individual travelers cannot improve their travel times by unilaterally changing routes. Stated differently [Wardrop 1952], user equilibrium can be defined as follows:

*The travel time between a specified origin and destination on all used routes is the same and is less than or equal to the travel time that would be experienced by a traveler on any unused route.*

**EXAMPLE 8.10    BASIC USER EQUILIBRIUM**

Two routes connect a city and a suburb. During the peak-hour morning commute, a total of 4500 vehicles travel from the suburb to the city. Route 1 has a 100-km/h speed limit and is 10 km in length; route 2 is 5 km in length with a 75-km/h speed limit. Studies show that the total travel time on route 1 increases two minutes for every additional 500 vehicles added. Minutes of travel time on route 2 increase with the square of the number of vehicles, expressed in thousands of vehicles per hour. Determine user-equilibrium travel times.

*SOLUTION*

Determining free-flow travel times, in minutes, gives

$$Route\ 1\text{: }10 \text{ km}/(100 \text{ km/h}) \times 60 \text{ min/h} = 6 \text{ min}$$
$$Route\ 2\text{: }5 \text{ km}/(75 \text{ km/h}) \times 60 \text{ min/h} = 4 \text{ min}$$

With these data, the performance functions can be written as

$$t_1 = 6 + 4x_1$$
$$t_2 = 4 + x_2^2$$

*where*

> $t_1, t_2$ = average travel times on routes 1 and 2 in minutes, and
> $x_1, x_2$ = traffic flow on routes 1 and 2 in thousands of vehicles per hour.

Also, the basic flow conservation identity is

$$q = x_1 + x_2 = 4.5$$

where $q$ = total traffic flow between the origin and destination in thousands of vehicles per hour.

**Figure 8.7** Illustration of performance curves for Example 8.10.

With Wardrop's definition of user equilibrium, it is known that the travel times on all used routes are equal. However, the first order of business is to determine whether or not both routes are used. Figure 8.7 gives a graphic representation of the two performance functions. Note that because route 2 has a lower free-flow travel time, any total origin-to-destination traffic flow less than $q'$ (in Fig. 8.7) will result in only route 2 being used, because the travel time on route 1 would be greater even if only one vehicle used it. At flows of $q'$ and above, route 2 is sufficiently congested, and its travel time sufficiently high, that route 1 becomes a viable alternative.

To check if the problem's flow of 4500 vehicles per hour exceeds $q'$, the following test is conducted:

1. Assume that all traffic flow is on route 1. Substituting traffic flows of 4.5 and 0 into the performance functions gives $t_1(4.5) = 24$ min and $t_2(0) = 4$ min.
2. Assume that all traffic flow is on route 2, giving $t_1(0) = 6$ min and $t_2(4.5) = 24.25$ min.

Thus, because $t_1(4.5) > t_2(0)$ and $t_2(4.5) > t_1(0)$, both routes will be used. If $t_1(0)$ had been greater than $t_2(4.5)$, the 4500 vehicles would have been less than $q'$ in Fig. 8.7, and only route 2 would have been used.

With both routes used, Wardrop's user-equilibrium definition gives

$$t_1 = t_2$$

or

$$6 + 4x_1 = 4 + x_2^2$$

From flow conservation, $x_1 + x_2 = 4.5$, so substituting, we get

$$6 + 4(4.5 - x_2) = 4 + x_2^2$$
$$x_2 = 2.899 \quad \text{or} \quad 2899 \text{ veh/h}$$
$$x_1 = 4.5 - x_2 = 4.5 - 2.899$$
$$= 1.601 \quad \text{or} \quad 1601 \text{ veh/h}$$

which gives average route travel times of

$$t_1 = 6 + 4(1.601) = \underline{\underline{12.4 \text{ min}}}$$

$$t_2 = 4 + (2.899)^2 = \underline{\underline{12.4 \text{ min}}}$$

**EXAMPLE 8.11**   **USER EQUILIBRIUM—EFFECT OF CAPACITY AND TRAFFIC REDUCTION**

Peak-hour traffic demand between an origin-destination pair is initially 3500 vehicles. The two routes connecting the pair have performance functions $t_1 = 2 + 3(x_1/c_1)$ and $t_2 = 4 + 2(x_2/c_2)$, where the $t$'s are travel times in minutes, the $x$'s are the peak-hour traffic volumes expressed in thousands, and the $c$'s are the peak-hour route capacities expressed in thousands of vehicles per hour. Initially, the capacities of routes 1 and 2 are 2500 and 4000 veh/h, respectively. A reconstruction project reduces capacity on route 2 to 2000 veh/h. Assuming user equilibrium before and during reconstruction, what reduction in total peak-hour origin-destination traffic flow is needed to ensure that total travel times (summation of all $x_a t_a$'s, where $a$ denotes route) during reconstruction are equal to those before reconstruction?

*SOLUTION*

First, focusing on the roads before reconstruction, a check to see if both routes are used gives (using performance functions)

$$t_1(3.5) = 6.2 \text{ min}, \qquad t_2(0) = 4 \text{ min}$$
$$t_1(0) = 2 \text{ min}, \qquad t_2(3.5) = 5.75 \text{ min}$$

which, because $t_1(3.5) > t_2(0)$ and $t_2(3.5) > t_1(0)$, indicates that both routes are used. Setting route travel times equal and substituting performance functions gives

$$2 + \frac{3}{2.5}(x_1) = 4 + \frac{2}{4}(x_2)$$

From conservation of flow, $x_2 = 3.5 - x_1$, so that

$$2 + 1.2x_1 = 4 + 0.5(3.5 - x_1)$$

Solving gives $x_1 = 2.206$ and $x_2 = 3.5 - 2.206 = 1.294$. For travel times,

$$t_1 = 2 + 1.2(2.206) = 4.647 \text{ min}$$
$$t_2 = 4 + 0.5(1.294) = 4.647 \text{ min}$$

The total peak-hour travel time before reconstruction will simply be the average route travel time multiplied by the number of vehicles:

Total travel time = 4.647(3500) = 16,264.5 veh-min

During reconstruction, the performance function of route 1 is unchanged, but the performance function of route 2 is altered because of the reduction in capacity to

$$t_2 = 4 + \frac{2}{2}(x_2) = 4 + x_2$$

If it is assumed that both routes are used, $t_1 = t_2$. Also, it is known that the total travel time is

$$t_1(q) = t_2(q)$$
$$= 16{,}264.5 \text{ veh-min}$$

Using the performance function of route 2, we find

$$(4 + x_2)(q) = 16.2645 \text{ (using thousands of vehicles)}$$

$$q = \frac{16.2645}{4 + x_2}$$

From $t_1 = t_2$, and $x_1 = q - x_2$ (flow conservation),

$$2 + 1.2x_1 = 4 + x_2$$
$$2 + 1.2(q - x_2) = 4 + x_2$$
$$q = 1.67 + 1.83x_2$$

Equating the two expressions for $q$ gives

$$1.67 + 1.83x_2 = \frac{16.2645}{4 + x_2}$$

$$1.83x_2^2 + 8.99x_2 - 9.5845 = 0$$

which gives $x_2 = 0.901$, $q = 1.67 + 1.83(0.901) = 3.319$, and $x_1 = 3.319 - 0.901 = 2.418$. Because flow exists on both routes, the earlier assumption that both routes would be used is valid, and a reduction of <u>181 vehicles</u> (3500 − 3319) in peak-hour flow is needed to ensure equality of total travel times.

---

**EXAMPLE 8.12    USER EQUILIBRIUM—EFFECT OF CAPACITY REDUCTION ON TOTAL TRAVEL TIME**

Two highways serve a busy corridor with a traffic demand that is fixed at 6000 vehicles during the peak hour. The performance functions for the two routes are $t_1 = 4 + 5(x_1/c_1)$ and $t_2 = 3 + 7(x_2/c_2)$, where $t$'s are in minutes and flows ($x$'s) and capacities ($c$'s) are in thousands of vehicles per hour. Initially, the capacities of routes 1 and 2 are 4400 veh/h and 5200 veh/h, respectively. If a highway reconstruction project cuts the capacity of route 2 to 2200 veh/h, how many additional vehicle hours of travel time will be added in the corridor assuming that user-equilibrium conditions hold?

*SOLUTION*

To determine the initial number of vehicle hours, first check to see if both routes are used:

$$t_1(6) = 10.82 \text{ min}, \qquad t_2(0) = 3 \text{ min}$$
$$t_1(0) = 4 \text{ min}, \qquad t_2(6) = 11.08 \text{ min}$$

Both routes are used, because $t_2(6) > t_1(0)$ and $t_1(6) > t_2(0)$. At user equilibrium, $t_1 = t_2$, so substituting performance functions gives

$$4 + \frac{5}{4.4}(x_1) = 3 + \frac{7}{5.2}(x_2)$$

With flow conservation, $x_2 = 6 - x_1$, so that

$$4 + 1.136(x_1) = 3 + 1.346(6 - x_1)$$
$$x_1 = 2.85$$

and

$$x_2 = 6 - 2.85$$
$$= 3.15$$

The total travel time in hours is $(t_1 x_1 + t_2 x_2)/60$ or, by substituting,

$$\frac{\left\{\left[4 + 1.136(2.85)\right]2850 + \left[3 + 1.346(3.15)\right]3150\right\}}{60} = 723.88 \ \text{veh-h}$$

For the reduced-capacity case, the route usage check is

$$t_1(6) = 10.82 \ \text{min}, \qquad t_2(0) = 3 \ \text{min}$$
$$t_1(0) = 4 \ \text{min}, \qquad t_2(6) = 22.09 \ \text{min}$$

Again, both routes are used $[t_2(6) > t_1(0)$ and $t_1(6) > t_2(0)]$. Equating performance functions (because travel times are equal) and using flow conservation, $x_2 = 6 - x_1$,

$$4 + \frac{5}{4.4}(x_1) = 3 + \frac{7}{2.2}(x_2)$$
$$4 + 1.136x_1 = 3 + 3.182(6 - x_1)$$
$$x_1 = 4.19$$

and

$$x_2 = 6 - 4.19 = 1.81$$

which gives a total travel time of $(t_1 x_1 + t_2 x_2)/60$ or, by substituting,

$$\frac{\left\{\left[4 + 1.136(4.19)\right]4190 + \left[3 + 3.182(1.81)\right]1810\right\}}{60} = 875.97 \ \text{veh-h}$$

Thus the reduced capacity results in an additional <u>152.09 veh-h</u> (875.97 − 723.88) of travel time.

### 8.6.3 Mathematical Programming Approach to User Equilibrium

Equating travel time on all used routes is a straightforward approach to user equilibrium, but can become cumbersome when many alternative routes are involved. The approach used to resolve this computational obstacle is to formulate the user equilibrium problem as a mathematical program. Specifically, user-equilibrium route flows can be obtained by minimizing the following function [Sheffi 1985]:

$$\min \; S(x) = \sum_n \int_0^{x_n} t_n(w) \, dw \qquad (8.8)$$

*where*

$n$ = a specific route, and

$t_n(w)$ = performance function corresponding to route $n$ ($w$ denotes flow, $x_n$'s).

This function is subject to the constraints that the flow on all routes is greater than or equal to zero ($x_n \geq 0$) and that flow conservation holds (the flow on all routes between an origin and destination sums to the total number of vehicles, $q$, traveling between the origin and destination, $q = \sum_n x_n$).

Formulating the user equilibrium problem as a mathematical program allows an equilibrium solution to very complex highway networks (many origins and destinations) to be readily undertaken by computer. The reader is referred to Sheffi [1985] for an application of user-equilibrium principles to such a network.

---

**EXAMPLE 8.13    USER EQUILIBRIUM – MATHEMATICAL PROGRAMMING SOLUTION**

Solve Example 8.10 by formulating user equilibrium problem as a mathematical program.

*SOLUTION*

From Example 8.10, the performance functions are

$$t_1 = 6 + 4x_1$$
$$t_2 = 4 + x_2^2$$

Substituting these into Eq. 8.8 gives

$$\min S(x) = \int_0^{x_1} (6 + 4w) \, dw + \int_0^{x_2} (4 + w^2) \, dw$$

The problem can be viewed in terms of $x_2$ only by noting that flow conservation implies $x_1 = 4.5 - x_2$. Substituting yields

$$S(x) = \int_0^{4.5 - x_2} (6 + 4w) \, dw + \int_0^{x_2} (4 + w^2) \, dw$$

$$= 6w + 2w^2 \Big|_0^{4.5-x_2} + 4w + \frac{w^3}{3} \Big|_0^{x_2}$$

$$= 27 - 6x_2 + 40.5 - 18x_2 + 2x_2^2 + 4x_2 + \frac{x_2^3}{3}$$

To arrive at a minimum, the first derivative is set to zero, giving

$$\frac{dS(x)}{dx_2} = x_2^2 + 4x_2 - 20 = 0$$

which gives $x_2 =$ <u>2899 veh/h</u>, the same value as found in Example 8.10. It can readily be shown that all other flows and travel times will also be the same as those computed in Example 8.10.

## 8.6.4  System Optimization

From an idealistic point of view, one can visualize a single route choice strategy that results in the lowest possible number of total vehicle hours of travel for some specified origin-destination traffic flow. Such strategy is known as a system-optimal route choice and is based on the choice rule that travelers will behave such that total system travel time will be minimized, even though travelers may be able to decrease their own individual travel times by unilaterally changing routes. From this definition it is clear that system-optimal flows are not stable, because there will always be a temptation for travelers to switch to non–system-optimal routes in order to improve their travel times. Thus system-optimal flows are generally not a realistic representation of actual traffic. Nevertheless, system-optimal flows often provide useful comparisons with the more realistic user-equilibrium traffic forecasts.

The system-optimal route choice rule is made operational by the following mathematical program:

$$\min S(x) = \sum_n x_n t_n (x_n) \tag{8.9}$$

This program is subject to the constraints of flow conservation ( $q = \sum_n x_n$ ) and nonnegativity ($x_n \geq 0$).

## EXAMPLE 8.14   SYSTEM OPTIMIZATION

Determine the system-optimal travel time for the situation described in Example 8.10.

### SOLUTION

Using Eq. 8.9 and substituting the performance functions for routes 1 and 2 yields

$$S(x) = x_1\left(6 + 4x_1\right) + x_2\left(4 + x_2^2\right)$$
$$= 6x_1 + 4x_1^2 + 4x_2 + x_2^3$$

From flow conservation, $x_1 = 4.5 - x_2$; therefore,

$$S(x) = 6\left(4.5 - x_2\right) + 4\left(4.5 - x_2\right)^2 + 4x_2 + x_2^3$$
$$= x_2^3 + 4x_2^2 - 38x_2 + 108$$

To find the minimum, the first derivative is set to zero, giving

$$\frac{dS(x)}{dx_2} = 3x_2^2 + 8x_2 - 38 = 0$$

which gives $x_2 = 2.467$ and $x_1 = 4.5 - 2.467 = 2.033$. For system-optimal travel times,

$$t_1 = 6 + 4\left(2.033\right) = 14.13 \ \text{min}$$
$$t_2 = 4 + \left(2.467\right)^2 = 10.08 \ \text{min}$$

which are not user-equilibrium travel times, because $t_1$ is not equal to $t_2$. In Example 8.10, the total user-equilibrium travel time is computed as 930 veh-h [4500(12.4)/60]. For the system-optimal total travel time [$(t_1x_1 + t_2x_2)/60$],

$$\frac{\left[2033\left(14.13\right) + 2467\left(10.08\right)\right]}{60} = 893.2 \ \text{veh-h}$$

Therefore, the system-optimal solution results in a systemwide travel time savings of 36.8 veh-h.

## EXAMPLE 8.15   COMPARISON OF USER-EQUILIBRIUM AND SYSTEM-OPTIMAL SOLUTIONS

Two roads begin at a gate entrance to a park and take different scenic routes to a single main attraction in the park. The park manager knows that 4000 vehicles arrive during the peak hour, and he distributes these vehicles among the two routes so that an equal number of vehicles take each route. The performance functions for the routes are $t_1 = 10 + x_1$ and $t_2 = 5 + 3x_2$, with the $x$'s expressed in thousands of vehicles per hour and the $t$'s in minutes. How many vehicle-hours would have been saved had the park manager distributed the vehicular traffic so as to achieve a system-optimal solution?

## SOLUTION

For the number of vehicle hours, assuming an equal distribution of traffic among the two routes,

$$Route\ 1: \quad \frac{x_1 t_1}{60} = \frac{2000\left[10 + (2)\right]}{60} = 400 \ \text{veh-h}$$

$$Route\ 2:\quad \frac{x_2 t_2}{60} = \frac{2000[5+3(2)]}{60} = 366.67 \ \text{veh-h}$$

for a total of 766.67 veh-h. With the system-optimal traffic distribution, the performance functions are substituted into Eq. 8.9, giving

$$S(x) = (10 + x_1)\,x_1 + (5 + 3x_2)x_2$$

With flow conservation, $x_1 = 4.0 - x_2$, so that

$$S(x) = 4x_2^2 - 13x_2 + 56$$

Setting the first derivative equal to zero,

$$\frac{dS(x)}{dx_2} = 8x_2 - 13 = 0$$

gives $x_2 = 1.625$ and $x_1 = 4 - 1.625 = 2.375$. The total travel times are

$$Route\ 1:\quad \frac{x_1 t_1}{60} = \frac{2375[10+2.375]}{60} = 489.84 \ \text{veh-h}$$

$$Route\ 2:\quad \frac{x_2 t_2}{60} = \frac{1625[5+3(1.625)]}{60} = 267.45 \ \text{veh-h}$$

which gives a total system travel time of 757.27 veh-h or a savings of <u>9.38 veh-h</u> (766.67 − 757.29) over the equal distribution of traffic to the two routes.

## EXAMPLE 8.16    SYSTEM OPTIMAL SOLUTION—MINIMIZING PERSON-HOURS

During the peak hour, an urban freeway segment has a traffic flow of 4000 veh/h (2000 vehicles with one occupant and 2000 vehicles with two occupants). The freeway has five lanes, four of which are unrestricted (open to all vehicles regardless of vehicle occupancy) and one that is restricted for use by vehicles with two occupants. The performance functions for the length of this freeway segment are $t_u = 4 + 0.5x_u$ for the unrestricted lanes (all four combined) and $t_r = 4 + 2x_r$ for the restricted lane ($t$'s are in minutes and $x$'s in thousands of vehicles per hour). Determine the distribution of traffic among the lanes such that the total number of person hours is minimized, and compare the savings in person hours relative to a user equilibrium solution (assume that compliance is perfect and that no single-occupant vehicles use the restricted lane).

### SOLUTION

As stated in the problem, the 2000 single-occupant vehicles must use the unrestricted lanes. Begin by determining the distribution of traffic that will minimize total person hours. Using the subscripts $r$ for restricted lane, $u1$ for single-occupant vehicles using the unrestricted lanes, and $u2$ for two-occupant vehicles using the unrestricted lanes, total person hours can be written as

$$S(x) = x_r t_r \times 2 + x_{u2} t_u \times 2 + x_{u1} t_u \times 1$$

*where*

$x_r$ = flow on the restricted lane (two-occupant vehicles only),

$t_r$ = travel time on the restricted lane,

$x_{u2}$ = flow of two-occupant vehicles on the unrestricted lanes,

$t_u$ = travel time on the unrestricted lanes, and

$x_{u1}$ = flow of single-occupant vehicles on the unrestricted lanes.

It is given that $t_u = 4 + 0.5x_u$, where $x_u = x_{u1} + x_{u2}$. And because $x_{u1} = 2.0$, $t_u = 4 + 0.5(2.0 + x_{u2})$. Substituting gives

$$S(x) = 2x_r(4 + 2x_r) + 2x_{u2}(4 + 0.5(2.0 + x_{u2})) + 2(4 + 0.5(2.0 + x_{u2}))$$
$$= 8x_r + 4x_r^2 + 10x_{u2} + x_{u2}^2 + 10 + x_{u2}$$

The total number of two-occupant vehicles is 2000, so $x_r + x_{u2} = 2.0$. Substituting,

$$S(x) = 8(2 - x_{u2}) + 4(2 - x_{u2})^2 + 10x_{u2} + x_{u2}^2 + 10 + x_{u2} = 5x_{u2}^2 - 13x_{u2} + 42$$

Taking the first derivative,

$$\frac{dS(x)}{dx_{u2}} = 10x_{u2} - 13 = 0$$

which gives $x_{u2} = 1.3$, and so $x_r = 2.0 - 1.3 = 0.7$. With this, total person hours is [with $t_r = 4 + 2(0.7) = 5.4$ and $t_u = 4 + 0.5(3.3) = 5.65$]

$2[5.4(700)] + 2[5.65(1300)] + 2000(5.65) = 33,550$ person-min or <u>559.167 person-h</u>

For the user-equilibrium solution, with 2000 vehicles on the unrestricted lanes, $t_u$ can be written as

$$t_u = 4 + 0.5(2.0 + x_{u2}) = 5 + 0.5x_{u2}$$

To check if both two-occupant lane choices are used by two-occupant vehicles, note that when $x_{u2} = 2$ and $x_r = 0$, $t_u = 6$ and $t_r = 4$. And when $x_{u2} = 0$ and $x_r = 2$, $t_u = 5$ and $t_r = 8$, so both lane choices might be used by two-occupant vehicles. Equating travel times ($t_u = t_r$) gives

$$5 + 0.5x_{u2} = 4 - 2x_r$$

and with $x_r = 2 - x_{u2}$,

$$5 + 0.5x_{u2} = 4 - 2(2 - x_{u2})$$

Solving gives $x_{u2} = 1.2$ and $x_r = 2 - 1.2 = 0.8$. This produces user-equilibrium travel times $t_u = t_r = 5.6$. Total person hours for the user-equilibrium solution is then

$2[5.6(2000)] + 5.6(2000) = 33,600$ person-min or <u>560 person-h</u>

So the savings is <u>0.833 person-h</u> $(560 - 559.167)$ when person-hours are minimized relative to the user-equilibrium solution.

## 8.7    TRAFFIC FORECASTING IN PRACTICE

With the basic procedures outlined in the previous sections of this chapter, a traffic forecasting model similar to those used in practice can be developed and implemented. Although there are many subtleties in the process that are beyond the scope of this book, the basic procedure is as follows (referring to the example network shown in Fig. 8.8):

1.  The geographic region being studied is segmented into nearly homogeneous areas (see Fig. 8.8) based on similarities in land use, socioeconomic conditions, and so on. These areas are referred to as traffic analysis zones (TAZs) and are used to determine the origins and destinations of trips (as used in the highway route choice models described in Section 8.6). The choice of the number of TAZs (often simply referred to as zones) presents a trade-off between accuracy (smaller TAZs provide more detailed forecasts) and ease of implementation (larger TAZs require less data and are easier to incorporate in the overall model system). A single point is usually chosen within the TAZ as the assumed origin/destination point of all TAZ trips. This point is referred to as the centroid (see example network in Fig. 8.8).

2.  The highway network is defined to include the relevant highway segments. Highway segments are linked by using nodes (see example network in Fig. 8.8), which are usually placed at intersections or other points where highway capacity could change. Nodes permit traffic to travel from one highway segment to the next. The highway network is a representation of the actual street network and carries traffic flow between TAZs. As was the case with the size of the TAZs, a very large and detailed highway network can provide very detailed forecasts but also requires large amounts of data, and thus smaller networks are often used. Defining the highway network includes detailed information on each highway segment's performance function (see Section 8.6.1) so that traffic flows can be computed (usually by assuming user equilibrium route choice). The performance function often used in practice to relate traffic flow with travel time was originally developed by the U.S. Bureau of Public Roads and takes the form:

$$t_n = t_{fn} \left[ 1 + \alpha (x_n/x_{cn})^\beta \right] \tag{8.10}$$

*where*

$t_n$ = travel time on highway segment (route) $n$, usually in minutes,

$t_{fn}$ = free-flow travel time on highway segment (route) $n$, usually in minutes,

$x_n$ = traffic flow on highway segment (route) $n$, usually in veh/h,

$x_{cn}$ = capacity of highway segment (route) $n$, usually in veh/h, and

$\alpha, \beta$ = model parameters that usually vary with respect to the capacity and speed limit of the highway segment (route). Typical values of $\alpha$ and $\beta$ are shown in Table 8.1.

Access links are used to connect the highway network with the centroids of the TAZs (see Fig. 8.8). These links also have highway performance functions associated with them so that access/egress from the highway network to the centroids can be approximated.

3. Trip generation models and mode/destination choice models are then used to determine the number of vehicles traveling between all TAZs during a specified time period (usually the peak hour). The resulting vehicle trips are used to create an origin-destination matrix that gives the total number of vehicle trips going between each TAZ combination during the analysis period. Although most traffic forecasting models developed in practice have hundreds of TAZs, for illustrative purposes, Table 8.2 gives a sample vehicle origin-destination matrix with just the five TAZs shown in Fig. 8.8. If TAZs 1 and 2 are in the center of the city, the origin-destination matrix shown in Table 8.2 is what one might expect during the morning peak hour—with higher vehicular flows going from the outskirts of the region (zones 3, 4 and 5) to the city center (zones 1 and 2) and substantially lower vehicular flows going from zones 1 and 2 to zones 3, 4 and 5. As an example, Table 8.2 indicates that 3,386 vehicle trips go from TAZ 5 to TAZ 1 while only 213 vehicle trips go from TAZ 1 to TAZ 5.

4. With the vehicle origin-destination trip matrix, traffic flows on each highway segment are determined, usually by assuming that user-equilibrium holds. This is achieved by solving Eq. 8.8, which requires a computer to solve the mathematical program of a network of any reasonable realistic size. An example of such a user equilibrium computer program can be obtained from: http://www.wiley.com/college/mannering.

**Table 8.1** Typical Values of $\alpha$ and $\beta$ for Bureau of Public Roads Highway Performance Function (see [Mannering et al. 1989])

| Route speed limit (km/h) | Route capacity, $x_{cn}$ (veh/h) | Performance function parameters | |
|:---:|:---:|:---:|:---:|
| | | $\alpha$ | $\beta$ |
| < 50 | < 250 | 0.7312 | 3.6596 |
| < 50 | 251-499 | 0.6128 | 3.5038 |
| < 50 | 500-749 | 0.8774 | 4.4613 |
| < 50 | 750-999 | 0.6846 | 5.1644 |
| < 50 | 1000+ | 1.1465 | 4.4239 |
| 51-65 | < 499 | 0.6190 | 3.6544 |
| 51-65 | 500-749 | 0.6662 | 4.9432 |
| 51-65 | 750-999 | 0.6222 | 5.1409 |
| 51-65 | 1000+ | 1.0300 | 5.5226 |
| 66-80 | < 750 | 0.6609 | 5.0906 |
| 66-80 | 750-999 | 0.5423 | 5.7894 |
| 66-80 | 1000+ | 1.0091 | 6.5856 |
| > 80 | < 750 | 0.8776 | 4.9287 |
| > 80 | 750-999 | 0.7699 | 5.3443 |
| > 80 | 1000+ | 1.1491 | 6.8677 |

**Figure 8.8** Example highway network.

**Table 8.2** Example of a peak-hour vehicle origin-destination trips with the five traffic analysis zones (TAZs) shown in Fig. 8.8

| Origin traffic analysis zone | Destination traffic analysis zone | | | | | Total origin trips |
|---|---|---|---|---|---|---|
| | 1 | 2 | 3 | 4 | 5 | |
| 1 | – | 1783 | 386 | 245 | 213 | 2627 |
| 2 | 2378 | – | 546 | 197 | 101 | 3222 |
| 3 | 4412 | 2232 | – | 745 | 343 | 7732 |
| 4 | 1399 | 1201 | 822 | – | 212 | 3634 |
| 5 | 3386 | 2866 | 1202 | 345 | – | 7799 |
| Total destination trips | 11575 | 8082 | 2956 | 1532 | 869 | |

## 8.8 THE TRADITIONAL FOUR-STEP PROCESS

The approach to travel demand and traffic forecasting presented in this chapter provides an excellent exposition of the principles underlying the problem. In practice, however, the most widely used approach to travel demand and traffic forecasting is a four-step procedure: trip generation, mode choice, destination choice, and route choice (also referred to as traffic assignment). This differs from the procedure presented in this book, which is a three-step procedure: trip generation, joint mode/destination choice, and route choice. The additional step (added by separating mode and destination choices) can make the estimation of the model less complex and was quite popular decades ago, before advances in computer estimation software became widely available. In splitting mode and destination choices, a logit formulation is still most often used for the mode choice decision. However, the destination choice is often modeled by using a gravity model [see Meyer and Miller 2001]. The gravity model is based on the gravitational modeling principles covered in physics (the gravitational forces of planets) where the likelihood of a trip going to a destination is a function of the distance from the trip origin and some measure of attractiveness (the equivalent of mass in gravitational theory) of the destination. To implement the gravity model for trip distribution, the basic gravitational equation from Newtonian physics is appropriately modified. From physics, the basic gravity equation is

$$F_{ab} = \frac{M_a M_b}{D_{ab}^2}$$

(8.11)

*where*

$F_{ab}$ = gravitational force between bodies $a$ and $b$,
$M_a$ = mass of body $a$,
$M_b$ = mass of body $b$, and
$D_{ab}$ = distance between bodies $a$ and $b$.

For trip distribution, Eq. 8.11 is modified as

$$T'_{ab} = T'_a \frac{A_b f_{ab} K_{ab}}{\sum_{\forall b} A_b f_{ab} K_{ab}} \tag{8.12}$$

*where*

$T'_{ab}$ = total number of trips from TAZ $a$ to TAZ $b$,

$T'_a$ = total number of trips from TAZ $a$,

$A_b$ = total number of trips attracted to TAZ $b$,

$f_{ab}$ = distance/travel cost "friction factor," and

$K_{ab}$ = estimated parameter to ensure results balance.

In this equation, the term for the number of trips from a TAZ ($T'_a$) is determined from regression techniques as described in Section 8.4, and the number of trips to a TAZ ($A_b$) is also determined using regression techniques. These values (trips aggregated for an entire zone) produce the origin trips or "from" trips ($T'_a$'s), which gives the last row in Table 8.2 (these from trips are sometimes referred to as trip productions), and the destination or "to" trips ($A_b$'s) which give the last column in Table 8.2 (these to trips are sometimes referred to as trip attractions). Given these data, the intent of Eq. 8.12 is to fill in the remaining cells of Table 8.2 (given the last row and column). The other terms used to do this are the friction factor ($f_{ab}$), which accounts for the accessibility (cost in terms of average travel time and/or distance) between TAZs $a$ and $b$, and the parameters, $K_{ab}$'s, which are solved iteratively to ensure that the total trips produced and attracted balance (see Table 8.2).

## 8.9   THE CURRENT STATE OF TRAVEL DEMAND AND TRAFFIC FORECASTING

Travel demand and traffic forecasting models, and specifically the four-step process (trip generation, trip distribution, mode choice, and route choice), have been used for more than 50 years. While they have given reasonable forecasts given the limits of the profession's understanding of travel behavior and the limits of computational tools, in the last two decades the weaknesses of this modeling approach have become obvious. Currently, the profession is shifting from the traditional four-step process to models that start at the individual traveler level and look at travel generation as an outgrowth of activity involvement (shopping, recreation, and work). Also critical in such an approach is the concept of tours, which are trips that sequentially link multiple activities (from home to exercise class, to shopping, and back home). While more complex models of traveler behavior have been appearing in the academic literature for decades, models more sophisticated than the simple four-step process have only recently begun to appear in practice. To be sure, the transition from traditional four-step models to tour-based and activity-based models presents many challenges. Included among these challenges are significant increases in required data (to be able to predict individual activity patterns) and limitations of current computer technology. However, more detailed and accurate models for travel and traffic forecasting will allow analysts to determine the impacts of many new transportation policies relating to parking controls, toll roads, congestion pricing, vehicle occupancy restrictions, reductions in energy consumption and emissions, and other emerging transportation policies.

## APPENDIX 8A    LEAST SQUARES ESTIMATION

Least squares regression is a popular method for developing mathematical relationships from empirical data. As mentioned earlier, it is a method that is well suited to the estimation of trip generation models. To illustrate the least squares approach, consider the hypothetical trip generation data presented in Table 8A.1, which could have been gathered from a typical survey of travelers.

To begin formalizing a mathematical expression, note that the objective is to predict the number of shopping trips made on a Saturday for each household, $i$; this number is referred to as the dependent variable ($Y_i$). This prediction is to be a function of the number of people in household $i$ ($z_i$), which is referred to as the independent variable. A simple linear relationship between $Y_i$ and $z_i$ is

$$Y_i = b_0 + b_1 z_i \qquad (8A.1)$$

*where*

$$
\begin{aligned}
Y_i &= \text{number of shopping trips made by household } i, \\
b_0, b_1 &= \text{coefficients to be determined by estimation, and} \\
Z &= \text{number of people in household } i.
\end{aligned}
$$

Ideally, one wants to determine the $b$'s in Eq. 8A.1 that will give predictions of the number of shopping trips ($Y_i$'s) that are as close as possible to the actual observed number of shopping trips ($Y_i$'s, as shown in Table 8A.1). The difference or deviation between the observed and predicted number of shopping trips can be expressed mathematically as

$$\text{Deviation} = Y_i - (b_0 + b_1 z_i) \qquad (8A.2)$$

**Table 8A.1** Example of Shopping Trip Generation Data

| Household number, $i$ | Number of shopping trips made all day Saturday, $Y_i$ | People in household $i$, $z_i$ |
|:---:|:---:|:---:|
| 1 | 3 | 4 |
| 2 | 1 | 2 |
| 3 | 1 | 3 |
| 4 | 5 | 4 |
| 5 | 3 | 2 |
| 6 | 2 | 4 |
| 7 | 6 | 8 |
| 8 | 4 | 6 |
| 9 | 5 | 6 |
| 10 | 2 | 2 |

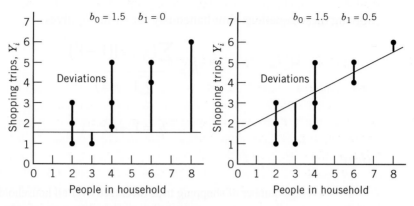

**Figure 8A.1** Illustration of deviations.

Such deviations are illustrated graphically in Fig. 8A.1 for two groups of $b_0$ and $b_1$ values. In the first illustration in this figure, $b_0 = 1.5$ and $b_1 = 0$, which implies that the number of household members does not affect the number of shopping trips made. The second illustration has $b_0 = 1.5$ and $b_1 = 0.5$ and, as can readily be seen, the deviations (differences between the points representing the observed number of shopping trips and the line representing the equation $b_0 + b_1 z_i$) are reduced relative to the first illustration. These two illustrations suggest the need for some method of determining the values of $b_0$ and $b_1$ that produce the smallest possible deviations relative to observed data. Such a method can be solved by a mathematical program whose objective is to minimize the sum of the square of deviations, or

$$\min \; S\left(b_0 + b_1\right) = \sum_i \left(Y_i - b_0 - b_1 z_i\right)^2 \tag{8A.3}$$

The minimization is accomplished by setting partial derivatives equal to zero:

$$\frac{\partial S}{\partial b_0} = -2\sum_i \left(Y_i - b_0 - b_1 z_i\right) = 0 \tag{8A.4}$$

$$\frac{\partial S}{\partial b_1} = -2\sum_i z_i \left(Y_i - b_0 - b_1 z_i\right) = 0 \tag{8A.5}$$

Solving these equations using gives

$$\sum_i Y_i - nb_0 - b_1\sum_i z_i = 0 \tag{8A.6}$$

$$\sum_i z_i Y_i - b_0\sum_i X_i - b_1\sum_i z_i^2 = 0 \tag{8A.7}$$

*where*

$n$ = number of households used to estimate the coefficients, and
Other terms are as defined previously.

Solving these equations simultaneously for $b_0$ and $b_1$ gives

$$b_1 = \frac{\sum_i (z_i - \overline{z})(Y_i - \overline{Y})}{\sum_i (z_i - \overline{z})^2} \tag{8A.8}$$

$$b_0 = \overline{Y} - b_1 \overline{z} \tag{8A.9}$$

*where*

$\overline{Y}$ = average number of shopping trips (averaged over all households, $n$),

$\overline{z}$ = average household income (averaged over all households, $n$), and

Other terms are as defined previously.

This approach to determining the values of estimable coefficients ($b$'s) is referred to as least squares regression, and it can be shown that for the data values given in Table 8A.1, the smallest deviations between the number of predicted and actual shopping trips is given by the equation

$$Y_i = 0.33 + 0.7z_i \tag{8A.10}$$

When many coefficient values ($b$'s) must be determined, a matrix representation of the least squares solution is appropriate:

$$B = (Z'Z)^{-1}Z'Y \tag{8A.11}$$

*where*

$B$ is an $n \times 1$ vector of coefficients (with $n$ being the number of households),

$Z$ is an $n \times k$ matrix of variables determining $Y$ (where $k$ is the number of variables, such as household income, number of people in the household, etc.),

$Z'$ is the $k \times n$ transpose matrix of $Z$, and

$Y$ is an $n \times 1$ vector of the dependent variable (number of shopping trips in this example).

For additional information on least squares regression, refer to Washington et al. [2011].

## APPENDIX 8B    MAXIMUM-LIKELIHOOD ESTIMATION

Maximum-likelihood estimation is used extensively in the statistical analysis of traffic data [Washington et al. 2011]. The idea underlying maximum-likelihood estimation is that different statistical distributions generate different samples, and any one sample is more likely to come from some distributions than from others.

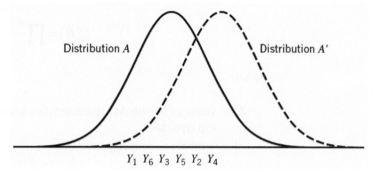

**Figure 8B.1** Illustration of randomly drawn numbers and possible source distributions.

To illustrate this, suppose there is a sample of six randomly drawn numbers, $Y_1, Y_2, \ldots, Y_6$, and there are two possible distributions that could generate these numbers, as shown in Fig. 8B.1.

It is clear from Fig. 8B.1 that distribution $A$ is much more likely to generate these six numbers than distribution $A'$. The objective of maximum-likelihood estimation is to estimate a coefficient vector, say $B$, that defines a distribution that is most likely to generate some observed data. To show how this is done, consider the Poisson regression of trip generation discussed in Section 8.4.2. The maximum-likelihood function can be written as a simple product of the probabilities of a Poisson distribution with coefficients $B$ generating observed household trip generation. This is, for a given trip type,

$$L(B) = \prod_i P(T_i) \tag{8B.1}$$

*where*

$$
\begin{aligned}
B &= \text{vector of estimable coefficients,} \\
L(B) &= \text{likelihood function,} \\
T_i &= \text{number of vehicle-based trips of a specific type (shopping, social/recreational,} \\
&\quad\ \text{etc.) in some specified time period by household } i, \text{ and} \\
P(T_i) &= \text{probability of household } i \text{ making } T \text{ trips.}
\end{aligned}
$$

Using the Poisson equation (Eq. 8.2), Eq. 8B.1 becomes

$$L(B) = \prod_i \frac{e^{-\lambda_i} \lambda_i^{T_i}}{T_i!} \tag{8B.2}$$

*where*

$\lambda_i$ = Poisson parameter for household $i$, which is equal to household $i$'s expected number of vehicle-based trips in some specified time period, $E[T_i]$, and

Other terms are as defined previously.

With $\lambda_i = e^{BZ_i}$ as in Eq. 8.3,

$$L(\boldsymbol{B}) = \prod_i \frac{e^{-e^{\boldsymbol{BZ}_i}}\left(e^{\boldsymbol{BZ}_i}\right)^{T_i}}{T_i!} \qquad\qquad \textbf{(8B.3)}$$

*where*

$\boldsymbol{Z}_i$ = vector of household $i$ characteristics determining trip generation for a given trip type, and

Other terms are as defined previously.

The problem then becomes one of finding the vector $\boldsymbol{B}$ that maximizes this function (maximizes the product of probabilities as shown in Eq. 8B.1). To do this, the natural logarithm is used to transform the likelihood function into a log-likelihood function (this does not affect the maximization process). In the Poisson regression case, this log transformation of Eq. 8B.3 gives

$$LL(\boldsymbol{B}) = \sum_i \left[ -e^{\boldsymbol{BZ}_i} + T_i\boldsymbol{BZ}_i - \ln(T_i!) \right] \qquad\qquad \textbf{(8B.4)}$$

*where*

$LL(\boldsymbol{B})$ = log-likelihood function, and

Other terms are as defined previously.

Maximization of this expression with respect to $\boldsymbol{B}$ is undertaken by setting the first derivative to zero such that

$$\frac{\partial LL(\boldsymbol{B})}{\partial \boldsymbol{B}} = \sum_i \left[ -e^{\boldsymbol{BZ}_i} + T_i \right] \boldsymbol{Z}_i = 0 \qquad\qquad \textbf{(8B.5)}$$

Equation 8B.5 can be solved numerically using standard software packages [Washington et al. 2011]. Using such a software package with the data in Table 8A.1, the estimated maximum-likelihood values of $\boldsymbol{B}$ give

$$\lambda_i = e^{\boldsymbol{BZ}_i} = e^{-0.206 + 0.2\,z_i} \qquad\qquad \textbf{(8B.6)}$$

*where*

$z_i$ = people in household $i$ (see Table 8A.1), and

Other terms are as defined previously.

Refer to Washington et al. [2011] for an extensive discussion of maximum-likelihood estimation.

## NOMENCLATURE FOR CHAPTER 8

$A_b$ — the number of trips attracted to traffic analysis zone $b$

$b_k$ — estimated coefficients

**$B$** — vector of estimable coefficients

$D_{ab}$ — the distance between bodies $a$ and $b$

$F_{ab}$ — the gravitational force between bodies $a$ and $b$

$f_{ab}$ — the distance/travel cost friction factor

$K_{ab}$ — estimated parameter for gravity model

$L(\cdot)$ — likelihood function

$LL(\cdot)$ — log-likelihood function

$M_a$ — the mass of body $a$

$P$ — probability of an alternative being selected

$P(T_i)$ — probability of $T_i$ trips (a non-negative integer) being generated by household $i$

$q$ — total origin-to-destination traffic flow

$S(\cdot)$ — mathematical objective function

$s$ — notation for the set of available alternatives

$T_i$ — number of household trips generated per unit time for household $i$

$T'_{ab}$ — total number of trips from traffic analysis zone $a$ to traffic analysis zone $b$

$T'_a$ — total number of trips from traffic analysis zone $a$

$t_n$ — travel time on route $n$

$t_{fn}$ — free-flow travel time on route $n$

$U$ — specifiable portion of an alternative's utility

$V$ — total alternative utility

$w$ — route flow operative for $x_n$

$x_n$ — traffic flow on route $n$

$x_{cn}$ — traffic flow capacity of route $n$,

$Y_i$ — dependent variable for household $i$

$z$ — household or alternative characteristic

**$Z_i$** — vector of household $i$'s characteristics

$\alpha$ — model parameter for highway performance function

$\beta$ — model parameter for highway performance function

$\lambda_i$ — Poisson parameter for household $i$

$\varepsilon$ — unspecifiable portion of an alternative's utility (assumed to be a random variable)

## REFERENCES

Mannering, F., D. Garrison, and B. Sebranke. "Generation and Assessment of Incident Management Strategies, Volume IV: Seattle-Area Incident Impact Analysis: Microcomputer Traffic Simulation Results." WA-RD 204.4 and TNW90-11vol4, 1989, available online at http://www.transnow.org/publication/final-reports/documents/TNW90-11vol4.pdf

Hamed, M., and F. Mannering. "Modeling Travelers' Post-Work Activity Involvement: Toward a New Methodology." *Transportation Science*, vol. 17, no. 4, 1993.

Mannering, F. "Poisson Analysis of Commuter Flexibility in Changing Route and Departure Times." *Transportation Research, Part B*, vol. 23, no. 1, 1989.

McFadden, D. "Econometric Models of Probabilistic Choice." In *Structural Analysis of Discrete Data with Econometric Applications*, edited by C. Manski and D. McFadden. Cambridge, MA: MIT Press, 1981.

Meyer, M., and E. Miller. *Urban Transportation Planning: A Decision-Oriented Approach*. New York: McGraw-Hill, 2001.

Sheffi, Y. *Urban Transportation Networks: Equilibrium Analysis with Mathematical Programming Models*. Englewood Cliffs, NJ: Prentice-Hall, 1985.

Wardrop, F. "Some Theoretical Aspects of Road Traffic Research." *Proceedings, Institution of Civil Engineers II*, vol. 1, 1952.

Washington, S., M. Karlaftis, and F. Mannering. *Statistical and Econometric Methods for Transportation Data Analysis*. Second Edition. Boca Raton, FL: Chapman & Hall/CRC, 2011.

# PROBLEMS

**8.1** A large retirement village has a total retail employment of 120. All 1600 of the households in this village consist of two nonworking family members with household income of $20,000. Assuming that shopping and social/recreational trip rates both peak during the same hour (for exposition purposes), predict the total number of peak-hour trips generated by this village using the trip generation models of Examples 8.1 and 8.2.

**8.2** Consider the retirement village described in Problem 8.1. Determine the amount of additional retail employment (in the village) necessary to reduce the total predicted number of peak-hour shopping trips to 200.

**8.3** A large residential area has 1400 households with an average household income of $40,000, an average household size of 4.8, and, on average, 1.5 working members. Using the model described in Example 8.2 (assuming it was estimated using zonal averages instead of individual households), predict the change in the number of peak-hour social/recreational trips if employment in the area increases by 25% and household income by 10%.

**8.4** Consider the Poisson trip generation model in Example 8.4. Suppose that a household has five members with an annual income of $150,000 and lies in a neighborhood with a retail employment of 320. What is the expected number of peak-hour shopping trips, and what is the probability that the household will make more than one peak-hour shopping trip?

**8.5** Consider a Poisson regression model for the number of social/recreational trips generated during a peak-hour period that is estimated by (see Eq. 8.3) $BZ_i$ = −0.75 + 0.025(household size) + 0.008(annual household income, in thousands of dollars) + 0.10(number of nonworking household members). Suppose a household has five members (three of whom work) and an annual income of $100,000. What is the expected number of peak-hour social/recreational trips, and what is the probability that the household will not make a peak-hour social/recreational trip?

**8.6** If small express buses leave the origin described in Example 8.5 and all are filled to their capacity of 20 travelers, how many work trip vehicles leave from origin to destination in Example 8.5 during the peak hour?

**8.7** Consider the conditions described in Example 8.5. If an energy crisis doubles the cost of the auto modes (drive-alone and shared-ride) and bus costs are not affected, how many workers will use each mode?

**8.8** It is known that 4000 automobile trips are generated in a large residential area from noon to 1:00 P.M. on Saturdays for shopping purposes. Four major shopping centers have the following characteristics:

| Shopping center | Distance from residential area (km) | Commercial floor space (thousands of m²) |
|---|---|---|
| 1 | 4 | 20 |
| 2 | 9 | 15 |
| 3 | 8 | 30 |
| 4 | 14 | 60 |

If a logit model is estimated with coefficients of −0.283 for distance and 0.172 for commercial space (in thousands of m²), how many shopping trips will be made to each of the four shopping centers?

**8.9** Consider the shopping trip situation described in Problem 8.8. Suppose that shopping center 3 goes out of business and shopping center 2 is expanded to 50,000 m² of commercial space. What would be the new distribution of the 4000 Saturday afternoon shopping trips?

**8.10** If shopping center 3 is closed (see Problem 8.9), how much commercial floor space is needed in shopping centers 1 and 2 to ensure that each of them has the same probability of being selected as shopping center 4?

**8.11** Consider the situation described in Example 8.7. If the construction of a new freeway lowers auto and transit travel times to shopping center 2 by 20%, determine the new distribution of shopping trips by destination and mode.

**8.12** Consider the conditions described in Example 8.7. Heavily congested highways have caused travel times to shopping center 2 to increase by 4 min for both auto and transit modes (travel times to shopping center 1 are not affected). In order for shopping center 2 to attract as many total trips (auto and transit) as it did before the congestion, how much commercial floor space must it add (given that the total number of departing shopping trips remains at 900)?

**8.13** A total of 725 auto-mode social/recreational trips are made from an origin (residential area) during the peak hour. A logit model estimation is made, and three factors were found to influence the destination choice: (1) population at the destination, in thousands (coefficient = 0.17); (2) distance from origin to destination, in kilometers (coefficient = −0.14); and (3) square meters of amusement floor space (movie theaters, video game centers, etc.), in thousands (coefficient = 1.0). Four possible destinations have the following characteristics:

| | Population (thousands) | Distance from origin (km) | Amusement space (thousands of m$^2$) |
|---|---|---|---|
| Destination 1 | 15.5 | 12 | 0.5 |
| Destination 2 | 6.0 | 8 | 1.0 |
| Destination 3 | 0.8 | 3 | 0.8 |
| Destination 4 | 5.0 | 11 | 1.5 |

Determine the distribution of trips among possible destinations.

**8.14** Consider the situation described in Problem 8.13. If a new 1,500-m$^2$ arcade center is built at destination 3, determine the distribution of the 725 peak-hour social/recreational trips.

**8.15** Consider the situation described in Problem 8.13. If the total number of trips remains constant, determine the amount of amusement floor space that must be added to destination 2 to attract an additional 50 social/recreational trips.

**8.16** Note that with the situation described in Example 8.8, 26.6% [(110 + 23)/500] of all social/recreational trips are to destination 1. If the total number of trips remains constant, how much additional amusement floor space would have to be added to destination 1 to have it capture 35.0% of the total social/recreational trips?

**8.17** Two routes connect an origin and a destination, and the flow is 15,000 veh/h. Route 1 has a performance function $t_1 = 4 + 3x_1$, and route 2 has a function of $t_2 = b + 6x_2$, with the $x$'s expressed in thousands of vehicles per hour and the $t$'s in minutes.

**(a)** If the user equilibrium flow on route 1 is 9780 veh/h, determine the free-flow travel time on route 2 ($b$) and equilibrium travel times.

**(b)** If population declines reduce the number of travelers at the origin, and the total origin-destination flow is

reduced to 7000 veh/h, determine user equilibrium travel times and flows.

**8.18** An origin-destination pair is connected by a route with a performance function $t_1 = 8 + x_1$, and another with a function $t_2 = 1 + 2x_2$ (with $x$'s in thousands of vehicles per hour and $t$'s in minutes). If the total origin-destination flow is 4000 veh/h, determine user equilibrium and system-optimal route travel times, total travel time (in vehicle-minutes), and route flows.

**8.19** Because of the great increase in vehicle-hours caused by the reconstruction project in Example 8.12, the state transportation department decides to regulate the flow of traffic on the two routes (until reconstruction is complete) to achieve a system-optimal solution. How many vehicle-hours will be saved during each peak-hour period if this strategy is implemented and travelers are not permitted to achieve a user equilibrium solution?

**8.20** For Example 8.11, what reduction in peak-hour traffic demand is needed to ensure an equality of total vehicle travel time (in vehicle-minutes) assuming a system-optimal solution before and during reconstruction?

**8.21** Two routes connect an origin and a destination. Routes 1 and 2 have performance functions $t_1 = 2 + x_1$ and $t_2 = 1 + x_2$, where the $t$'s are in minutes and the $x$'s are in thousands of vehicles per hour. The travel times on the routes are known to be in user equilibrium. If an observation for route 1 finds that the gaps between 30% of the vehicles are less than 6 seconds, estimate the volume and average travel times for the two routes. (*Hint*: Assume a Poisson distribution of vehicle arrivals, as discussed in Chapter 5.)

**8.22** Three routes connect an origin and a destination with performance functions $t_1 = 8 + 0.5x_1$, $t_2 = 1 + 2x_2$, and $t_3 = 3 + 0.75x_3$, with the $x$'s expressed in thousands of vehicles per hour and the $t$'s expressed in minutes. If the peak-hour traffic demand is 3400 vehicles, determine user equilibrium traffic flows.

**8.23** Two routes connect a suburban area and a city, with route travel times (in minutes) given by the expressions $t_1 = 6 + 8(x_1/c_1)$ and $t_2 = 10 + 3(x_2/c_2)$, where the $x$'s are expressed in thousands of vehicles per hour and the $c$'s are the route capacities in thousands of vehicles per hour. Initially, the capacities of routes 1 and 2 are 4000 and 2000 veh/h, respectively. A reconstruction project on route 1 reduces the capacity to 3000 veh/h, but total traffic demand is unaffected. Observational studies note a 35.28-second increase in average travel time on route 1 and a 68.5% increase in flow on route 2 after reconstruction begins. User equilibrium conditions exist before and during reconstruction. If both routes are always used, determine equilibrium flows and travel times before and after reconstruction begins.

**8.24** Three routes connect an origin and destination with performance functions $t_1 = 2 + 0.5x_1$, $t_2 = 1 + x_2$, and $t_3 = 4 + 0.2x_3$ (with $t$'s in minutes and $x$'s in thousands of vehicles per hour). Determine user equilibrium flows if the total origin-to-destination demand is (a) 10,000 veh/h and (b) 5000 veh/h.

**8.25** For the routes described in Problem 8.24, what is the minimum origin-to-destination traffic demand (in vehicles per hour) that will ensure that all routes are used (assuming user equilibrium conditions)?

**8.26** A multilane highway has two northbound lanes. Each lane has a capacity of 1500 vehicles per hour. Currently, northbound traffic consists of 3100 vehicles with 1 occupant, 600 vehicles with 2 occupants, 400 vehicles with 3 occupants, and 20 buses with 50 occupants each. The highway's performance function is $t = t_0[1 + 1.15(x/c)^{6.87}]$, where $t$ is in minutes, $t_0$ is equal to 15 minutes, and $x$ and $c$ are volumes and capacities in vehicles per hour. An additional lane is being added (with 1500 veh/h capacity). What will the total person hours of travel be if the lane is (a) open to all traffic, (b) open to vehicles with 2 or more occupants only, and (c) open to vehicles with 3 or more occupants only? (Assume that all qualified higher-occupancy vehicles use only the new lane, no unqualified vehicles use the new lane, and there is no mode shift.)

**8.27** Consider the new lane addition in Problem 8.26. First, suppose 500 one-occupant vehicle travelers take 10 buses (50 on each bus), and the new lane is open to vehicles with two or more occupants. What would the total person hours be? Second, referring back to part (a) of Problem 8.26, what is the minimum mode shift from one-occupant vehicles to buses (with 50 persons each) needed to ensure that the person hours of travel time on the highway with the new lane (which is restricted to vehicles with two or more occupants) is as low as if all three lanes (the two existing lanes and the new lane) were open to all traffic? (Set up the equation, and solve to the nearest 100 one-occupant vehicles.)

**8.28** Two routes connect an origin and destination with performance functions $t_1 = 5 + 3x_1$ and $t_2 = 7 + x_2$, with $t$'s in minutes and $x$'s in thousands of vehicles per hour. Total origin-destination demand is 7000 vehicles in the peak hour. What are user equilibrium and system-optimal route flows and total travel times?

**8.29** Consider the conditions in Problem 8.28. What is the value of the derivative of the user equilibrium math program evaluated at the system-optimal solution with respect to $x_1$ (with $x_1$ equal to the system-optimal solution)?

**8.30** Two routes connect an origin and a destination. Their performance functions are $t_1 = 3 + 1.5(x_1/c_1)2$ and $t_2 = 5 + 4(x_2/c_2)$, with $t$'s in minutes and $x$'s and $c$'s being route flows and capacities, respectively. The origin-destination demand is 6000 vehicles per hour, and $c_1$ and $c_2$ are equal to 2000 and 1500 vehicles per hour, respectively. Proposed capacity improvements will increase $c_2$ by 1000 vehicles per hour. It is known that the routes are currently in user equilibrium, and it is estimated that each 1-minute reduction in route travel time will attract an additional 500 vehicles per hour (from latent travel demand and mode shifts). What will the user equilibrium flows and total hourly origin-destination demand be after the capacity improvement?

**8.31** Two routes connect an origin-destination pair with performance functions $t_1 = 5 + 4x_1$ and $t_2 = 7 + 2x_2$, with $t$'s in minutes and $x$'s in thousands of vehicles per hour. Assuming both routes are used, can user equilibrium and system-optimal solutions be equal at some feasible value of total origin-destination demand ($q$)? (Prove your answer.)

**8.32** Three routes connect an origin-destination pair with performance functions $t_1 = 5 + 1.5x_1$, $t_2 = 12 + 3x_2$, and $t_3 = 2 + 0.2 x_3^2$ (with $t$'s in minutes and $x$'s in thousands of vehicles per hour). Determine user equilibrium flows if $q = 4300$ veh/h.

**8.33** Two routes connect an origin-destination pair with performance functions $t_1 = 6 + 4x_1$ and $t_2 = 2 + 0.5x_2^2$ (with $t$'s in minutes and $x$'s in thousands of vehicles per hour). The origin-destination demand is 4000 veh/h at a travel time of 2 minutes, but for each additional minute beyond these 2 minutes, 100 fewer vehicles depart. Determine user equilibrium route flows and total vehicle travel time.

**8.34** A freeway has six lanes, four of which are unrestricted (open to all vehicle traffic), and two of which are restricted lanes that can be used only by vehicles with two or more occupants. The performance function for the highway is $t = 12 + (2/NL)x$ (with $t$ in minutes, $NL$ being the number of lanes, and $x$ in thousands of vehicles). During the peak hour, 3000 vehicles with one occupant and 4000 vehicles with two occupants depart for the destination. Determine the distribution of traffic between restricted and unrestricted lanes such that total person hours are minimized.

**8.35** Two routes connect an origin-destination pair with performance functions $t_1 = 5 + (x_1/2)^2$ and $t_2 = 7 + (x_2/4)^2$ (with $t$'s in minutes and $x$'s in thousands of vehicles per hour). It is known that at user equilibrium,

75% of the origin-destination demand takes route 1. What percentage would take route 1 if a system-optimal solution were achieved, and how much travel time would be saved?

**8.36** Two routes connect an origin-destination pair with performance functions $t_1 = 5 + 3.5x_1$ and $t_2 = 1 + 0.5\,x_2^2$ (with $t$'s in minutes and $x$'s in thousands of vehicles per hour). It is known that at $x_2 = 3$, the difference between the first derivatives of the system-optimal and user equilibrium math programs, evaluated with respect to $x_2$, is 7 $[dS(x)_{SO}/dx_2 - dS(x)_{UE}/dx_2 = 7]$. Determine the difference in total vehicle travel times (in vehicle minutes) between user equilibrium and system-optimal solutions.

## Multiple Choice Problems (Multiple Sections)

**8.37** You are conducting a trip generation study based on Poisson regression. You estimate the following coefficients for a peak-hour shopping-trip generation model.

$$BZ_i = \quad -0.30 + 0.04(\text{household size})$$
$$+ 0.005(\text{annual household income in thousands of dollars})$$
$$- 0.12(\text{employment in the household's neighborhood in hundreds})$$

For a household with five members, an annual income of $95,000, and in a neighborhood with an employment of 250, what is the probability of the household making three or more peak-hour trips?

a) 0.071
b) 0.095
c) 0.905
d) 0.024

**8.38** A work-mode-choice model is developed from data acquired in the field in order to determine the probabilities of individual travelers selecting various modes. The mode choices include automobile drive-alone (DL), automobile shared-ride (SR), and bus (B). The utility functions are estimated as:

$$U_{DL} = 2.6 - 0.3(\text{cost}_{DL}) - 0.02(\text{travel time}_{DL})$$
$$U_{SR} = 0.7 - 0.3(\text{cost}_{SR}) - 0.04(\text{travel time}_{SR})$$
$$U_B = -0.3(\text{cost}_B) - 0.01(\text{travel time}_B)$$

where cost is in dollars and time is in minutes. The cost of driving an automobile is $5.50 with a travel time of 21 minutes, while the bus fare is $1.25 with a travel time of 27 minutes. How many people will use the shared-ride mode from a community of 4500 workers, assuming the shared-ride option always consists of three individuals sharing costs equally?

a) 866
b) 2805
c) 828
d) 314

**8.39** In a 1-hour period, 1100 vehicle-based shopping trips leave a large residential area for two shopping plazas. A joint shopping-trip mode-destination choice logit model is estimated, providing the following coefficients:

| Variable | Auto coefficient | Bus coefficient |
|---|---|---|
| Constant | 0.25 | 0.0 |
| Travel time in minutes | −0.4 | −0.5 |
| Commercial floor space (in thousands of m$^2$) | 0.13 | 0.13 |

Initial travel times to shopping plazas 1 and 2 are as follows:

| | By auto | By bus |
|---|---|---|
| Travel time to shopping plaza 1 (in minutes) | 15 | 18 |
| Travel time to shopping plaza 2 (in minutes) | 16 | 19 |

If shopping plaza 2 has 30,000 m$^2$ of commercial floor space and shopping plaza 1 has 25,000 m$^2$, determine the number of bus trips to shopping plaza 2.

a) 10
b) 18
c) 597
d) 21

**8.40** Two highways connect an origin-destination pair, having performance functions $t_1 = 5 + 4(x_1/c_1)$ and $t_2 = 4 + 5(x_2/c_2)$, for routes 1 and 2, respectively. For the performance functions, the $t$'s are in minutes, and the flows ($x$'s) and capacities ($c$'s) are in thousands of vehicles per hour. The total traffic demand for the two highways is 4000 vehicles during the peak hour. Initially, the capacities of routes 1 and 2 are 3500 veh/h and 4600 veh/h, respectively. A construction project is planned that will cut the capacity of route 2 by 2100 veh/h. How many additional vehicle-hours of travel time will be added to the system assuming user-equilibrium conditions hold?

a)  55.3
b)  503.0
c)  77.6
d)  525.3

**8.41** A study showed that during the peak-hour commute on two routes connecting a suburb with a large city, there are a total of 5500 vehicles that make the trip. Route 1 is 11 kilometers long with a 105-km/h speed limit and route 2 is 6 kilometers long with a speed limit of 80 km/h. The study also found that the total travel time on route 2 increases with the square of the number of vehicles, while the route 1 travel time increases two minutes for every 500 additional vehicles added. Determine the system-optimal total travel time (in veh-h).

a)  29.4
b)  1128.6
c)  1343.3
d)  1568.4

**8.42** Two routes connect an origin-destination pair, with 2500 and 2000 vehicles traveling on routes 1 and 2 during the peak hour, respectively. The route performance functions are $t_1 = 12 + x_1$ and $t_2 = 7 + 2x_2$, with the $x$'s expressed in thousands of vehicles per hour and the $t$'s in minutes. If vehicles could be assigned to the two routes such as to achieve a system-optimal solution, how many vehicle-hours of travel time could be saved?

a)  965.83
b)  970.83
c)  333.33
d)  5.56

# Index